T0245417

CAMBRIDGE LIBRARY COLLECTION

Books of enduring scholarly value

Travel and Exploration

The history of travel writing dates back to the Bible, Caesar, the Vikings and the Crusaders, and its many themes include war, trade, science and recreation. Explorers from Columbus to Cook charted lands not previously visited by Western travellers, and were followed by merchants, missionaries, and colonists, who wrote accounts of their experiences. The development of steam power in the nineteenth century provided opportunities for increasing numbers of 'ordinary' people to travel further, more economically, and more safely, and resulted in great enthusiasm for travel writing among the reading public. Works included in this series range from first-hand descriptions of previously unrecorded places, to literary accounts of the strange habits of foreigners, to examples of the burgeoning numbers of guidebooks produced to satisfy the needs of a new kind of traveller - the tourist.

Narrative of a Voyage to the Pacific and Beering's Strait

Frederick William Beechey (1796–1856), naval officer and hydrographer, was born into a family of artists, joined the Navy at a young age and went on to travel the world to survey coastlines and oceans. He published several accounts of his expeditions to destinations including the Arctic and Africa. This two-volume work, first published in 1831, describes his voyage as commander of the *Blossom* in 1825–1828. The ship's mission was to support the exploration of the North-West Passage by travelling eastwards via the Bering Strait to meet the explorers Sir John Franklin and Sir Edward Parry who were travelling west from the North Atlantic. Volume 1 records Beechey's outward journey via Cape Horn, his visits to Pitcairn Island, where he met the last surviving *Bounty* mutineer and documented his story (retold in Chapters 3 and 4), Tahiti, and Hawai'i, and his first season exploring the Bering Strait.

Cambridge University Press has long been a pioneer in the reissuing of out-of-print titles from its own backlist, producing digital reprints of books that are still sought after by scholars and students but could not be reprinted economically using traditional technology. The Cambridge Library Collection extends this activity to a wider range of books which are still of importance to researchers and professionals, either for the source material they contain, or as landmarks in the history of their academic discipline.

Drawing from the world-renowned collections in the Cambridge University Library, and guided by the advice of experts in each subject area, Cambridge University Press is using state-of-the-art scanning machines in its own Printing House to capture the content of each book selected for inclusion. The files are processed to give a consistently clear, crisp image, and the books finished to the high quality standard for which the Press is recognised around the world. The latest print-on-demand technology ensures that the books will remain available indefinitely, and that orders for single or multiple copies can quickly be supplied.

The Cambridge Library Collection will bring back to life books of enduring scholarly value (including out-of-copyright works originally issued by other publishers) across a wide range of disciplines in the humanities and social sciences and in science and technology.

Narrative of a Voyage to the Pacific and Beering's Strait

Performed in His Majesty's Ship Blossom, in the years 1825, 26, 27, 28

VOLUME 1

FREDERICK WILLIAM BEECHEY

CAMBRIDGE UNIVERSITY PRESS

CAMBRIDGE UNIVERSITY PRESS

Cambridge, New York, Melbourne, Madrid, Cape Town,
Singapore, São Paolo, Delhi, Tokyo, Mexico City

Published in the United States of America by Cambridge University Press, New York

www.cambridge.org
Information on this title: www.cambridge.org/9781108031035

© in this compilation Cambridge University Press 2011

This edition first published 1831
This digitally printed version 2011

ISBN 978-1-108-03103-5 Paperback

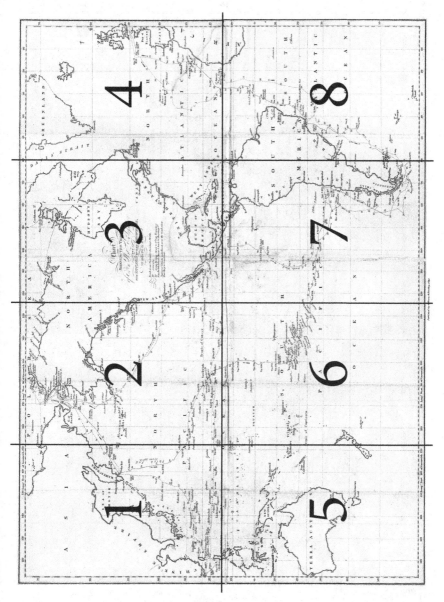

Key to following pages.

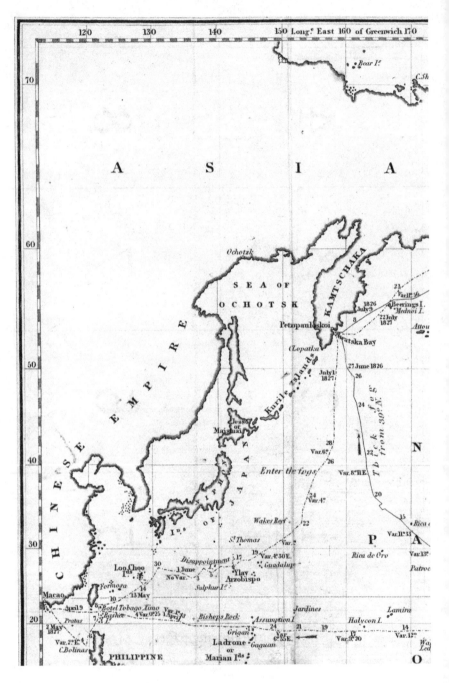

1

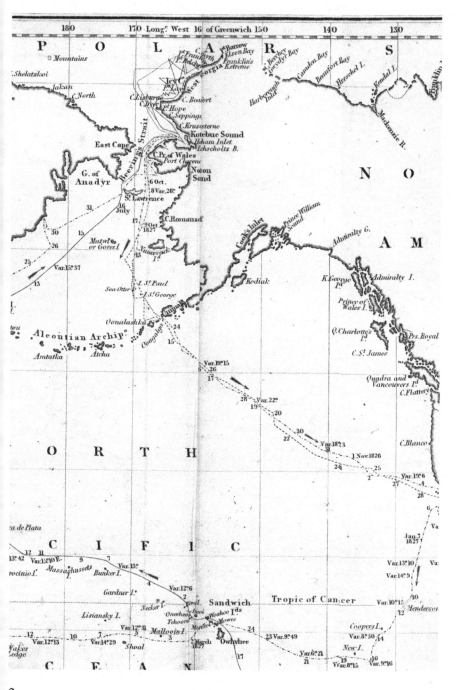

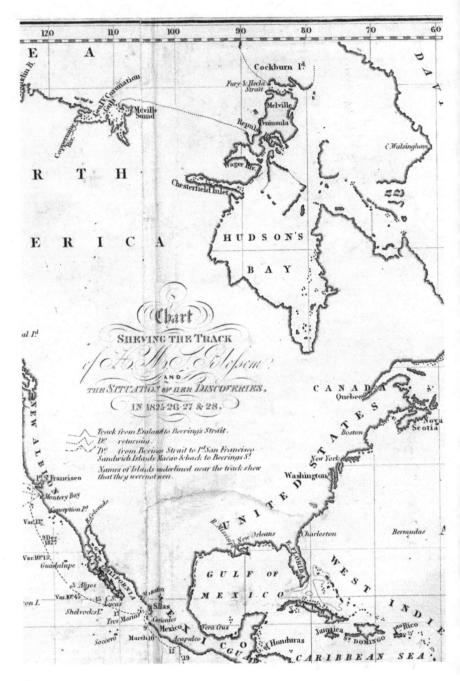

GREENLAND

ICELAND

DAVIS' STRAIT

C. Farewell

NEWFOUND LAND

N O R T H

Sep. 24
1828
Var. 33° W. 23 May 1825

22

26

20

Var. 39° W.

28

Oporto

PORTUGAL

SPAIN

Bay of Biscay

LONDON
Portsmouth

IRELAND
ENGLAND

Azores or
Flores
Western Is.
Pico
Terceira
S.t Michael

Lisbon

Var. 27°

30

Str. of
Gibraltar

15

17

A T L A N T I C

12

Madeira

Pto Santo

Var. 29°

Mogadore

AFRICA

10

3

Salvages

Canary
Is.
Ferro

Teneriffe

Gr.
Canary

6

Var. 21°

7 June

C. Blanco

1 Sep.

10

Var. 18°

9

O C E A N

30

S.t Antonio
Bravo
Cape
Verd Is.

Var 12° W.

S.t Jaao

C. Verd

4

PHILIPPINE
I^ds.
Marian I^ds.
Guahan
Rada
Chain
10
C A R O L I N E
Scarboro
Ra
0
NEW
GUINEA
Solomons
I^s.
10
New
Hebrid
20
New
Caledonia
T
TERRA   AUSTRALIS
30
Port
Jackson
40
VanDiemen's
Land
N
E
W
Z
E
50

120   130   140   150 Long^e. East 160 of Greenwich 170

5

C E A N   Var.8°.47 E.

21   19 Var.8°.15   Var.9°.16

lack
un

15

Manuel Rodriguez

I ds

Var.7°.20 E. 13

ough
ange

Fanning I.

Christmas I.

n

EQUATOR

Var.7°.15 9

Var.5°.16 7

5

Var.7° E.

Var.5° E. Caroline I.

Maarquesas

Flint I. May 1 1826
Var.8° E.

Navigators   S   O   U   T

29

Orange I. K George's I. Disappointment I.

Friendly I ds

Scilly I s Society 27 5 Pallicers

des   Ulieta I da Chain I. Melville I.

Fidies   Otaheite Crocker I. Resan Cap I.
Maitea Common I.
Pr. W m Henry I. Whit Sunday I.
Byam Martin I. Henle I.

Wateeo   Q. Charlotte I. Clermont de Tonnere
Barrow I.
Crescent I. Staysort I. Hood I.

Tropic of Capricorn   Mangeea   Blighs Lagoon Gambiers I s
Osnaburg I. 25 Oeno
Cockburn I. Elizabeth Var.6°
Crescent 22 1 Dec.
Pitcairn I. Ducies I.

Oparo

P   A   C   I   F

N Z

O   C   E

Chatham I.

180   170   Long t West 50 of Greenwich 150   140   130

Published by Coll.

6

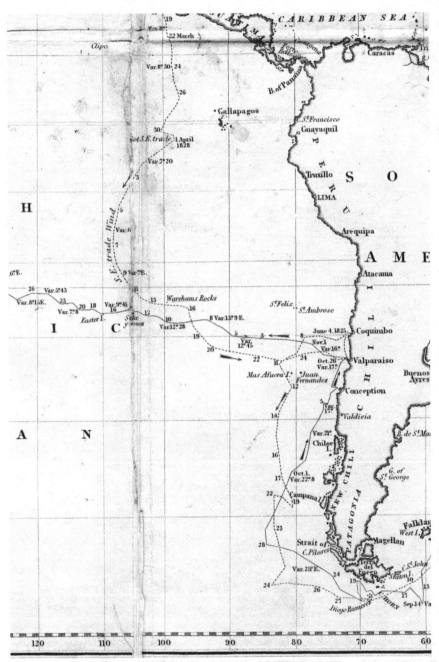

CARIBBEAN SEA

Caracas

Clipo

22 March

Var. 8° 30′ 24

26

Gallapagos

B. of Panama

S.t Francisco

Guayaquil

H

30

Got S.E. trade 1 April
1828
Var. 7° 20

3

LIMA

Truxillo

P
E
R
U

S
O

Var. 6

5

7

Arequipa

A M E

9 Var. 7° E.

Atacama

6.° E.

26   Var. 5° 43

11

Var. 8° 15′ E.

23
Var. 7° 8

20 18
16

Var. 9° 41

13

Warehams Rocks

16

S.t Felix

S.t Ambrose

I

C

Easter I.

S.ta
y'ones

12

10
Var. 12° 28

19

8 Var. 13° 9 E.

5
Var.
12°45

3

8

June 4. 1825

Coquimbo

Nov.1
Var. 16°

20

22

11

24

Oct. 26
Var. 17°

Valparaiso

Buenos
Ayres

Mas Afuera I.

Juan
Fernandes

12

5

Var.
17°

Conception

Valdivia

C

A

N

14

16

Var. 21°

Chiloe

B. de S.t Ma

17

Oct.1.
Var. 22° 8

G. of
S.t George

22

23

Campana I.

19

N
E
W

C
H
I
L
I

P
A
T
A
G
O
N
I
A

Falklan
West I.

28

Strait of
C.Pilares

Magellan

Var. 23° E.

24

Terra
del
Fuego

C.S.t John

24

26

19

'taten I.

50

13

27

Diego Ramirez

 HORY

17

Sep 14 Va

120   110   100   90   80   70   60

:lburn &Bentley,1831.

7

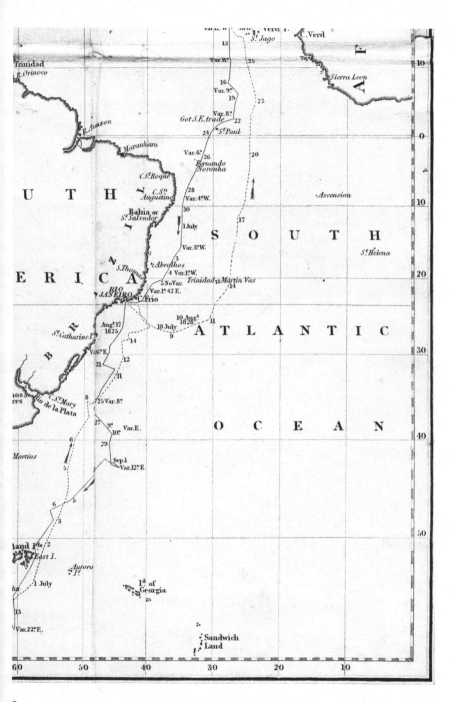

S. Jago
C. Verd
Var. W.   Verd I.

13
Var. II°   25
Trinidad
r. Orinoco
Sierra Leon

16
Var. 9°
19
23

AFR
Var. 8°
Got S.E. trade   22
24 St Paul
0

Amazon
Var. 6°   26
Maranham
Fernando Noronha
20

C. St Roque
C. Sn Augustine
28
Var. 4°.W.
Ascension
10

30
Bahia or
St Salvador
1.July
S   O   U   T   H

Var.3°.W.
St Helena

ERICA
S.Thomé
Abrolhos
4 Var.1°.W.
20
Var.1° 42.E.
5 No Var.
Trinidad Martin Vas
RIO JANEIRO
14
C.Frio

B
Aug.17
1825
10 Aug.   11
A   T   L   A   N   T   I   C
St Catharine I.
18.July   18,26°
18.July
9

R
14
Var.6°.E.
12
30

21
11

os
res
Rio de la Plata
St Mary
8
25 Var.8°
O   C   E   A   N

27   9°
10°
Var.E.
40
Martias
6
29

5
Sep.1
Var.12°.E.

6   5

3
50

land I.ds   2
East I.

Aurora
I.ds
1 July

Id of Georgia
13

Var.22°.E.
Sandwich Land

60   50   40   30   20   10

NARRATIVE

OF A

VOYAGE TO THE PACIFIC

AND BEERING'S STRAIT,

TO CO-OPERATE WITH

THE POLAR EXPEDITIONS:

PERFORMED IN

HIS MAJESTY'S SHIP BLOSSOM,

UNDER THE COMMAND OF

CAPTAIN F. W. BEECHEY, R.N.

F.R.S. &c.

IN THE YEARS 1825, 26, 27, 28.

PUBLISHED BY AUTHORITY OF THE LORDS COMMISSIONERS OF
THE ADMIRALTY.

IN TWO VOLUMES.

VOL. I.

LONDON:

HENRY COLBURN AND RICHARD BENTLEY,

NEW BURLINGTON STREET

1831.

TO THE KING.

In availing myself of Your Majesty's gracious
permission to dedicate this work to Your Majesty,
I feel that I am performing a most pleasing duty.

The claims of Your Majesty's family on the gratitude of the nation, for the efficient patronage they
have afforded to maritime discovery, require merely
to be alluded to, to ensure the attention of every
well-wisher to his country.

Under a less powerful Sovereign than your Royal
Father, the voyages of Cook and Vancouver, in all
probability, would never have been projected, and
could hardly have prospered; while it is certain
that the expeditions of Parry and Franklin owed
their chief distinction to the enlightened encouragement of His late Majesty.

But these great enterprises—so productive of
national renown—so extensively useful in diffusing
the blessings of civilization over distant and savage

lands—and so eminently beneficial to the cause of science and of commerce, could never have been successfully accomplished, had not the character of the Navy been habitually maintained at, perhaps, the highest level which human exertion is capable of reaching.

To produce this generous spirit, however, and to preserve it entire when once created, there was required, on the part of the Royal Family, some signal example of personal sacrifice to the popular service of the country. And although it would be very presumptuous in any one to pretend to estimate the advantages which the profession has derived, in our own days, from Your Majesty having condescended to become one of its working members, there can be no doubt, that in all future times, the British Navy will retain the salutary impression, and cherish the remembrance of this high honour.

May it please Your Majesty,
Your Majesty's
most dutiful servant,
most grateful
and most faithful subject,

FREDERICK WILLIAM BEECHEY.

INTRODUCTION.

THE discovery of a north-west passage to the Pacific had for some years occupied the attention of the British government and of the public at large, and several brilliant attempts had been made both by sea and by land to ascertain the practicability of its navigation, which, though conducted with a zeal and perseverance that will transmit them to the latest posterity, had, from insurmountable difficulties, failed of success.

In 1824, His late Majesty having commanded that another attempt should be made by way of Prince Regent's Inlet, an expedition was equipped — the last that sailed upon this interesting service — and the command again conferred upon Captain Parry, whose exploits have so deservedly earned him the approbation of his country. At the same time Captain Franklin, undaunted by his former perilous expedition, and by the magnitude of the contemplated undertaking, having, with the promptness and perseverance peculiar to his character, proposed to connect his brilliant discoveries at the mouth of the Coppermine River with the furthest known point on the western side of America, by descending the Mackenzie River, and, with the assistance of his intrepid associate, Dr. Richardson, by coasting the northern shore in opposite directions towards the two previously discovered points, His late Majesty was also pleased to command that this expedition should be simultaneously undertaken.

From the nature of these services it was nearly impossible that either of these expeditions could arrive at the open sea in Beering's Strait, without having nearly, if not wholly, exhausted their resources ; and Captain Franklin's party being, in addition, destitute of a conveyance to a place whence it could return to Europe To obviate these anticipated difficulties, his Majesty's government determined upon sending a ship to Beering's Strait to await the arrival of the two expeditions.

As this vessel would traverse, in her route, a portion of the globe hitherto little explored, and as a considerable period must necessarily elapse before her presence would be required in the north, it was intended to employ her in surveying and exploring such parts of the Pacific as were within her reach, and were of the most consequence to navigation.

The vessel selected for this service was his Majesty's ship Blossom, of twenty-six guns, but on this occasion mounting only sixteen ; and on the 12th of January, 1825, I had the honour of being appointed to the command of her. The following officers, most of them men distinguished for their abilities, were placed under my orders, viz.—

Lieutenant,	George Peard.	Naturalist, . .	George T. Lay.
Ditto, .	{ Edw. Belcher, Supernumerary, and Assistant Surveyor.	Assistant Surgeon,	Thomas Neilson.
		Clerks, . .	{ John Evans, Chas. H. Osmer.

Ditto, .	John Wainwright.	*Volunteers,* 1st *Class,*	{ John Crawley,	
Master, .	Thomas Elson.		{ John Hockley.	
Surgeon, .	Alex. Collie.	*Ditto,* . 2d *Class,*	{ J. Clarke Barlow,	
Purser, .	George Marsh.		{ Charles Lewis.	
Admiralty Mates,	{ J. F. Gould,* William Smyth, James Wolfe.	*Gunner,* . .	John Richardson.	
		Boatswain, . .	James Clarkson.	
		Carpenter, . .	Thos. Garrett.	
Midshipmen,	{ John Rendall, Richard B. Beechey.			

To these were added such a number of seamen, marines, and boys, as, with the exception of the supernumeraries, would form a complement of a hundred and ten persons; but in consequence of the weakness of our crew when collected, I was permitted to discharge ten of the most inefficient; a reduction which, without sensibly diminishing the strength of our crew, materially increased the duration of our stock of provisions, and in the sequel proved of the most happy consequence.

The ship was partially strengthened, and otherwise adapted to the service, by increasing her stowage. A boat was supplied, to be used as a tender, and for this purpose she was made as large as the space on the deck would allow. She was rigged as a schooner, decked, and fitted in the most complete manner, and reflected great credit upon Mr. Peake, the master-shipwright of Woolwich dock-yard, who modelled and built her.

To the usual allowance of provision was added a variety of anti-scorbutics. Cloth, beads, cutlery, and other articles of traffic, were put on board ; and two fowling-pieces, embossed with silver, and fitted in the most complete manner, were supplied as presents to the kings of the Society and Sandwich Islands.

* This valuable young officer was obliged to quit the ship at Rio Janeiro on account of his health.

The College of Surgeons sent bottles of spirits for the preservation of specimens, and the Horticultural Society enhanced our extra stores with a box of seeds properly prepared for keeping.

The seamen were furnished with two suits of clothes gratis, and were allowed the further privilege of having six months' wages in advance.

In the equipment of all the expeditions of this nature it has been the good fortune of the officers engaged in them to meet with the utmost courtesy and attention to their wishes from the departments which have the power so materially to contribute to their comfort ; and I take this opportunity of expressing my sincere thanks to Sir G. Cockburn and the other Lords Commissioners of the Admiralty, to Sir Thomas Byam Martin, and the Commissioners of the Navy and Victualling Boards, for the readiness with which they at all times complied with my requests.

Being in every respect ready, on the 19th May I received the following instructions from the Lords Commissioners of the Admiralty :—

> " By the Commissioners for executing the office of Lord High Admiral of the United Kingdom of Great Britain and Ireland, &c. &c.

"Whereas it is our intention that his majesty's sloop Blossom, under your command, should be at Beering's Strait in the Autumn of 1826, and, contingently, in that of 1827, for the purpose of affording such assistance as may be required, either by Captain Parry or Captain Franklin, should one or both of those officers make their appearance in that neighbourhood. You are hereby required and directed to put to sea in the said sloop, so soon as in

every respect ready, and observe the following instructions for your guidance:—

"You are to proceed with all convenient expedition to Rio Janeiro, where you are to complete your
provisions and water; after which you are to make
the best of your way round Cape Horn, and endeavour to make Easter Island; from whence you are
to take your departure, steering for the Society
Islands, and passing near the spot where Gomez
Island appears in the charts, in order to ascertain
whether such island has any existence; and, in like
manner, whether Ducie's and Elizabeth Islands be
not one and the same. You will then proceed to
Pitcairn's Island at the south-eastern extremity of
the groupe of the Society Islands, or, as they are
sometimes called, the Georgian Islands, where you
will commence a survey of this groupe, proceeding
north-westerly to Otaheite. In the execution of
this survey it may be found most advisable to
anchor, if practicable, every evening, under one of
the islands, in order that the situation of the ship
may, by these means, be more secure, and that you
may be certain that none of them are passed by you
unobserved. If, however, you should experience
any difficulty in pursuing the route herein pointed
out, from the prevailing winds, you will make the
best of your way to Otaheite, and proceed from
thence in your survey to Pitcairn's Island.

"During your stay among these or any other of
the islands of the Pacific which you may visit, you
are to use every possible endeavour to preserve an
amicable intercourse with the natives, and to caution
your officers and ship's company to avoid giving
offence or engaging in disputes with them; and you
are to show them on all occasions every act of kind-

ness that may be in your power, taking care that when any purchases, by barter or otherwise, are made, an officer of the ship may always be present to prevent disputes: and you are particularly to impress on the minds of your officers and men the necessity of being extremely guarded in their intercourse with the females of those places, so as to avoid exciting the jealousy of the men.

" Having completed the survey of this groupe of islands, if you find that your time will admit of it, you are to direct your course to the Navigator's Islands, settling in your way thither the true position of Suwarrow's Islands; from whence, in your progress to the northward, you will touch at Owhyhee, to deliver the despatches and packages addressed by the Foreign Office for his Majesty's consul at that island, and to procure refreshments and water.

" You are, however, to be particularly careful not to prolong your stay at any of those islands, so as to retard your arrival at the appointed rendezvous in Beering's Strait later than the 10th July, 1826; which period, together with the rendezvous, has been fixed by Captain Franklin and yourself, by a memorandum, a copy of which is annexed, and we desire and direct you to pay particular attention to the various matters contained therein.

" You are to remain at the said appointed rendezvous until the end of October, or to as late a period as the season will admit, without incurring the risk of being obliged to winter there, provided you shall hear nothing of Captain Franklin or his party; but in the event of his joining, you are to receive him and his party on board, and convey him either to Kamtschatka, the Sandwich Islands, Panama, or to China, as he may determine, in order to procure a further conveyance to England. If, however, you should receive certain intelligence of Captain Parry having passed through Beering's Strait into the Pacific, you are in that case to proceed with the Blossom round Cape Horn, and bring Captain Franklin and his party to England ; touching at Callao, and such other ports on the western coast of South America as you may deem proper for refreshments, intelligence, &c.

" In the former event, namely, of your leaving Beering's Strait with Captain Franklin, but without having obtained any intelligence of Captain Parry, you are to complete your water and provisions at the place to which you convey Captain Franklin ; or in the event of your hearing nothing either of Captains Franklin or Parry, previous to the season obliging you to leave Beering's Strait in 1826, you are to proceed to such place as you may deem most eligible and convenient for completing your provisions and water; taking care in either of the last-mentioned cases to be again in Beering's Strait by the 1st August, 1827, calling in your way thither again at Owhyhee, at which place Captain Parry has been directed to give the preference of touching in his way homeward, for the purpose of affording you intelligence of him.

" If you should find that Captain Parry has passed, or should he pass after joining you, and that you have heard nothing of Captain Franklin, you are, nevertheless, to proceed to, or remain at (as the case may be) Beering's Strait, in the autumn of 1827, as already directed, following in all respects the directions already given for your conduct in the autumn of 1826.

" In order that you may be put in full possession of that part of our instructions to Captain Parry which relates to his arrival in Beering's Strait, we enclose you herewith an extract from them, as also a copy of a ' Memorandum,' drawn up by Captain Parry, and dated ' Hecla, Davis' Strait, June, 1824 ;' to both of which we desire to call your particular attention, in order that you may govern your proceedings accordingly.

" Having remained in Beering's Strait as late in the autumn of 1827 as the season will admit, and without risking the chance of being obliged to winter on account of the ice, you are to proceed to England by the route before directed ; reporting to our secretary your arrival, and transmitting the journals of yourself and officers for our information.

" In the prosecution of your voyage out, and during your stay in the Pacific, you are to be particular in noticing the *differences of longitude* given by your chronometers, from any one place to another, which you may visit in succession.

" As we have appointed Mr. Tradescant Lay as naturalist on the voyage, and some of your officers are acquainted with certain branches of natural history, it is expected that your visits to the numerous islands of the Pacific will afford the means of collecting rare and curious specimens in the several departments of this branch of science. You are to cause it to be understood that two specimens, *at least*, of each article are to be reserved for the public museums ; after which the naturalist and officers will be at liberty to collect for themselves. You will pay every attention in your power to the preservation of the various specimens of natural history, and on your arrival in England transmit them to this office ; and if, on your arrival at any place in the course of your voyage, you should meet with

a safe conveyance to England, you are to avail yourself of it to send home any despatches you may
have, accompanied by journals, charts, drawings,
&c., and such specimens of natural history as may
have been collected.   And you will, on each of your
visits to Owhyhee, deliver to his Majesty's consul at
that place duplicates of all your previous collections
and documents, to be transmitted by him, by the
first safe opportunity, to England.

" In the event of England becoming involved in
hostilities with any other power during your absence, you are, nevertheless, clearly to understand
that you are not on any account to commit any
hostile act whatsoever; the vessel you command being sent out only for the purpose of discovery and
science, and it being the practice of all civilized nations to consider vessels so employed as excluded
from the operations of war : and, confiding in this
general feeling, we should trust that you would receive every assistance from the ships or subjects of
any foreign power which you may fall in with.

" On your return home you will proceed to Spithead, informing our secretary of your arrival.

" Given under our hands, the 11th of May, 1825.

" MELVILLE.
WM. JOHNSTONE HOPE.
G. COCKBURN.
G. CLERK.
W. R. K. DOUGLAS.

" To Frederick William Beechey, Esq.
Commander of his Majesty's Sloop Blossom, at Spithead.

" By command of their lordships.

" J. W. CROKER."

MEMORANDUM ACCOMPANYING THE
INSTRUCTIONS.

" We deem it advisable that the ship should be in Beering's Strait by the 10th of July, and that she should remain at some appointed rendezvous until the end of October, or to as late a period as the season will admit, without incurring the risk of being obliged to winter there.

" At present we know of but one place on the eastern shore of the strait which we can recommend as a rendezvous for both parties, viz. Kotzebue's Sound; there it appears the ship may remain with all winds. Desirable as it is to take up a more northerly position than this, in order that the voyage of Captain Franklin's party in open boats may be shortened; yet, admitting the possibility of deep inlets on the coast, it is evident that the boats of Captain Franklin would have more difficulty in searching for the ship in them than in proceeding at once to the above-mentioned sound; and the certainty of finding the ship at a fixed point would be more satisfactory to Captain Franklin.

" In order, however, to lessen as far as possible the difficulties of the land party (still preserving the fixed rendezvous), it is recommended that a party, well armed, and having a supply of provisions and fuel, shall be left at Chamisso Island with a boat; or, if it be necessary, the defences of the island may be strengthened by the two forecastle guns, which, with a strong boat's crew, will be sufficient to protect the only landing-place in the island against any force the natives can bring, should they be hostile.*

" Leaving this party at the rendezvous, the Blossom may proceed to examine the coast, assisted by her decked launch, keeping in-shore of her; and signals can then be regularly placed on every conspicuous cape or height, according to the mode agreed upon, for the purpose of directing Captain Franklin's attention to bottles containing written information, which will be buried at each station.

" In this manner it is proposed, circumstances permitting, to navigate from Kotzebue's Sound northward, and then eastward as far as the state of the ice will allow, following up every opening, and never quitting the main shore. The distance to which the ship can proceed to the eastward will be limited by the lateness of the season, and the necessity of avoiding the hazard of being beset in the ice and obliged to winter.

" Fog-signals and night-lights will of course be established between the launch and the ship; and should the launch part company with the ship, it will proceed to the last formed signal station, and there await the junction of the ship; but if she does not arrive there in five days, the launch is to prosecute the voyage along shore, in search of Captain Franklin, but not to go so far as to put the certainty of returning to Chamisso Island by the 30th of September at any risk, by which date the ship will also have arrived there; and Captain Franklin will proceed to the same place should he not have met either the ship or launch before.

* This erroneous idea was suggested by Captain Kotzebue's account of the island, arising no doubt from a bad translation.

" During the time the Blossom remains in Kotzebue Sound, a party will be directed to proceed inland on a north course, if practicable, in order that should the coast of the Polar Sea be within reasonable distance, signals may be erected upon the heights for Captain Franklin, whose party may by this means be spared a long journey round the N. W. promontory of America. At this and every other station where information is deposited of Captain Beechey, it is advisable that a request in the Russian language be also placed, that this information be not taken away, or the signals disturbed.

" Since the transmission of the above, Captain Franklin has received his instructions from Earl Bathurst, the contents of which have been made known to Captain Beechey, and the only addition which we think necessary to make is, that in the event of Captain Franklin arriving at an early period at Icy Cape, or at the N. W. extremity of America, or in the longitude of Icy Cape (161° 42' W.) and returning the same season to his former winter quarters, he will, in the above-mentioned meridian, erect a signal, and bury a bottle containing the information of his having done so for Captain Beechey's guidance.

(Signed) " JOHN FRANKLIN, Captain.
F. W. BEECHEY, Commander, His
Majesty's Sloop Blossom.
Woolwich, 10*th February*, 1825."

After the receipt of these instructions, I took an early opportunity of communicating to the officers under my command the sentiments of their lordships, contained in the twelfth paragraph. How satisfactorily these expectations were fulfilled, must appear from the manner in which their lordships have marked their approbation of their conduct. As commander of the expedition, however, I am happy of an opportunity of again bearing testimony to their diligence, and of expressing my thanks for the assistance I derived during the voyage from their exertions. They are especially due to my first lieutenant, Mr. Peard, upon whom much additional duty devolved, in consequence of my attention being in some measure devoted to other objects of the expedition: to Lieutenant Belcher and Mr. Elson, the master, for their indefatigable attention to the minor branches of surveying; and to the former, again, for his assistance in geological re-searches: to Lieutenant Wainwright for his astro-nomical observations; to Mr. Collie, for his unre-mitting attention to natural history, meteorology, and geology; to Mr. James Wolfe, for his attendance at the observatory and the construction of charts; and, lastly, to Messrs. Smyth and Richard Beechey, for the devotion of their leisure time to drawing.

On the return of the expedition to England, the journals and papers of the officers were placed in my hands by the Admiralty, with directions to pub-lish an account of the voyage. I found those of Messrs. Collie and Belcher to contain much useful information on the above-mentioned branches of science, and in other respects I have derived much assistance from their remarks, and also from those in the journals of Messrs. Evans, Smyth, and Beechey. I have in general noticed these obligations in the course of my narrative : but as this could not always be done without inconvenience to the reader, I take this opportunity of more fully expressing my acknowledgments.

In the compilation I have endeavoured to combine information useful to the philosopher with remarks that I trust may prove advantageous to the seaman, and to convey to the general reader the impressions produced upon my mind at the moment of each occurrence. How far I have succeeded in acquitting myself of the task my duty compelled me to undertake, I must leave to the public to decide, and shall conclude with expressing a hope that my very early entry into the service may be taken in extenuation of any faults they may discover.

The collections of botanical and other specimens of natural history have been reserved for separate volumes, being far too numerous to form part of an appendix to the present narrative. His Majesty's government having liberally appropriated a sum of money to their publication, I hope, with the assistance of several eminent gentlemen, who have kindly and generously offered to describe them, shortly to be able to present them to the public, illustrated by engravings by the first artists. The botany, of which the first number has already been published, is in the hands of Dr. Hooker, professor of Botany, at Glasgow, who in addition to having devoted the whole of his time to our collection, has borne with the numerous difficulties and disappointments which have attended the progress of the publication of this branch of natural history, and my thanks on this account are the more especially due to him in particular. The department which he has so kindly undertaken will extend to ten numbers 4to.; making, in the whole, about 500 pages, and 100 plates of plants, wholly new, or such as have been hitherto imperfectly described.

The other branches of natural history are under the care of Messrs. N. A. Vigors, Edward Bennett, J. E. Gray, Richard Owen, Dr. Richardson, R. N., and Mr. T. Lay, the naturalist to the expedition, and the geology of Professor Buckland and Captain Belcher, R. N.; to all of whom I must express my warmest thanks, for their cordial assistance, and for the ready and handsome manner in which they have taken upon themselves the task of describing and of superintending the delineation of the various specimens. Their contributions will form another 4to. volume of species entirely new, or, as before, of such as have been imperfectly described. The public in general are not aware how much is due to these gentlemen, without whose zeal and aid they would be deprived of much useful knowledge; for, notwithstanding the liberal assistance of his Majesty's government, there is so little encouragement for works of the above-mentioned description, that they could not be published unless the contributions were gratuitously offered to the publishers.

CONTENTS

OF THE FIRST VOLUMF.

CHAPTER IX.

CHAPTER X.

CHAPTER XI.

CHAPTER XII.

DIRECTIONS TO THE BINDER FOR INSERTING THE PLATES AND CHARTS.

VOL. I.

VOYAGE

PACIFIC AND BEERING'S STRAIT.

CHAPTER I.

Departure from England—Teneriffe—Sun eclipsed—Fernanda Norhona—Make the Coast of Brazil—Rio Janeiro—Passage round Cape Horn—Conception—Valparaiso.

On the 19th of May we weighed from Spithead, and the following afternoon took our parting view of the Devonshire coast, and steered out of the Channel with a fair wind. For several days afterwards our progress was impeded by boisterous weather, for which the approach to the Bay of Biscay has long been proverbial. We however escaped tolerably well, and favourable breezes soon succeeding, we advanced to the southward.

CHAP.
I.
May,
1825.

On the 30th we ascertained, by running over the spot in a fine clear day, that a reef of rocks, named the Eight Stones, did not exist in the situation which it has for a number of years occupied in our charts: the next morning we passed the Desertas, and on the 1st of June were off Teneriffe.

June.

VOL. I.   B

As I purposed touching at Santa Cruz, we immediately hauled up for the land, and it was a fortunate circumstance that we did so, for so strong a
current set to the southward during the night, that
had we trusted to our reckoning, the port would
have been passed, and there would have been much
difficulty in regaining it. I mention the circumstance with a view of bringing into notice the great
southerly set that usually attends the passage of
ships from Cape Finisterre southward.   From this
cape to Point Naga, our error in that direction, or
more correctly S. 33° W., was not less than ninety
miles.   I do not stop to inquire into the cause of
this great tendency of the water to the equator,
which might probably be traced to the remote
effect of the trade-wind, but merely mention the
fact as a guide to persons who may pursue the same
route.

We approached the island on a fine sunny day,
but from a quarter that was highly unfavourable
for a view of the lofty Peak, which was almost hid
from us by intervening mountains.   At four o'clock
we came to an anchor in the roads of Santa Cruz,
and there found His Majesty's ship Wellesley, Captain, now Admiral Sir G. E. Hamond, Bart., on her
way to Rio Janeiro, with his Excellency Sir Charles
Stuart, the British Ambassador to the court of Brazil.   As soon as we had exchanged salutes with
the fort, we landed to procure the supplies the ship
required, with all despatch; and met with much
assistance and civility from Mr. Dupland, who was
acting in the absence of the Consul.

Santa Cruz, at the time of our arrival, was under
the government of Don Ysidore Uriarti, who very

obligingly allowed me to pitch a tent in one of the forts for the purpose of making observations, and placed a guard of soldiers to keep watch over the instruments. In Santa Cruz there is very little to interest a stranger: when he has paraded some inferior gardens which perpetuate the memory of the Marquis de Brancifort, cast his eye round the interior of the great church of San Francisco, where a flag that once belonged to Lord Nelson will not be allowed to escape his attention, and scanned a monument erected to the Virgin Mary de la Candelaria, the patroness-saint of the island, he has seen all that can offer an inducement to expose himself to a dusty walk on a hot day, which he will be sure to find in the month of June in this scattered town. The Plaza Reale will amuse those persons who wish to indulge their criticism on the manner and costume of the inhabitants, who assemble there in the evening to smoke their cigars, and enjoy the luxurious freshness of the air.

At Laguna the capital, visiters will find a better town, a more fertile country, a climate several degrees cooler than that of Santa Cruz, and every species of produce more abundant and forward than at the port; and though the road is bad, few will regret having encountered its difficulties. The celebrated Peak of Teyde is the great object of curiosity which engages the attention of travellers to the Canary groupe, and we experienced much mortification at not having it in our power to ascend it. To have added our mite toward the determination of its altitude by barometrical measurement, was a consideration not overlooked; but, circumstanced as we were, it was not of sufficient importance to jus-

tify the detention of the ship; and we were obliged
to console ourselves with the hope that we should
shortly visit places less known, and where our time,
consequently, would be more usefully employed.

Teneriffe is an island which lies in the track of
all outward-bound ships from Europe, and most
voyagers have touched at it : being the first object
of interest they meet, their zeal is naturally more
excited there, than at any subsequent period of
their voyage : it is consequently better described
than almost any other island in the Atlantic, and
nothing is now left for a casual visiter, but to go over
the ground of his predecessors for his own gratification or improvement. My observations for the determination of the latitude and longitude of the place,
&c. were made in the Saluting Battery, but they
are omitted here, as I purpose, throughout these volumes, to avoid, as far as possible, the insertion of
figures and calculations, which, by the majority of
readers, are considered interruptions to the narrative,
and are interesting only to a few. On the 3rd, His
Majesty's ship Wellesley sailed for Rio Janeiro with
His Excellency Sir Charles Stuart ; and on the 5th,
having procured what supplies we required, we
weighed, and shaped a course for the same place.

From our anchorage we had been daily tantalized
with a glimpse only of the very summit of the
Peak, peeping over a nearer range of mountains, and
the hazy state of the weather on the day of our departure made us fearful we should pass on without
beholding any more of it ; but towards sunset,
when we had reached some miles from the coast, we
were most agreeably disappointed by a fair view of
this gigantic cone. The sun set behind it ; and as

his beams withdrew, the mountain was thrown forward, until it appeared not half its real distance. Then followed a succession of tints, from the glowing colours of a tropical sky, to the sombre purple of the deepest valleys; varying in intensity with every intermediate range, until a landscape was produced, which, for beauty of outline, and brilliancy of colour, is rarely surpassed; and we acknowledged ourselves amply repaid for our days of suspense. Night soon closed upon the view; and, directing our compass to a well-known headland, we took our last look at the island, which was the only one of the Canary groupe we had seen: not on account of our distance from them, but owing to that mass of clouds which "navigators behold incessantly piled over this Archipelago." The breeze was fair, and we rolled on, from day to day, with our awnings spread; passing rapidly over the ground with a fresh trade-wind, and daily increasing the heat and humidity of our atmosphere: amused, occasionally by day, with shoals of flying-fish starting from our path, followed by their rapacious pursuers; and by night, with the phosphoric flashing of the sea, and the gradual rising of constellations not visible in our native country.

Toward the termination of the trade, the wind veered gradually to the eastward, and became fresh, until noon of the 15th, when it suddenly ceased, and the sea, foaming like breakers, beneath a black thunder-cloud, warned us to take in our lighter sails. We were presently taken aback with a violent gust of wind from the southward, and from that time lost the north-east trade. As we approached its limit, the atmosphere gradually became more charged with humidity, and the sky thickened with dark clouds, which, latterly, moved heavily in all directions, pouring down torrents of rain.

On the 16th, the sun was eclipsed ; and we made many observations to determine the moment of conjunction. In doing this, my attention was arrested by a very unusual appearance. It consisted of a luminous haze about the moon, as if the light had been transmitted through an intervening atmosphere. I made a sketch of it very soon afterwards,* of which I was very glad, as a similar phenomenon, I found, had been observed by M. Dolland in another eclipse ; and as the subject has since received much interest from the circumstance of Aldebaran, and Jupiter and his satellites, having been seen projected upon the disc of the moon. About the time of the greatest obscuration, Leslie's photometer stood at 27°, exactly half what it afterwards showed. Between the intervals of observation, we amused ourselves with making experiments with a burning glass upon differently coloured cloths, in imitation of those recorded in the Memoirs of the Astronomical Society, and which will convey to the general reader a more intelligible idea of the decrease of intensity in the sun's rays at the time of the greatest obscuration, than the observations with the photometer, as well as of the readiness with which some colours ignite in comparison with others : for instance,

Black	Blue	Scarlet,	Pea-Green
burned instantly ;	required $3^s,7$,	$15^s,7$:	would not ignite.

---

* See the plate.

After the eclipse, and when the sun was shining bright,

Black	Blue	Scarlet,	Pea-Green,	Yellow,
burned instantly;	instantly;	2ˢ;	7ˢ,8;	4ˢ,3.

The results are the mean of several observations; and the intervals, the number of seconds between the rays being brought to a focus on the cloth, and its ignition.

After losing the trade-wind, we went through the usual ordeal of baffling winds and calms, with oppressively hot moist weather, and heavy rains; and then, on the 19th, in latitude 5° 30′ N., got the south-east trade, with which we pursued our course towards the equator, and crossed it on the 24th, in longitude 30° 2′ West, much further from the meridian of Greenwich than choice would have dictated. Some anxiety was in consequence felt lest the current, which here ran to the westward at the rate of thirty miles a day, should sweep the ship so far to leeward, as to prevent her weathering Cape St. Roque, the north-eastern promontory of the Brazilian coast, which would materially protract the passage, by making it necessary to return to the variable winds about the equator in order to regain the easting, as it is almost impossible to make way against the rapid current which sets past Cape St. Roque.

During the forenoon of the 26th, we observed an unusual number of birds. To our companions, the tropic bird, shearwaters, and Mother Carey's chickens, were added gannets and boatswains: they were conjectured to be the forerunners of land; and, at three o'clock, the island of Fernanda Norhona was seen from the deck, bearing southwest, twelve leagues. When we had neared this island within six leagues, there was an irregular sea; but we had no soundings at 351 fathoms' depth. Our observations reduced to the Peak, placed it eighteen miles to the eastward of its position in the East India Directory. Some squally weather, which occasionally broke the ship off her course, increased our anxiety; but we kept clean full, to pass as quickly as possible the current, which here runs with great rapidity.

On the 29th we had the satisfaction to find ourselves to the southward of the promontory, and that it would not be necessary to make a tack. The wind, however, led us in with the coast of Brazil, which was seen on the morning of the 8th. The same evening we passed the shoal off Cape St. Thomas—a danger which until very lately was erroneously placed upon the charts, and not sufficiently marked to warn ships of the peril of approaching it.* Thence, our course was for Cape Frio, a headland which all vessels bound to Rio Janeiro should, on several accounts, endeavour to make. In fine weather the south-east winds blow home to the cape, and gradually fall into either the land or sea breeze, according to the time of day, though the prevailing wind off it is from the north-east: with either of these winds, a ship can proceed to her port. The southerly monsoon, which, while it blows, materially facilitates the navigation along the coast to the

* A merchant-vessel on her way from Rio Janeiro to Bahia, when about ten miles from the land, struck upon this shoal, and beat over it, fortunately with the loss of her rudder only. She afterwards stood for five hours along the shoal, to the eastward, and her master stated that the sea broke upon it out of sight of land.

northward, scarcely affects the wind close in with
the cape. The greatest interruptions to which they
are liable are from the pamperos, which in the winter blow with great violence from the river Plate,
sweep past Rio Janeiro, extend to the before-
mentioned cape, and often beyond it, to a considerable offing. It was during the influence of one of
these gales that we approached Cape Frio, and had
no sooner opened the land on the western side of
the promontory, than we were met by a long rolling
swell from the south-west, gusts of wind, and unsettled weather; and at noon encountered a violent
squall, attended by thunder and lightning, which
obliged us to take in every sail on the instant. Towards sunset the weather cleared up, and we saw
Cape Frio, N. W. by W., very distant.

Calms and baffling winds succeeded this boisterous weather, so that on the morning of the 11th we
were still distant from our port; and the daylight
was gone, and with it the sea-breeze, before we could
reach a place to drop our anchor. It, however, sometimes happens, fortunately for those who are late in
making the entrance of the harbour, that in the interval between the sea and land breezes, gusts blow
off the eastern shore, and ships, by taking advantage
of them, and at the same time by keeping close over
on that side, may succeed in entering the port. This
was our case; and at nine at night we anchored
among the British squadron, under the command of
Rear-Admiral Sir George Eyre, who was the following morning saluted with thirteen guns—a compliment which would have been paid by the ships to
the authorities of the place, had it not been suspend-

ed in consequence of his Imperial Highness requiring certain forms on the occasion, with which his
Britannic Majesty's government did not think it
right to comply.

The ship being in want of caulking, and the rigging of a refit, previous to encountering the boisterous latitude of Cape Horn, these repairs were immediately commenced, and the few stores expended
on the passage were replaced. While these services
were going forward, and observations were in progress for determining the geographical position of
the port, and for other scientific purposes, excursions
were made to the various places of interest in which
Rio Janeiro abounds : — Bota-Fogo, Braganza, the
Falls of Tejuca, and the lofty Corcovado, were successively visited, and afforded amusement to the naturalist, the traveller, and the artist. Few places are
more worthy the description that has been given of
them by various authors, than those above mentioned ; and they have been so frequently described
that they are familiar to every reader, and, as well as
the picturesque scenery of Rio Janeiro itself, are
quite proverbial. Indeed there is little left in the
vicinity of this magnificent port, of which the description will possess the merit of novelty.

The observations which were made during our
stay in Rio Janeiro will be found in the Appendix
to the quarto edition. It may, however, be interesting to insert here the height of the Peak of Corcovado, a singularly shaped mass of granite which over-
looks the placid waters of Bota-Fogo, as the measurements hitherto given are at variance with each other,
and as it is a subject which has caused many discus-
sions among the good people who live in its vicinity.

Our first measurement was with barometers, which, calculated by Mr. Daniel's new formula, gave the base of the flag-staff on the Peak, above half-tide   ·   .   2308 feet.
The next, by trigonometrical measurement, gave   2306

On my return to the same place three years afterwards, I repeated the observations, which gave the height as follows :—

By barometrical measurement   .   .   .   2291½* feet.
By trigonometrical measurement   .   .   ·   2305¼†
The Sugar Loaf by the first base in 1825 was   1286
       by the second base in 1828 was   1299‡

The astronomical observations were made at an observatory erected in Mr. May's garden at Gloria, an indulgence for which I feel particularly indebted to that gentleman, as well as for other civilities which I received from him during my stay at the place.

On the 13th of August we sailed from Rio Janeiro for the Pacific: a passage interesting from the difficulties which sometimes attend it, and from its possessing the peculiarity of producing the greatest change of climate in the shortest space of time. The day after we left the port, we encountered a

---

* This differs sixteen feet from the first result, which may partly be owing to the barometers, on this occasion, not being in such good order as at first: the amount, however, is so small as almost to need no apology, particularly as the observations were made on days as opposite as possible to each other — the first in drizzling rain, the last on a clear sunshining day—whereby the formula was put to the severest trial.

† In this operation I was assisted by the late Captain Henry Forster, R.N. an officer well known to the scientific world, with whom I had the pleasure to become acquainted at this place.

‡ The difference in these measurements is, no doubt, owing to there being no object on the summit of the hill sufficiently defined for the purpose of observation, and it is almost impossible to ascend it.

dangerous thunder-storm, which commenced in the evening, and lasted till after midnight: during this time the sheet lightning was vivid and incessant, and the forked frequently passed between the masts. The wind varied so often, that it was with the greatest difficulty the sails were prevented coming aback ; and it blew so hard that it was necessary to lower the close-reefed topsails on the cap. Shortly after midnight, a vivid flash of lightning left five meteors upon the mast-heads and topsail yard-arms, but did no damage: they were of a bluish cast, burnt about a quarter of an hour, and then disappeared. The weather almost immediately afterwards moderated, and the thunder cloud passed away.

We had afterwards light and variable winds, with which we crept down to the southward, until the night of the 25th, when being nearly abreast of the River Plate, a succession of pamperos* began, and

continued until the 2nd of September, with their usual characteristics, of thunder and lightning, with hail and sunshine between. On the 9th, soundings were obtained in 75 fathoms off the Falkland Islands ; but no land was seen at the time, in consequence of misty weather. We here again experienced a short though heavy gale. As it was against us, we turned our proximity to the land to good account, by seeking shelter under its lee, strik-

* These are heavy gusts of wind which blow off the heated plains (or pampas) lying between the foot of the Cordillera Mountains and the sea. In the River Plate, and near the coast, they are very violent and dangerous, from the sudden manner in which they occur. Their force diminishes as the distance from the coast increases.

ing soundings upon a sandy bottom, from 50 to 80 fathoms, the depth increasing with the distance from the coast.   The weather moderated on the day following, and we saw the land, from S. 25° W. to S. 56° W., eight or nine leagues distant : the wind, at the same time, became favourable, and carried us past the Islands during the night.   The eastern point of these Islands (Cape St. Vincent), by such observations as we were able to make, appears to be correctly placed in the charts.   The position I have assigned to it will be seen in the table at the end of the work.

From the Falkland Islands we stood to the southward; and after two short gales from the westward, made Cape Horn on the 16th, bearing N. 40° W. six or seven leagues.   This was quite an unexpected event, as a course had been shaped the day before to pass it at a distance of seventy miles.   It appeared, however, by the noon observation, that a current had drifted the ship fifty miles to the northward in the twenty-four hours, a circumstance which might have been attended with very serious consequences had the weather been thick; and ships in passing the Strait le Maire will do well to be on their guard against a like occurrence.*   The view of this celebrated promontory, which has cost navigators, from the earliest period of its discovery to the present time, so much difficulty to double, was highly gratifying to all on board, and especially so to those who had never seen it before; yet it was a pleasure we would all willingly have exchanged

* For remarks on the currents, and observations on the winds, in the vicinity of Cape Horn, the reader is referred to the Nautical Remarks in the quarto edition.

for the advantage of being able to pursue an uninterrupted course along the shore of Tierra del Fuego, which the flattering prospect of the preceding
day led us to expect, and which, had it not been for
the northerly current, would have been effected with
ease. The disappointment was of course very great,
particularly as the wind at the moment was more
favourable for rounding the cape than it usually is.

In the evening, the Islands of Diego Ramirez
were seen on the weather bow; and nothing remained but to pursue the inner route, at the risk of
being caught upon a lee-shore with a gale of wind,
or stand back to the south-eastward, and lose in one
day what it would require perhaps a week to recover. We adopted the former alternative, and
passed the Islands as close as it was prudent in a
dark night, striking soundings in deep water upon
an uneven bottom.

The next morning, the small groupe of Ildefonzo
Islands was distant six miles on the lee-beam, and
the mainland of Tierra del Fuego appeared behind
it, in lofty ranges of mountains streaked with snow.
The cape mistaken for Cape Horn by Lord Anson
bore N. 49° E., and the promontory designated York
Minster by Captain Cook, W. by N. The coast was
bold, rocky, and much broken, and every here and
there deeply indented, as if purposely to afford a refuge from the pitiless gales which occasionally beat
upon it. The general appearance of the landscape
was any thing but exhilarating to persons recently
removed from the delightful scenery of Rio Janeiro;
and we were particularly struck with the contrast
between the romantic and luxurious scenery of that
place and the bleak coast before us, where the snow,

filling the valleys and fissures, gave the barren projections a darker hue and a more rugged outline than they in reality possessed.

As we drew in with the land, the water became discoloured, and specifically lighter than that in the offing, whence it was concluded that some rivers emptied themselves into the sea in the vicinity. In the evening it became necessary to stand off the coast; and we experienced the disadvantages of the offing, by getting into the stream of the easterly current, and by the increase of both wind and sea.* We stood to the westward again as soon as it could be done; and on the 26th were fifty leagues due west of Cape Pillar, a situation from which there is no difficulty in making the remainder of the passage.

We now, for a time at least, bade adieu to the shores of Tierra del Fuego, whose coast and climate we quitted with far more favourable impressions than those under which they were approached. This, I think, will be the case with every man-of-war that passes it, excepting the few that may be particularly unfortunate in their weather; for early navigation has stamped it with a character which will ever be coupled with its name, notwithstanding its terrors are gradually disappearing before the progressive improvement in navigation. It must be admitted we were much favoured: few persons, probably, who effect the passage, will have it in

---

* It is a curious fact, that on this day, at a distance of only fifty leagues from where we were, it blew a strong gale of wind, with a high sea, which washed away the bulwark of a fine brig, the Hellespont, commanded by Lieutenant Charles Parker, R. N., to whom I am indebted for this and other interesting information on the winds and currents encountered by him in his passage.

their power to say they were only a week from the
meridian of Cape Horn to a station fifty leagues due
west of Cape Pillar, and that during that time there
was more reason to complain of light winds and
calms, than the heavy gales which proverbially visit
these shores.

Navigators distinguish the passages round Cape
Horn by the *outer* and *inner*; some recommending
one, some the other; and doubtless both have their
advantages and disadvantages. It would be very
uninteresting here to discuss the merits of either, as
the question has been sufficiently considered elsewhere; and it would, in my opinion, be equally
useless, as very few persons follow the advice of
their predecessors in a matter of this nature, but
pursue that course which from circumstances may
seem most advantageous at the moment; and this
will ever be the case where such a difference of
opinion exists.   What I had to say on this subject
has been published in the Nautical Remarks to the
quarto edition.

In describing the passage round Cape Horn, I
have omitted to mention some particulars on the
days on which they occurred, in order that they
might not interrupt the narrative.   As we approached the Falkland Islands from Rio Janeiro,
some penguins were seen upon the water in latitude
47° S., at a distance of three hundred and forty
miles from the nearest land; a fact which either
proves the common opinion, that this species never
stray far from land, to be in error, or that some unknown land exists in the vicinity.   As their situation was not far from the parallel in which the long-
sought Ile Grande of La Roche was said to have

been seen, those who are wedded to the common
opinion above alluded to, may yet fancy such an
island has existence; although it is highly improbable that it should have escaped the observation,
not only of those who purposely went in search of
it, but of the numerous ships also which have of late
made the passage from the Atlantic to the Pacific.
Another opinion, not quite so general, (but which I
have heard repeatedly expressed with reference to
the coast of California), is, that of aquatic birds confining their flight within certain limits, so that a
person who has paid attention to the subject will
know by the birds that are about him, without seeing the land, what part of the coast he is off. My
own experience does not enable me to offer any
remarks on the subject, except in the instance of the
St. Lawrence Islands, in Beering's Strait, the vicinity to which is always indicated by the Crested
Auk *(alca crestatella)*. But the following fact may
be serviceable in adding weight to the opinion, provided it were not accidental; and if so, it may still
be useful in calling the attention of others to the
subject. Off the River Plate, we fell in with the
dusky albatross *(diomedia fulginosa)*, and as we proceeded southward, they became very numerous; but
on reaching the latitude of 51° S. they all quitted
us. We rounded the cape; and on regaining the
same parallel of 51° S. on the opposite side, they
again came round us, and accompanied the ship up
the Chili coast. The pintados were our constant
attendants the whole way.

From the time of our leaving England, the temperature of the surface of the sea had been registered
every two hours. Off Cape Horn, I caused it to be

tried every hour, under an impression that it might
apprise us of our approach to floating ice, when,
from the darkness of the night, or foggy weather,
it could not be seen ; a plan I would recommend
being adopted, as it may be useful, notwithstanding
its fallibility ; for though ice in detached masses,
when drifting fast with the wind, extends its influence
a very short way in the direction of its course : yet
on the other hand, its effect may be felt a considerable distance in its wake. We had only one warning of this nature, by a decrease of temperature of
four degrees, which lasted about an hour. The
temperature of the sea, at the greatest depth our
lines would reach, was not below 39°, 2. Off the
Falkland Islands, it was the same at 854 fathoms
as at 603 fathoms. The lowest temperature of the
air was 26°. The current, which at a distance from
the land runs fast to the eastward to the discomfiture of ships bound in the opposite direction, near
the coast to the westward of Cape Horn, at first
entirely ceased, and afterwards took a contrary
course. There is much reason to believe that it
continues this north-westerly course, and ultimately
falls into the northerly current so prevalent along
the coast of Chili.

The wind was now favourable for making progress to the northward. My instructions did not
direct me to proceed to any port on the coast of
Chili, but circumstances rendered it necessary to
put into one of them, and I selected Conception as
being the most desirable for our purpose.

The weather had for a long time been cloudy ;
but on this night a clear sky presented to our view
a comet of unusual magnitude and brilliancy, situ-

ated to the S. E. of the square formed by επσρ Ceti.
The head had a bluish cast, and increased in lustre
towards its nucleus, where indeed it was so bright,
that with our small telescopes it appeared to be a
star ; but this was evidently a deception, as Mr.
Herschell, who made some interesting and satisfactory observations on the same comet, found on turning his twenty feet reflector upon it, that the star-
like appearance of the nucleus was only an illusion.*
The tail extended between 9° and 10° of arc in a
N. W. direction, and gradually increased in width
from the nucleus till near its termination.   We
made a number of measurements to ascertain its
place, and continued them every night afterwards
on which the comet appeared ; but as its orbit has
been calculated from far more accurate observations,
and ours were necessarily made with stars unequally
affected by refraction, which involves a laborious
reduction, besides the abstruse calculation for determining its orbit, I have not given them a place.

On the following night we noticed distinctly the
bifurcation of the tail represented in the Memoirs
of the Astronomical Society. The branches were
of unequal length, and the lower one diverged from
the nucleus, at an angle of about 40°.

On the 6th we made the island of Mocha, on the
coast of Chili, a place once celebrated as a resort of the
Buccaneers, who anchored off it for the useful supplies
which in their days it furnished. Its condition was
then certainly very different from the present : several Indian chiefs and a numerous population resided
there, and it was well stocked with cattle, sheep,

* See Memoir Ast. Soc. vol. ii. p. 2.

hogs, and poultry. At present it is entirely deserted, except by horses and hogs, both of which,
Captain Hall states, are used as fresh stock by
whaling ships in the Pacific. The Indians appear
to have been generally very cordial with their visiters, exchanging the produce of the island for
cutlery and trinkets. They, however, apparently
without provocation, attacked Sir Francis Drake,
and wounded him and all his boat's crew. In 1690
the island was found deserted by Captain Strong,
and it has since remained uninhabited. The cause
of this is not known, though I was informed in
Chili, that it was in consequence of the frequent
depredations committed by vessels that touched at
the island.

We quitted Mocha, passed the Island of St.
Mary, which must not be approached on account
of sunken rocks, and anchored at Talcahuana, the
sea-port of Conception, on the 8th, fifty-six days
from Rio Janeiro. Here we found the British
squadron, under the command of Captain Maling,
from whom I received every assistance and attention. Our arrival off the port was on one of those
bright days of sunshine which characterize the
summer of the temperate zone on the western side
of America. The cliffs of Quiriquina, an island situated in the entrance of the harbour, were covered
with birds, curiously arranged in rows along the various strata; and on the rocks were numberless seals
basking in the sun, either making the shores re-echo
with their discordant noise, or so unmindful of all
that was passing, as to allow the birds to alight
upon them and peck their oily skin without offering any resistance.

The sea-port of Conception is a deep, commodious bay, well protected from northerly winds by the fertile little island above-mentioned, lying at its entrance: there is a passage on either side of it, but the eastern is the only one in use, the other being very narrow and intricate. The land on the eastern and western sides of the bay is high, well wooded, and on the latter very steep; on the former it slopes from the mountains toward the sea with gentle undulations. Several villages are situated along the shore on both sides, but principally on the eastern. Around these hamlets, some diminutive patches of a more lively green than the surrounding country, show the very limited extent to which cultivation is carried; of which we had further proof as we proceeded up the bay, by witnessing groups of both sexes up to their middle in the sea, collecting their daily subsistence from beds of choros and other shell-fish.

Talcahuana we found to be a miserable little town, extending along the beach, and up a once fertile valley; divided into streets and squares, but much dilapidated, dusty, and in some places overgrown with grass. A thousand inmates occupied these wretched dwellings, who acknowledged the supremacy of a governor, poor, but independant; and intrusted their spiritual concerns to the care of a patriot priest. In the principal square stood a church, in character with the rest of the buildings; and in front of it a belfry, which for some time past must have endangered the life of the bellman. His occupation, however, was less laborious than in other catholic countries, as it was here called into action but once in seven days; and was then at-

tended to only by the female part of the inhabitants.

It was painful to compare the present circumstances of this place with the prosperity that once
prevailed, and impossible to look upon the unhappy
inhabitants without feelings of pity at the state to
which they were reduced. The other villages in the
bay were in a very similar condition; and one,
Tombé, where there was formerly an extensive salt-
petre manufactory, was entirely deserted.

The day after my arrival, I accompanied the captains of the squadron, and Mr. Nugent the consul-
general, to Conception, pursuant to an invitation we
received from the Intendente to visit that city. Its
distance from Talcahuana is about three leagues.
The road, at first, leads over a steep hill to the eastward of the town, the summit of which commands
an excellent view of the natural advantages of defence which the peninsula of Talcahuana possesses,
and shows how formidable it might become under
judicious management. The royalists were not ignorant of this, and during the turbulent times of
emancipation, sought shelter amongst them, cut
ditches, and threw up temporary works of defence,
all of which are now nearly effaced by the heavy
rains that visit this country at particular periods of
the year. At the back of this range of hills, the
country is flat and occasionally swampy, and continues so, with very little interruption, to the Collé
de Chepé, a small eminence, whence a stranger obtains the first view of the river Bio Bio and the
city. The intendente met us about a mile outside
the town, and accompanied us to his residence,
where we experienced a most cordial and hospitable
reception.

Conception, during its prosperity, has been described by the able pens of Juan de Ulloa, La Perouse, and others; and since its misfortunes, by a well-known naval author, who has admirably pictured the ruin and desolation which the city at that time must have presented. Much of his description would have correctly applied to the time of our visit; but, generally speaking, there was a decided improvement in every department. The panic occasioned by the daring associates of the outlaw Benavides, Peneleo and Pinchero, was beginning to subside. These chiefs, unable to make head against the people when united, had of late confined their depredations to the immediate vicinity of their strong holds among the mountains: the peasants had returned to the cultivation of the soil; looms were active in various parts of the town; and dilapidations were gradually disappearing before cumbrous brickwork and masonry. Commerce was consequently beginning to revive; there were several merchant-vessels in the port; and the Quadra, once "silent as the dead," now resounded with the voices of muleteers conducting the exports and imports of the country.

The tranquil and improving condition of the state was further evinced by the equipment of an expedition against the island of Chiloe, which still maintained its allegiance to the mother country. The preparations appeared to give general satisfaction in Conception, and recruits were daily inlisting, and training in the Presidio. I peeped through the gate one morning, and saw these tyros in arms going through the ordeal of the awkward squad. They were half Indians, without shoes or stockings, and with heads like mushrooms. Their appearance, however, was immaterial: they were the troops on which the people placed their dependence, which the result of the expedition did not disappoint; and the effect upon their minds was equally exhilarating. Hitherto obliged to act on the defensive against a few piratical Indian chiefs, they now found themselves lending their troops to carry on a warfare in a distant province. Such was the prosperous state of affairs at the time of our arrival; and the highest expectations pervaded all classes of society.

The town of Conception occupies nearly a square mile of ground. It is situated on the north side of the river Bio Bio, and is distant from it about a quarter of a mile. Its site was chosen in 1763, about twelve years after the old city of Penco was destroyed by an earthquake, or rather by an inundation, occasioned by a tremendous reaction of the sea. Such a catastrophe, it might be supposed, would be sufficient to deter the inhabitants from again building on low ground; nevertheless, the present city is erected on a spot scarcely more elevated than the other, and the river, when high, washes the threshold of the nearest houses. It has no defences; and is also very badly situated in this respect, being commanded by a range of hills close behind it. Benavides was fully aware of this, and constructed a battery upon the eminence, which still bears his name: but the guns are spiked, and the fort is in ruins.

During the late incursions, we were told, that the mode of repelling an attack was to collect the inhabitants in the squares, and barricade the streets leading out of them, with whatever came first to hand: the musketry and the muzzles of the field-pieces were then thrust through these temporary bulwarks, and a fire opened upon the assailants. This was a sufficiently secure defence against the Indians, but it is easy to imagine what would have been the effect of a few well-placed cannon upon a crowd of persons so collected.

In the selection of the site of the new city, the advantage of the river Bio Bio was, no doubt, the great consideration ; and when inland navigation is as well understood in that country as in some others, it will be of the greatest importance, though its numerous shoals must occasion serious difficulties. Part of the produce of the interior is now brought down upon rafts, which, not being able to return, are broken up and sold for timber. There is a ferry-boat over the river for the accommodation of persons who wish to pass from Conception to the Indian country, and sufficiently large to carry cattle or horses. The natives cross in punts, but have so much difficulty in stemming the current and avoiding banks and shallows, that, though the extreme distance is only a mile, they are sometimes an hour and a half performing the passage. Although the Spaniards nominally possessed territory far to the southward of this river, yet it in reality formed their boundary, and until very lately it was unsafe for an European to venture far upon that side, on account of straggling parties of the Indians.* The mouth of the Bio Bio is circumscribed by banks, which have progressively risen, to 210 yards; and even this

* I have been informed that since this period (1825), the Intendente has a magnificent estate on that side of the river, that the Indians are quiet, and that Conception has undergone great improvement.

narrow stream is divided by a rock one-third of the way across it. If the plan of the entrance be correct in the chart annexed to La Perouse's Voyage, the formation of these banks has been very rapid, and has altered the channel of the river.

The population of Conception is about 6500 persons. The inhabitants, the labouring class at least, have a particularly healthy look. The men have hard features and strong sinewy limbs, and the women and children are fatter than would be agreeable to most persons: short stature, dark hair and eyes, and pretty Indian features, are the characteristics of their persons. They are subject to but few diseases; and for these they have their own remedies, consisting principally of medicinal herbs, with which the country abounds, and in the preparation of which they are well skilled. Fevers, occasioned by cold and dampness, are the most common complaints.

In the streets of Conception I did not see a single cripple, a very rare circumstance in Spanish towns; nor were we molested by beggars, beyond a few troublesome boys beseeching alms; and this arose more from impudence, and a determination to try their luck, than from any real necessity: in secret, however, there are not wanting persons who, if opportunity offered, would not only solicit charity, but enforce their demand with a pistol or a stiletto. On meeting the Indians in an unfrequented part of the country, it is particularly necessary to be upon your guard; for these half-civilized barbarians are generally intoxicated, and care very little about insulting or maltreating strangers even in the heart of the town, much less when alone in the country. A regiment of Araucaneans is embodied in the army of the state, and quartered in the town: they retain their own weapons, and continue their own tactics. A specimen of their extraordinary and barbarous warfare was exhibited at Conception during our stay.

Since the trade of Chili has been thrown open, a remarkable change in the costume of the inhabitants, and also in the furniture of their houses, has taken place; and an Englishman may now see with pride the inferior manufactures of his own country prized, to the exclusion of the costly gold and silver tissue stuffs of Spain, which, Perouse observes, were entailed in families like diamonds, and descended from the great-grandmother to the children of the third and fourth generation. Even the national musical instrument, the guitar, has fallen into neglect, and has been supplanted by the English piano-forte. It would have been better for the lower orders of society, of which a large portion of the population of Conception consists, if the use of this simple instrument had been retained; for it is well known, in foreign countries, how many hours of innocent mirth are beguiled in the happy circles it assembles around the cottage doors; and how many idle characters its fascination deters from indulging in less innocent occupations, to which the Chilians are equally prone with other nations; though I am by no means an advocate for its being prized to the extent it once was by the Portuguese, who, after a battle in which they were defeated, left 14,000 guitars upon the field.*

The entertainments most frequented in Concep-

* Mengiana, tom. i.

tion are cock-fighting and billiards. All classes of
society assemble at the pit, and if there be no fight,
will light their cigars, and chat whole hours away,
in the hope of a match being made up, and are dispersed only by the approach of night. The English
cocks are most esteemed, and are sometimes valued
at a hundred dollars a-piece (twenty guineas). The
Chilian spurs cut as well as thrust, and greatly
shorten the cruel exhibition. Some of the gover-
nors are said to have imposed a tax on these esta-
blishments for their own private advantage, but
without the authority of the laws.

Of the country round Conception I have little to
say, except that it has undergone a great change
since the days of its prosperity. In the parallel of
37° on the western side of a great continent, a luxuriant soil may be expected to produce an abundant
vegetation. This district has, in consequence, been
famous for its grain, vines, fruits, esculent roots, &c.,
and for its pasture lands, on which formerly were
reared immense herds of cattle, and horses of the
finest breed. But the effects of the disturbed state
of the country are as manifest here as in the different parts of the city. At present, as much arable
land as is absolutely necessary for the support of the
inhabitants is cleared, and no more; and even its
produce is but scantily enjoyed by the lower classes
on the coast, who are obliged to subsist almost entirely upon shell-fish. The soil, if attended to, will
give an abundant return: wheat, barley, Indian corn,
beans, pease, potatoes, and arrow-root; grapes, apples, pears, currants, strawberries, and olives, are the
common produce of the country. From the latter
a fine oil is extracted; but the fruit is too rank to

be eaten at table, except by the natives. The arrow-root is of a good quality, and very cheap. In the ravines and moist places, the panque *(gunnera scabra)* grows luxuriantly and strong: it is a very useful root, and serves for several purposes ; a pleasant and cooling drink is extracted from it, which is deemed beneficial in feverish complaints ; its root furnishes a liquid serviceable in tanning, and superior to any of the barks of South America ; when made into tarts, it is scarcely inferior to the rhubarb, for which it is sometimes mistaken ; and it is eaten in strips after dinner, with cheese and wine, &c. Several European shrubs and herbaceous plants grow here, but more luxuriantly than in our own country ; among these were hemlock, flax, chickweed, pimpernel, water-cresses, and a species of elder.

The wines which were formerly so much esteemed, and carried along the coast to the northward, are now greatly deteriorated, and in the sea-port much adulterated. There is a great variety of them, and in general they are very intoxicating. The only palatable kind I tasted was made from the vines on the estate of General Friere, and for which I was indebted to the liberality of the governor, as there was none to be purchased. This wine, though agreeable to the English palate, is not in such estimation with the Chilians as one that has a strong empyreumatic flavour. It acquires this in the process of heating, or rather of boiling, the fruit, which is done with a view to extract a larger proportion of the juice than could be obtained by the ordinary means, and to produce a mellowness which age only could otherwise give. Cici and mattee are still in use, though less so than formerly ; and indeed it appeared to me that the Chilians were fast getting rid of all their old customs, of which the drinking of mattee is one.

After passing a very pleasant time in the society of the Intendente, we took our leave, and returned to the port. Our occupations there were divided between astronomical observations, making a survey of the bays of Conception and St. Vincent, and equipping the ship for sea.

I had some hesitation in procuring coal for our sea stock of fuel; not that the article was become scarce, but on account of the enormous price to which the owner thought proper on this occasion to raise it. Captain Hall states, that when he was at this place, the Penco coal, which was the best, was sold for twelve shillings a ton, all expenses included; but the same quantity was now valued at nine dollars, besides the labour of digging and carrying. This arose from a report that some mines which had been recently discovered were about to be worked, which would occasion a great and permanent demand for the material. The coal is of a very inferior quality, and fit only for the forge. Hitherto, however, experiment has been made only upon that which is near the surface: when the mines are worked, if they ever be, a better quality, in all probability, will be obtained. Talcahuana and Penco are, I believe, the only places where coal has yet been discovered near Conception. Were this article of a good quality and reasonable, there would be a great demand for it at Valparaiso, and among the several squadrons upon the station; and it would probably be well worth the experiment of the owner to search a little deeper in the earth, and ascertain the nature of the lower strata. These veins occur in red sandstone formation, and do not appear at the surface to be very extensive, or to promise any very large supply of fuel. This observation applies only to that part of the coast which lies in the vicinity of Conception and the port, a large proportion of which is composed of diluvial depositions.

We are informed by a visiter to this country, that limestone is found at Conception, and is used by the inhabitants for whitewashing their houses; but this is evidently a mistake, as the natives collect shells, and calcine them for that purpose; besides, in no part of the bay or vicinity of Conception could we perceive limestone, or even hear of its existence. A gentleman pointed out a place to the northward of Tomé Bay, where, he said, it occurred; but, on examination, only clay-slate, chert, and greenstone were found.

As the geology of Conception will appear in another place, I shall merely observe here, that in the secondary sand-stone a variety of petrifactions occur, of wood, shells, and bones, formed by an infiltration of siliceous and calcareous matter. The little island of Quiriquina presents alternate horizontal strata of pebbles, sand-stone, and petrified substances, principally of wood, and vertebral and other bones of the whale. On the opposite shore a fossil nautilus was found, which measured three feet in diameter. Upon the beach, in several parts of the bay, there are ridges of magnetic iron-sand which the waves have thrown up: they are seen adhering together, apparently by mutual attraction.

The abundance of shell-fish in Conception entices a great many birds within the bay. The shore is occasionally thronged with them and the shags sometimes fly in an unbroken line of two miles and more in length. The quebranta huessos, the black-backed gull, a species of tern, and two or three species of pelican, one of which pursues its food in a very entertaining manner. It first soars to a great height, and then suddenly darts into the sea, splashing the water in all directions: in a few seconds, it emerges and resumes its lofty flight until again attracted by its prey, when it plunges into the sea as before; and thus the flock, for these birds are gregarious, ranges over the whole bay, performing all its motions in concert and with a surprising rapidity. The penguin is also here, and a very large species of duck, the female of which has a callosity on the shoulder of each wing, and is very excellent eating; a species of colymbus with lobed toes; curlews, sea-pies, horned plovers, a beautiful species of chaverey, with iridescent plumage; the oyster-eater, or razor-bill, and sanderlings; turkey buzzards, the condor, several species of hawks, owls, black-birds, and wood-pigeons, the latter of which are very large and good to eat; a very beautiful species of duck, frequenting the marshes and lakes between Talcahuana and Conception; partridges, a species of woodpecker, a dark-brown fringilla, with a beautiful scarlet breast, a species of loxia, turdus, hirundo, ampelis, not remarkable for their plumage, and numerous flocks of green parrots, which the Chilian Spaniard, who eats almost every kind of bird, has no objection to place upon his table. The domesticated fowls are the same here as in Europe. The reptiles are few, and not venomous: small lizards are extremely common on the rocks, and among the trees. There are one or two species of snakes; a large one resem-

bling the common English adder is frequent, and a
small green snake was caught by one of the officers.

The fish are not very numerous, only coming into
the bay with a particular wind. The number of
whales which guard the entrance, and the shoals of
seals, grampuses, and porpoises, which crowd the
bay, must destroy a great many. Shell-fish are an
exception to this scarcity, and being very large,
form no small portion of the food of those inhabitants who live on the borders of the bay. Besides
the choros, a large muscle, and locas (*concho lepus*),
mentioned by Ulloa, there are several other small
species which are more esteemed than the large
choros, a number of razor, and some venus-shells.
Large sea-eggs are highly prized, and, like the
others, eaten raw. The smaller shell-fish are, various
sorts of limpets, turbos, neritas, murex, and some
others : there are also a great many crabs.

In the survey of the Bay of Conception a shoal
was discovered by Lieutenant Belcher on the Penco
side, which is probably that upon which a vessel
struck some time previous, but which the boats of
the squadron could not afterwards find. It was
necessary to make some alteration in the position of
the Belen bank, from the manner in which it is laid
down in the Spanish charts, and the shoal said to
occur off the sandy point of Quiriquina does not in
fact exist. The western entrance was thoroughly
examined, and found to be quite safe, though very
narrow, and should only be used in all cases of
difficulty in weathering Paxaros Ninos, with a northerly wind. The bay of St. Vincent does not appear
to me to afford security to any vessel of more than
a hundred tons with a strong westerly wind ; and I

would advise no large ship to put in there under such circumstances, if she could possibly avoid it. Further information on the subject will be found among the Nautical Remarks.

Conception, as a place of refreshment, in every way answered our expectations : fresh beef, poultry, good water, vegetables, and wood are to be had : they happened to be dear at the time of our visit, but no doubt, if the country remains tranquil, they will be both cheap and more abundant.

On the 20th our operations were completed ; but a strong northerly wind prevented our putting to sea, and we anchored under the little island of Quiriquina. This is a very secure stopping place, and, in the winter season, a better anchorage to refit a ship at, than that off Talcahuana. It is small, and a ship must lie very close to the shore. After two days of contrary wind, we put to sea on the 24th, and three days afterwards anchored at Valparaiso, in the hope of receiving some supplies which we could not procure at Conception ; but being disappointed in their arrival, on the 29th we weighed, and took our final leave of the coast.

CHAPTER II.

Leave the Coast of Chili—Visit Sala-y-Gomez—Easter Island—
Hostile Reception there—Description of the Inhabitants, Island,
&c.—Enter the dangerous Archipelago—Davis' Island—Elizabeth or Henderson's Island, its singular Formation—See Pitcairn Island.

ON leaving Valparaiso, my intention was, if possible, to pass within sight of Juan Fernandez, in
order to determine its position; but finding the wind
would not allow us to approach sufficiently near
even to see it, we kept away for the island of Sala-y-
Gomez, and with the view of making this part of
the voyage useful, the ship's course was directed
between the tracks of Vancouver and Malespina on
the south side, and many other navigators on the
north, who, engaged in pursuits similar to our own,
had run down the parallels of 27° and 28° S. in
search of the land discovered by Davis. These parallels, during the summer months, are subject to
light and variable airs; and we, in consequence,
made very slow progress, particularly as we approached the meridian of the island, where it became necessary to adopt the precaution of lying-to
every night, that the object of our search might not
be passed unobserved.

CHAP.
II.

Nov
1825.

D 2

When the nights were clear, we continued our observations on the comet. On the 30th the coma had increased to the enormous length of 24°; the nucleus was larger and more brilliant than before; and the ray, before mentioned as forming part of the coma, was more distinct, and apparently at a greater angle with it than when first seen.

The day after we quitted the coast of Chili, all the birds left us; even the pintados, which had been our constant attendants for upwards of 5000 miles, deserted us on this occasion. We afterwards saw very little on the wing, I believe nothing, except a wandering albatross, until we approached the island of Sala-y-Gomez.

In the Pacific, in particular, the navigator should not be inattentive to the presence or absence of birds, as they will generally be found in the vicinity of islands, and especially of such as are uninhabited and of coral formation. On the 14th, several tropic birds, boatswains, and gannets, flew round the ship, and were hailed as an omen which did not deceive us, for at daylight, on the following morning, the island of Sala-y-Gomez was seen from the masthead, bearing N. N. W., fifteen miles distant.

We shortly closed with this isolated spot, and found its extent much less than has been stated. It is, indeed, scarcely more than a heap of rugged stones, which the elements appear to have thrown together, and in a gale of wind would not be distinguished amidst the spray. The rocks, except such parts as have been selected for roosting places by the sea-gulls, are of a dark-brown colour. Upon a small flat spot there was a moss-like vegetation, and near it a few logs of wood, or planks, which the imagination might convert into the remains of some miserable vessel whose timbers had there found a resting-place. Though several vessels have been missing in these seas, we have no intelligence of any having been wrecked here. Sala-y-Gomez, when he discovered the island, imagined he found the frame of a vessel upon it, and in all probability the wood which we saw was the same; but whether it was so or not, our curiosity and desire to land were fully awakened, though we were disappointed by the high breakers which rolled over every part of the shore.

We remained some time under the lee of the island, narrowly scrutinizing it with our telescopes, but without adding to our information. During this time the ship was surrounded by sharks and bonitos, but none were taken, nor were our fishermen more fortunate at the bottom. The feathery tribe,* disturbed from their roost, came fearlessly around us; we shot several, and in the stomach of a pelican a volcanic pebble was found, which some of us conjectured to have been gathered upon the island, and thence inferred its particular formation.

Sala-y-Gomez, when first seen, has the appearance of three rocks: its direction is N.W. and S.E.; and it is something less than half a mile in length, and a fifth of a mile in width. Some sunken rocks lie off the N.E. and S.E. points: in other directions the island may be approached within a quarter of a mile. N. 50° W. $\frac{3}{4}$ mile there are soundings; in 46 fathoms sand and coral; and N. 33° W. $1\frac{1}{2}$ mile, 140 fathoms gray sand. A reference to the geographical table will show the position of the island, and

---

* Phæton ethereus, Pelicanus leucocephalus, sterna stolida, and a small dove-coloured tern.

I shall here only remark, that Captain Kotzebue's latitude is nine miles in error, which perhaps may be a typographical mistake.

From hence we bore away to the westward, with the intention of passing near the situation of an island named Washington and Coffin, reported to have been discovered by an American ship. At sunset we were within four leagues of the spot, with a perfectly clear sky and horizon, but could see nothing of it; nor had we any indication of land in the immediate vicinity, but rather the contrary, as the birds which had followed us from Sala-y-Gomez had quitted the ship some time before. As the night was fine, and the moon gave sufficient light to discover in time any danger that might lie in the route of the ship, the course was continued toward Easter Island, and daylight appeared without any thing being seen. Had such an island been in existence, and answered the description of that upon which Davis was so near losing his vessel, geographers would not have been long in reconciling their opinions on the subject of his discovery; as, in all probability, they would have waived their objection to its distance from Copiapo, in consideration of its identity.

The subject of this supposed discovery has been often discussed; and where the data are so unsatisfactory as to allow one party to choose the Islands of Felix and Ambrose for the land in question, and the other, Easter Island, two places nearly 1600 miles apart, they are not likely to be speedily reconciled, unless two islands exactly answering the description given by Davis, and situated in the proper latitude, shall be found. Such persons as are curi-

ously disposed on this subject will find it ably treated by the late Captain Burney, R. N., in his account
of the Buccaneers.

Without entering into a question which presents
so many difficulties, I shall merely observe, that,
considering the rapid current that exists in the vicinity of the Galapagos, and extends, though with
diminished force, throughout the trade wind, the
error in Davis's reckoning is not more than might
have happened to any dull sailing vessel circumstanced as his was. To substantiate this, I shall
advert to four instances out of many others which
might be named. In a short run from Juan Fernandez to Easter Island, Behrens, who was with
Roggewein, was drifted 318 geographical miles to
the westward of his supposed situation. The Blossom, in passing over the same ground, in the short
space of eighteen days experienced a set of 270
miles ; and on her passage from Acapulco to Valparaiso of 401 miles : and again M. La Perouse, on his
arrival at the Sandwich Islands from Conception,
touching at Easter Island on his way, found a similar error of 300 miles in the course of that passage.
It is fair to presume that the passage of Davis from
the Galapagos to Easter Island was longer than that
of either of the abovementioned vessels ; and consequently it is but reasonable to allow him a greater
error, particularly as the first part of his route was
through a much stronger current. But taking the
error in the Blossom's reckoning as a fair amount,
and applying it to the distance given by Wafer,
there will remain only 204 miles unaccounted for
between it and the real position of Easter Island,
which from the foregoing considerations, added to

the manner in which reckonings were formerly kept, does not appear to me to exceed the limit that might reasonably be ascribed to those causes.

M. La Perouse was of opinion that the islands of Felix and Ambrose were those under discussion, and in order to reconcile their distance from Copiapo with that given by Wafer, he has imputed to him the mistake of a figure in his text, without considering that it would have been next to impossible for Davis to have pursued a direct course from the Galapagos to those islands, (especially at the season in which his voyage was made), but on the contrary that he would be compelled to make a circuit which would have brought him much nearer to Easter Island; and that Davis acquainted Dampier with the situation of his discovery, which agreed with that contained in Wafer's account. The alteration in a figure, it must be admitted, is rather arbitrary, as it has nothing to support it but the circumstance of the number of islands being the same. A mistake certainly might have occurred, but in the admission of it either party may claim it as an advantage by interpreting the presumed error in a way which would support their own opinions.

At four o'clock in the afternoon of the 16th of November, Easter Island was seen from the mast head, bearing N. 78° W. (Mag.) fourteen or sixteen leagues, and we were consequently very nearly in the situation of the long looked for, small, sandy island, which, had it existed within reasonable limits of its supposed place, could not have escaped our observation. Nothing of it however was seen, nor had we any indication of the vicinity of such a spot as we proceeded, though we must have actually passed over the place assigned to it. Easter Island had at first the appearance of being divided into two, rather flat at the top, with rounded capes; the north-eastern of which is distinguished by two hillocks. To avoid over-running the distance, the ship was hove to at night, and at daylight on the following morning we bore up for the northern shore of the island. I preferred that side, as it had been but partially examined by Captain Cook, and not at all by M. La Perouse.

As we approached, we observed numerous small craters rising above the low land, and near the N. E. extremity, one of considerable extent, with a deep chasm in its eastern side. None of these were in action, nor indeed did they appear to have been so for a very long time, as, with the exception of the one above-mentioned, they were covered with verdure. The N. E. promontory, already noticed as having two small hillocks upon it, was composed of horizontal strata, apparently of volcanic origin; and near it, some patches of earth, sloping down to the cliff, were supposed to consist of red scoriae. The hills, and exposed parts of the earth, were overgrown with a short burnt-up grass, which gave the surface a monotonous and arid aspect; but the valleys were well cultivated, and showed that the island required only a due proportion of moisture and labour to produce a luxuriant vegetation.

Passing along the northern shore, we saw several of those extensive habitations which M. La Perouse has described, situated in a valley surrounded by groves of banana trees and other patches of cultivation. The larger huts were placed near the wood, and the smaller ones close together outside them.

Nearer the sea-shore, which here forms a bay, was a morai, surmounted by four images standing upon a long low platform, precisely answering the description and representation of one given by Perouse; and also an immense enclosure of stones, and several large piles, which, as well as the images, were capped with something white, a circumstance noticed both by Captain Cook and M. Perouse.

The greatest attention appeared to be paid to the cultivation of the soil. Such places as were not immediately exposed to the scorching rays of the sun were laid out in oblong strips, taking the direction of the ravines; and furrows were ploughed at right angles to them, for the purpose of intercepting the streams of water in their descent. Near the middle of the small bay just mentioned, there was an extinguished crater, the side of which, fronting the sea, had fallen in. The natives, availing themselves of this natural reservoir for moisture, in which other parts of the island are so deficient, had cultivated the soil in its centre, and reared a grove of banana-trees, which, as we passed, had a very pleasing effect. The natives lighted fires, and followed the ship along the coast, their numbers increasing at every step. Some had a white cloth thrown loosely over their shoulders, but by far the greater number were naked, with the exception of the maro.

When the ship had arrived off the N.W. point of the island, she was hove to for the purpose of taking observations; and a boat was lowered to examine the bays, and obtain soundings near the shore. Immediately she put off, the natives collected about the place where they supposed she would land. The sea broke heavily upon the rocks, and some of them

apprehending the boat would be damaged, waved
their cloaks to caution her against making the attempt to land: while others, eager to reach her,
plunged into the sea, and so surrounded her that she
was obliged to put about to get rid of them. They
all showed a friendly disposition, and we began
to hope that they had forgotten the unpardonable
conduct of the American master, who carried several of the islanders away by force to colonize Masa-
fuera.

Immediately the noon observation was obtained,
we ran along the western side of the island, towards
the bay in which Cook and Perouse had both anchored. The natives, as before, followed along the
coast, and lighted fires in different directions, the
largest of which was opposite the landing-place.
With a view to ascertain the feeling of the inhabitants, and, if possible, to establish an amicable intercourse with them, I desired Lieutenant Peard to
proceed with two boats to the shore, and by presents
and kindness to endeavour to conciliate the people
and to bring off what fruit and vegetables he could.
Lieutenant Wainwright was directed to accompany
him; and though I did not apprehend any hostility,
yet, as a precautionary measure, I armed the boats,
and placed two marines in each. Their strength
was further increased by several of the officers, and
the naturalist. Thus equipped, they rowed for the
landing-place in Cook's Bay, while the ship remained
at a short distance. The islanders were collected in
great numbers, and were seen running to and fro,
exhibiting symptoms of expectation and delight.
Some few, however, were observed throwing large
stones at a mark behind a bank erected near the beach.

As the boats approached, the anxiety of the natives was manifested by shouts, which overpowered
the voices of the officers : and our boats, before they
gained the beach, were surrounded by hundreds of
swimmers, clinging to the gunwale, the stern, and
the rudder, until they became unmanageable. They
all appeared to be friendly disposed, and none came
empty-handed. Bananas, yams, potatoes, sugar-cane,
nets, idols, &c. were offered for sale, and some were
even thrown into the boat, leaving their visiters to
make what return they chose. Among the swimmers there were a great many females, who were
equally or more anxious to get into the boats than
the men, and made use of every persuasion to induce
the crew to admit them. But to have acceded to
their entreaties would have encumbered the party,
and subjected them to depredations. As it was, the
boats were so weighed down by persons clinging to
them, that for personal safety the crew were compelled to have recourse to sticks to keep them off, at
which none of the natives took offence, but regained
their position the instant the attention of the persons
in the boat was called to some other object. Just
within the gunwale there were many small things
which were highly prized by the swimmers ; and
the boats being brought low in the water by the
crowd hanging to them, many of these articles were
stolen, notwithstanding the most vigilant attention
on the part of the crew, who had no means of recovering them, the marauders darting into the water,
and diving the moment they committed a theft.
The women were no less active in these piracies than
the men ; for if they were not the actual plun-
derers, they procured the opportunity for others,

by engrossing the attention of the seamen by their caresses and ludicrous gestures.

In proceeding to the landing-place, the boats had to pass a small isolated rock which rose several feet above the water. As many females as could possibly find room crowded upon this eminence, pressing together so closely, that the rock appeared to be a mass of living beings. Of these Nereids three or four would shoot off at a time into the water, and swim with the expertness of fish to the boats to try their influence on their visiters. One of them, a very young girl, and less accustomed to the water than her companions, was taken upon the shoulders of an elderly man, conjectured to be her father, and was, by him, recommended to the attention of one of the officers, who, in compassion, allowed her a seat in his boat. She was young, and exceedingly pretty ; her features were small and well made, her eyes dark, and her hair black, long, and flowing ; her colour, deep brunette. She was tattooed in arches upon the forehead, and, like the greater part of her countrywomen, from the waist downward to the knee in narrow compact blue lines, which at a short distance had the appearance of breeches. Her only covering was a small triangular maro, made of grass and rushes ; but this diminutive screen not agreeing with her ideas of propriety in the novel situation in which she found herself, she remedied the defect by unceremoniously appropriating to that use a part of one of the officers' apparel, and then commenced a song not altogether inharmonious. Far from being jealous of her situation, she aided all her countrywomen who aspired to the same seat of honour with herself, by dragging them out of the water by the hair of the head; but unkind as it might appear to interfere to prevent this, it was necessary to do so, or the boats would have been filled and unmanageable.

As our party passed, the assemblage of females on the rock commenced a song, similar to that chaunted by the lady in the boat; and accompanied it by extending their arms over their heads, beating their breasts, and performing a variety of gestures, which showed that our visit was acceptable, at least to that part of the community. When the boats were within a wading distance of the shore, they were closely encompassed by the natives; each bringing something in his hand, however small, and almost every one importuning for an equivalent in return. All those in the water were naked, and only here and there, on the shore, a thin cloak of the native cloth was to be seen. Some had their faces painted black, some red; others black and white, or red and white, in the ludicrous manner practised by our clowns; and two demon-like monsters were painted entirely black. It is not easy to imagine the picture that was presented by this motley crowd, unrestrained by any authority or consideration for their visiters, all hallooing to the extent of their lungs, and pressing upon the boats with all sorts of grimaces and gestures.

It was found impossible to land where it was at first intended: the boats, therefore, rowed a little to the northward, followed by the multitude, and there effected a disembarkation, aided by some of the natives, who helped the party over the rocks with one hand, while they picked their pockets with the other. It was no easy matter to penetrate the dense

multitude, and much less practicable to pursue a
thief through the labyrinth of figures that thronged
around. The articles stolen were consequently as
irretrievably lost here, as they were before in the
hands of the divers. It is extremely difficult, on
such occasions, to decide which is the best line of
conduct to adopt : whether to follow Captain Cook's
rigid maxim of never permitting a theft when clearly
ascertained to go unpunished ; or to act as Perouse
did with the inhabitants of Easter Island, and suffer
every thing to be stolen without resistance or remonstrance. Perhaps the happy medium of shutting
the eyes to those it is not necessary to observe, and
punishing severely such as it is imperative to notice,
will prove the wisest policy.

Among the foremost of the crowd were two men,
crowned with pelican's feathers, who, if they were
not chiefs, assumed an authority as such, and with
the two demons above mentioned attempted to clear
the way by striking at the feet of the mob ; careful,
however, so to direct their blows, that they should
not take effect. Without their assistance, it would
have been almost impossible to land : the mob cared
very little for threats : a musket presented at them
had no effect beyond the moment it was levelled,
and was less efficacious than some water thrown
upon the bystanders by those persons who wished
to forward the views of our party.

The gentleman who disembarked first, and from
that circumstance probably was considered a person
of distinction, was escorted to the top of the bank
and seated upon a large block of lava, which was the
prescribed limit to the party's advance. An endeavour was then made to form a ring about him ; but

it was very difficult, on account of the Islanders crowding to the place all in expectation of receiving something. The applicants were impatient, noisy, and urgent: they presented their bags, which they had carefully emptied for the purpose, and signified their desire that they should be filled: they practised every artifice, and stole what they could in the most careless and open manner: some went even farther, and accompanied their demands by threats. About this time one of the natives, probably a chief, with a cloak and head-dress of feathers, was observed from the ship hastening from the huts to the landing-place, attended by several persons with short clubs. This hostile appearance, followed by the blowing of the conch-shell, a sound which Cook observes he never knew to portend good, kept our glasses for a while riveted to the spot. To this chief it is supposed, for it was impossible to distinguish amongst the crowd, Mr. Peard made a handsome present, with which he was very well pleased, and no apprehension of hostilities was entertained. It happened, however, that the presents were expended and this officer was returning to the boat for a fresh supply, when the natives, probably mistaking his intentions, became exceedingly clamorous, and the confusion was further increased by a marine endeavouring to regain his cap, which had been snatched from his head. The natives took advantage of the confusion, and redoubled their endeavours to pilfer, which our party were at last obliged to repel by threats, and sometimes by force. At length they became so audacious that there was no longer any doubt of their intentions, or that a system of open plunder had commenced; which, with the appear-

ance of clubs and sticks, and the departure of the
women, induced Mr. Peard, very judiciously, to
order his party into the boats. This seemed to be
the signal for an assault : the chief who had received
the present threw a large stone, which struck Mr.
Peard forcibly upon the back, and was immediately
followed by a shower of missiles which darkened the
air. The natives in the water and about the boats
instantly withdrew to their comrades, who had run
behind a bank out of the reach of the muskets,
which former experience alone could have taught
them to fear, for none had yet been fired by us.

The stones, each of which weighed about a pound,
fell incredibly thick, and with such precision that
several of the seamen were knocked down under the
thwarts of the boat, and every person was more or
less wounded, except the female to whom Lieutenant Wainwright had given protection, who, as if
aware of the skilfulness of her countrymen, sat un-
concerned upon the gunwale, until one of the officers,
with more consideration for her safety than she herself possessed, pushed her overboard, and she swam
ashore. A blank cartridge was at first fired over the
heads of the crowd; but forbearance, which with
savages is generally mistaken for cowardice or inability, only augmented their fury. The showers of
stones were if possible increased, until the personal
safety of all rendered it necessary to resort to severe
measures. The chief, still urging the islanders on,
very deservedly, and perhaps fortunately, fell a victim to the first shot that was fired in defence. Terrified by this example, the natives kept closer under
their bulwark; and though they continued to throw
stones, and occasioned considerable difficulty in ex-

tricating the boats, their attacks were not so effectual as before, nor sufficient to prevent the embarkation of the crew, all of whom were got on board.

Several dangerous contusions were received in the affair, but fortunately no lives were lost on our part; and it was the opinion of the officer commanding the party, that the treacherous chief was the only victim on that of the islanders, though some of the officers thought they observed another man fall. Considering the manner in which the party were surrounded, and the imminent risk to which they were exposed, it is extraordinary that so few of the natives suffered; and the greatest credit is due to the officers and crews of both boats for their forbearance on the occasion.

After this unfortunate and unexpected termination to our interview, I determined upon quitting the island, as nothing of importance was to be gained by remaining, which could be put in competition with the probable loss of lives that might attend an attempt at reconciliation. The disappointment it occasioned was great to us, who had promised ourselves much novelty and enjoyment; but the loss to the public is trifling, as the island has been very well described by Roggewein, Cook, Perouse, Kotzebue, and others, and the people appeared, in all material points, the same now as these authors have painted them. With regard to supplies, nothing was to be gained by staying; for after Cook had traversed the island, he came to the conclusion that few places afford less convenience for shipping. " As every thing must be raised by dint of labour, it cannot be supposed the inhabitants plant much more than is sufficient for themselves; and as they are few in number, they cannot have much to spare to supply the wants of strangers."

The population of Easter Island has been variously stated : Roggewein declares several thousands surrounded the boats: Cook reckoned it at six or seven hundred; Mr. Forster, who was with him, at nine hundred ; M. la Perouse, at two thousand: my officers estimated it at about fifteen hundred. If a mean of these be taken, it will leave 1260, which is, perhaps, near the truth ; for it may be presumed, that in an island of such limited extent, and which does not increase its productions or personal comforts, and where sexual intercourse is unrestrained, the population will remain much the same.

One of the authors of Roggewein's Voyage represents the inhabitants of this island as giants, which, if his assertion be true, makes it evident that, like the Patagonians, they have degenerated very rapidly. Cook remarks that he did not see a man that would measure 6 feet; and our estimate of the average height of the people was 5 feet $7\frac{1}{2}$ inches. They are a handsome race, the women in particular. The fine oval countenances and regular features of the men, the smooth, high-rounded foreheads, the rather small and somewhat sunken dark eye, and the even rows of ivory-white teeth, impressed us with the similarity of their features to the heads brought from New Zealand. The colour of their skin is lighter than that of the Malays. The general contour of the body is good : the limbs are not remarkable for muscularity, but formed more for activity than strength. The hair is jet black, and worn moderately short. One man of about fifty years of age, the only exception that was noticed, had his hair

over the forehead of a reddish-ash gray. The beards
of such as had any were black; but many had none,
or only a few hairs on the chin. None of the men
had whiskers, which seemed to be rather a subject
of regret with them, and they appeared envious of
such of our party as had them, who were obliged to
submit to the ordeal of having them stroked and
twisted about for the admiration and amusement of
their new acquaintances. Both sexes still retain the
hideous practice of perforating the lobes of the ears,
though the custom is not so general with the men
as formerly. The aperture, when distended, which
is done by a leaf rolled up and forced through it, is
about an inch and a quarter in diameter. The
lobe, deprived of its ear-ring, hangs dangling against
the neck, and has a very disagreeable appearance,
particularly when wet. It is sometimes so long as
to be greatly in the way; to obviate which, they pass
the lobe over the upper part of the ear, or more rarely,
fasten one lobe to the other, at the back of the head.
The lips, when closed, form nearly a line, showing
very little of the fleshy part, and giving a character
of resolution to the countenance. The nose is aquiline and well-proportioned; the eyes small and dark
brown or black; the chin small and rather prominent;
and the tongue disproportionably large, and, on its
upper surface, of a diseased white appearance.

Tattooing or puncturing the skin is here practised
to a greater extent than formerly, especially by the
females, who have stained their skin in imitation of
blue breeches; copied, no doubt, from some of their
visiters, who frequently tuck up their trowsers to
the knee in passing through the water. The deception, which, at a short distance, completely deceives

the eye, is produced by a succession of small blue lines, beginning at the waist and extending downward to the knee. Besides this, some of them tattoo their foreheads in arched lines, as well as the edges of their ears, and the fleshy part of their lips. The males tattoo themselves in curved lines of a dark Berlin blue colour upon the upper part of the throat, beginning at the ear, and sloping round below the under jaw. The face is sometimes nearly covered with lines similar to those on the throat, or with an uninterrupted colouring, excepting two broad stripes on each side, at right angles to each other. Most of their lips were also stained. Others had different parts of their bodies variously marked, but in the greater number it was confined to a small space. All the lines were drawn with much taste, and carried in the direction of the muscle in a manner very similar to the New Zealanders. These people have had so little communication with Europeans, or have benefited so little by it, that we did not perceive any European cloth among them; and the cloth mulberry-tree, which grows upon their island, produces so small a supply, that part of the inhabitants necessarily go naked: the larger portion however wear a maro, made either of fine Indian cloth of a reddish colour, of a wild kind of parsley, or of a species of sea-weed.

Their weapons are short clubs of a flattened oval form, tapering toward the handle, and a little curved. The straw hats mentioned both by Cook and Perouse appeared to be no longer used. One man only had his head covered; and that with a tattered felt hat, which he must have obtained from some former visiters. A ramrod, which had probably been procured in the same way, was also seen among them. We noticed three boats hauled up on the shore to the northward of the landing-place, resembling the drawing in Perouse's Voyage, but the natives did not attempt to launch them.

Roggewein and Perouse were of opinion that these people lived together in communities, a whole village inhabiting one extensive hut, and that property was in common. The former idea was probably suggested by the very capacious dwellings which are scattered over the island; and the conjecture may be correct, though it is certain that there are a far greater number of small huts, sufficient to contain one family only; but with regard to the supposition that property is common, it seems very doubtful whether the land would be so carefully divided by rows of stones if that were the case. Some circumstances which occurred at the landing-place, during our visit, certainly favoured the presumption of its being so. One of the natives offered an image for sale, and being disappointed in the price he expected, refused to part with it; but a bystander, less scrupulous, snatched it from him without ceremony, and parted with it for the original offer without a word of remonstrance from his countryman. Others again threw their property into the boats, without demanding any immediate return; taking for granted, it may be presumed, that they would reap their reward when a distribution of the property obtained should take place. But this state of society is so unnatural that, however appearances may sanction the belief, I am disposed to doubt it. One strong fact in support of my opinion was the unceremonious manner in which the apparent proprietor of a piece of ground planted with potatoes drove away the mob, who, with very little consideration for the owner, were taking the crop out of the earth to barter with our party.

The Island, though situated nearer the Continent of America than any other of the Archipelago to which it belongs, has been less frequently visited; and unfortunately for its inhabitants, some of those visits have rather tended to retard than to advance its prosperity, or improve its moral condition; and they afford a striking example of the necessity of an extensive intercourse with mankind, before a limited community can emerge from barbarism to a state of civilization. One consolation for this privation is their exemption from those complaints by which some of the ill-fated natives of these seas have so dreadfully suffered.

The gigantic busts which excited the surprise of the first visiters to the island, have suffered so much, either from the effects of time, or maltreatment of the natives, that the existence of any of them at present is questionable. At first they were dispersed generally over the whole island: when Cook visited it there were but two on the western side near the landing-place: Kotzebue found only a square pedestal in the same place: and now a few heaps of rubbish only, occupy a spot where it is doubtful whether one of them was erected or not. When it is considered how great must have been the labour bestowed upon these images before they were hewn from the quarries with the rude stone implements of the Indians, and before such huge masses of rock could be transported to, and erected on, so many parts of the island, it is nearly positive that they

were actuated by religious motives in their construction ; and yet, if it were so, why were these objects
of adoration suffered to go to decay by succeeding
generations? Is it that the religious forms of the
islanders have changed, or that the aborigines have
died off, and been succeeded by a new race ? — Pitcairn Island affords a curious example of a race of
men settling upon an island, erecting stone images
upon its heights, and either becoming extinct or
having abandoned it ; and some circumstances connected with Easter Island occur, independent of
that above alluded to, in favour of the presumption
that the same thing may also have taken place there.
The most remarkable of these facts is, that the present generation are so nearly allied in language and
customs to many islands in the South Sea, as to
leave no doubt of their having migrated from some
of them, — and yet in none of these places are there
images of such extraordinary dimensions, or indeed
in any way resembling them. The Easter Islanders
have, besides, small wooden deities similar to those
used by the inhabitants of the other islands just
mentioned.

That there had been recent migrations from some of
the islands to the westward, about Roggewein's time,
may be inferred from the natives having recognised
the animals on board his ship, and from their having
hogs tattooed upon their arms and breasts ; whereas
there was not a quadruped upon the island at the
time, nor has any one except the rat ever been seen
there. Another curious fact connected with this
island is, that when it was first discovered it abounded in woods and forests, and palm branches were
presented as emblems of peace ; but fifty years after-

wards, when visited by Captain Cook, there were no traces of them left. The revolution that has taken place in La Dominica, one of the Marquesa Islands, affords another instance of this kind: when first visited by Mendana, in 1595, it exhibited an enchanting aspect: " vast plains displayed a smiling verdure, and divided hills, crowned with tufted woods," &c.: but in 1774 it was found by Captain Cook to have so completely altered its features, that Marchand ascribes the change to one of those great " convulsions of nature, which totally disfigure every part of the surface of the globe, over which its ravages extend." Easter Island is studded with volcanos, and an eruption may have driven the natives into the sea, or have so torn up the soil and vegetation, that they could no longer subsist upon it.

I cannot say a word on the success that has attended the humane efforts of the much-lamented Perouse, who planted many useful fruit-trees and seeds upon the island; but there is every reason to believe they have perished, or shared the fate of the vines at Otaheite, as they brought us no fruits or roots beyond what he found there on his arrival. Perhaps a tuft of trees in a sheltered spot at the back of Cook's Bay, which had the appearance of orange-trees, are the offspring of his benevolent care and attention. Cook had no opportunity of benefiting the islanders in this way ; but he planted in them a warm and friendly feeling towards strangers, and his usual rectitude and generous treatment taught them a lesson of which Perouse felt the good effects, and which possibly might have existed until now, but for the interference of a few unprincipled masters of vessels, who have unfortunately found their way to the island; and I fear these communications are more frequent than is generally supposed.

The island is 2000 miles from the coast of Chili, and 1500 from the nearest inhabited islands, Pitcairn Island excepted, which has been peopled by Europeans. A curious inquiry therefore suggests itself: in what manner has so small a place, and so distantly situated from any other, received its population? particularly as every thing favours the probability of its inhabitants having migrated from the westward, in opposition to the prevalent wind and current. Captain Cook obtained considerable knowledge upon this subject at Wateo; and I shall hereafter be able to offer something in support of the theory entertained by that celebrated navigator.

Cook and Perouse differ in a very trifling degree from each other, and also from us in the geographical position of Easter Island. The longitude by Cook is 109° 46′ 20″ W., and deducting 18′ 30″, in consequence of certain corrections made at Fetegu Island, leaves 109° 27′ 50″ W. That by Perouse, allowing the longitude of Conception to be 72° 56′ 30″ W., is 109° 32′ 10″ W.; and our own is 109° 24′ 54″ W. The island is of a triangular shape: its length is exactly nine miles from N. W. to S. E., nine and three quarters from W. N. W. to E. S. E., and thirteen from N. E. to S. W. The highest part of it is 1200 feet, and in clear weather it may be seen at sixteen or eighteen leagues distance. The geographical description by M. Bernizet, who was engineer in the Astrolabe, is exact: the views of the land are a little caricatured, but the angular measurements are perfectly correct. Further remarks on the coast and anchorage will be found in the Nautical Memoir.

We quitted Easter Island with a fresh N. E. wind, and bore away for the next island placed upon the chart. On the 19th, during a calm, some experiments were made on the temperature of the water at different depths. As the line was hauling in, a large sword-fish bit at the tin case which contained our thermometer, but, fortunately, he failed in carrying it off. On the 27th, in lat. 25° 36′ S., long. 115° 06′ W., many sea-birds were seen; but there was no other indication of land. From the time of our quitting Easter Island light and variable winds greatly retarded the progress of the ship, until the 24th, in lat. 26° 20′ S., and long. 116° 30′ W., when we got the regular trade-wind, and speedily gained the parallel of Ducie's Island, which it was my intention to pursue, that the island might by no possibility be passed. In the forenoon of the 28th we saw a great many gulls and tern; and at half-past three in the afternoon the island was descried right a-head. We stood on until sunset, and shortened sail within three or four miles to windward of it.

Ducie's Island is of coral formation, of an oval form, with a lagoon or lake, in the centre, which is partly inclosed by trees, and partly by low coral flats scarcely above the water's edge. The height of the soil upon the island is about twelve feet, above which the trees rise fourteen more, making its greatest elevation about twenty-six feet from the level of the sea. The lagoon appears to be deep, and has an entrance into it for a boat, when the water is sufficiently smooth to admit of passing over the bar. It is situated at the south-east extremity, to the right of two eminences that have the appearance of sand-hills. The island lies in a north-east and south-west direction, — is one mile and three quarters long, and one mile wide. No living things, birds excepted, were seen upon the island; but its environs appeared to abound in fish, and sharks were very numerous. The water was so clear over the coral, that the bottom was distinctly seen when no soundings could be had with thirty fathoms of line; in twenty-four fathoms, the shape of the rocks at the bottom was clearly distinguished. The coral-lines were of various colours, principally white, sulphur, and lilac, and formed into all manner of shapes, giving a lively and variegated appearance to the bottom; but they soon lost their colour after being detached.

By the soundings round this little island it appeared, for a certain distance, to take the shape of a truncated cone having its base downwards. The north-eastern and south-western extremities are furnished with points which project under water with less inclination than the sides of the island, and break the sea before it can reach the barrier to the little lagoon formed within. It is singular that these buttresses are opposed to the only two quarters whence their structure has to apprehend danger; that on the north-east, from the constant action of the trade-wind, and that on the other extremity, from the long rolling swell from the south-west, so prevalent in these latitudes; and it is worthy of observation that this barrier, which has the most powerful enemy to oppose, is carried out much farther, and with less abruptness, than the other.

The sand-mounds raised upon the barrier are confined to the eastern and north-western sides of the lagoon, the south-western part being left low, and

CHAP.
II.

Nov.
1825.

broken by a channel of water. On the rocky surface of the causeway, between the lake and the sea,
lies a stratum of dark rounded particles, probably
coral, and above it another, apparently composed of
decayed vegetable substances. A variety of evergreen trees take root in this bank, and form a canopy almost impenetrable to the sun's rays, and present to the eye a grove of the liveliest green.

As soon as we had finished our observations on
Ducie's Island, and completed a plan of it, we made
sail to the westward. The island soon neared the
horizon, and when seven miles distant ceased to be
visible from the deck. For several days afterwards
the winds were so light, that we made but slow
progress; and as we lay-to every night, in order
that nothing might be passed in the dark, our daily
run was trifling. On the 30th, we saw a great number of white tern, which at sun-set directed their
flight to the N. W. At noon on the 2d of December, flocks of gulls and tern indicated the vicinity of
land, which a few hours afterwards was seen from
the mast-head at a considerable distance. At daylight on the 3rd, we closed with its south-western
end, and despatched two boats to make the circuit
of the island, while the ship ranged its northern
shore at a short distance, and waited for them off a
sandy bay at its north-west extremity.

Dec.

We found that the island differed essentially from
all others in its vicinity, and belonged to a peculiar
formation, very few instances of which are in existence. Wateo and Savage Islands, discovered by
Captain Cook, are of this number, and perhaps also
Malden Island, visited by Lord Byron in the Blonde.
The island is five miles in length, and one in breadth,

and has a flat surface nearly eighty feet above the sea. On all sides, except the north, it is bounded by perpendicular cliffs about fifty feet high, composed entirely of dead coral, more or less porous, honeycombed at the surface, and hardening into a compact calcareous substance within, possessing the fracture of secondary limestone, and has a species of millepore interspersed through it. These cliffs are considerably undermined by the action of the waves, and some of them appear on the eve of precipitating their superincumbent weight into the sea; those which are less injured in this way present no alternate ridges or indication of the different levels which the sea might have occupied at different periods, but a smooth surface, as if the island, which there is every probability has been raised by volcanic agency, had been forced up by one great subterraneous convulsion. The dead coral, of which the higher part of the island consists, is nearly circumscribed by ledges of living coral, which project beyond each other at different depths; on the northern side of the island the first of these had an easy slope from the beach to a distance of about fifty yards, when it terminated abruptly about three fathoms under water. The next ledge had a greater descent, and extended to two hundred yards from the beach, with twenty-five fathoms water over it, and there ended as abruptly as the former, a short distance beyond which no bottom could be gained with 200 fathoms of line. Numerous *echini* live upon these ledges, and a variety of richly coloured fish play over their surface, while some cray-fish inhabit the deeper sinuosities. The sea rolls in successive breakers over these ledges of coral, and renders landing upon them extremely difficult. It may, however, be effected by anchoring the boat, and veering her close into the surf, and then, watching the opportunity, by jumping upon the ledge, and hastening to the shore before the succeeding roller approaches. In doing this great caution must be observed, as the reef is full of holes and caverns, and the rugged way is strewed with sea-eggs, which inflict very painful wounds ; and if a person fall into one of these hollows, his life will be greatly endangered by the points of coral catching his clothes and detaining him under water. The beach, which appears at a distance to be composed of a beautiful white sand, is wholly made up of small broken portions of the different species and varieties of coral, intermixed with shells of testaceous and crustaceous animals.

Insignificant as this island is in height, compared with others, it is extremely difficult to gain the summit, in consequence of the thickly interlacing shrubs which grow upon it, and form so dense a covering, that it is impossible to see the cavities in the rock beneath. They are at the same time too fragile to afford any support, and the traveller often sinks into the cavity up to his shoulder before his feet reach the bottom. The soil is a black mould of little depth, wholly formed of decayed vegetable matter, through which points of coral every now and then project.

The largest tree upon the island is the pandanus, though there is another tree very common, nearly of the same size, the wood of which has a great resemblance to common ash, and possesses the same properties. We remarked also a species of budleia, which was nearly as large and as common, bearing fruit. It affords but little wood, and has a reddish bark of considerable astringency : several species of this genus are to be met with among the Society Islands. There is likewise a long slender plant with a stem about an inch in diameter, bearing a beautiful pink flower, of the class and order hexandria monogynia. We saw no esculent roots, and, with the exception of the pandanus, no tree that bore fruit fit to eat.

This island, which on our charts bears the name of Elizabeth, ought properly to be called Henderson's Island, as it was first named by Captain Henderson of the Hercules of Calcutta. Both these vessels visited it, and each supposing it was a new discovery, claimed the merit of it on her arrival the next day at Pitcairn Island, these two places lying close together. But the Hercules preceded the former several months. To neither of these vessels, however, is the discovery of the land in question to be attributed, as it was first seen by the crew of the Essex, an American whaler, who accidentally fell in with it after the loss of their vessel. Two of her seamen, preferring the chance of finding subsistence on this desolate spot to risking their lives in an open boat across the wide expanse which lies between it and the coast of Chili, were, at their own desire, left behind. They were afterwards taken off by an English whaler that heard of their disaster at Valparaiso from their surviving shipmates.*

* The extraordinary fate of the Essex has been recorded in a pamphlet published in New York by the mate of that vessel, but of the veracity of which every person must consult his own judgment. As all my readers may not be in possession of it, I shall briefly state that it describes the Essex to have been in the act of

It appears from their narrative that the island possessed no spring; and that the two men procured a supply of water at a small pool which received the drainings from the upper part of the island, and was just sufficient for their daily consumption.

In the evening we bore away to the westward, and at one o'clock in the afternoon of the 4th of December we saw Pitcairn Island bearing S.W. by W. $\frac{1}{2}$ W. at a considerable distance.

catching whales, when one of these animals became enraged, and attacked the vessel by swimming against it with all its strength. The steersman, it is said, endeavoured to evade the shock by managing the helm, but in vain. The third blow stove in the bows of the ship, and she went down in a very short time, even before some of the boats that were away had time to get on board. Such of the crew as were in the ship contrived to save themselves in the boats that were near, and were soon joined by their astonished shipmates, who could not account for the sudden disappearance of their vessel; but found themselves unprovided with every thing necessary for a sea-voyage, and several thousand miles from any place whence they could hope for relief. The boats, after the catastrophe, determined to proceed to Chili, touching at Ducie's Island in their way. They steered to the southward, and, after considerable sufferings, landed upon an island which they supposed to be that above mentioned, but which was, in fact, Elizabeth Island. Not being able to procure any water here, they continued their voyage to the coast of Chili, where two boats out of the three arrived, but with only three or four persons in them. The third was never heard of; but it is not improbable that the wreck of a boat and four skeletons which were seen on Ducie's Island by a merchant vessel were her remains and that of her crew. Had these unfortunate persons been aware of the situation of Pitcairn Island, which is only ninety miles from Elizabeth Island, and to leeward of it, all their lives might have been saved.

CHAPTER III.

Pitcairn Island—Adams and Natives come off to the Ship—Adams'
    Account of the mutiny of the Bounty—Lieutenant Bligh sent
    adrift in the Launch—Mutineers proceed to Tobouai—Hostile
    Reception there—Proceed to Otaheite—Return to Tobouai—
    Again quit it, and return to Otaheite—Christian determines to
    proceed to Pitcairn Island—Lands there—Fate of the Ship—
    Insurrection among the blacks—Murder of Christian and four of
    the Mutineers—Adams dangerously wounded—Fate of the
    remaining Number.

CHAP.
III.

Dec.
1825.

THE interest which was excited by the announce-
ment of Pitcairn Island from the mast-head brought
every person upon deck, and produced a train of
reflections that momentarily increased our anxiety
to communicate with its inhabitants; to see and
partake of the pleasures of their little domestic circle; and to learn from them the particulars of every
transaction connected with the fate of the Bounty:
but in consequence of the approach of night this
gratification was deferred until the next morning,
when, as we were steering for the side of the island
on which Captain Carteret has marked soundings,
in the hope of being able to anchor the ship, we had
the pleasure to see a boat under sail hastening toward us. At first the complete equipment of this
boat raised a doubt as to its being the property of

the islanders, for we expected to see only a well-provided canoe in their possession, and we therefore concluded that the boat must belong to some whale-ship on the opposite side ; but we were soon agreeably undeceived by the singular appearance of her crew, which consisted of old Adams and all the young men of the island.

Before they ventured to take hold of the ship, they inquired if they might come on board, and upon permission being granted, they sprang up the side and shook every officer by the hand with undisguised feelings of gratification.

The activity of the young men outstripped that of old Adams, who was consequently almost the last to greet us. He was in his sixty-fifth year, and was unusually strong and active for his age, notwithstanding the inconvenience of considerable corpulency. He was dressed in a sailor's shirt and trousers and a low-crowned hat, which he instinctively held in his hand until desired to put it on. He still retained his sailor's gait, doffing his hat and smoothing down his bald forehead whenever he was addressed by the officers.

It was the first time he had been on board a ship of war since the mutiny, and his mind naturally reverted to scenes that could not fail to produce a temporary embarrassment, heightened, perhaps, by the familiarity with which he found himself addressed by persons of a class with those whom he had been accustomed to obey. Apprehension for his safety formed no part of his thoughts : he had received too many demonstrations of the good feeling that existed towards him, both on the part of the British government and of individuals, to entertain any

alarm on that head ; and as every person endeavoured to set his mind at rest, he very soon made
himself at home.*

The young men, ten in number, were tall, robust,
and healthy, with good-natured countenances, which
would any where have procured them a friendly
reception ; and with a simplicity of manner and a
fear of doing wrong which at once prevented the
possibility of giving offence. Unacquainted with
the world, they asked a number of questions which
would have applied better to persons with whom
they had been intimate, and who had left them but
a short time before, than to perfect strangers ; and
inquired after ships and people we had never heard
of. Their dress, made up of the presents which had
been given them by the masters and seamen of merchant ships, was a perfect caricature. Some had on
long black coats without any other article of dress
except trousers, some shirts without coats, and others
waistcoats without either ; none had shoes or stockings, and only two possessed hats, neither of which
seemed likely to hang long together.

They were as anxious to gratify their curiosity
about the decks, as we were to learn from them the
state of the colony, and the particulars of the fate of
the mutineers who had settled upon the island,
which had been variously related by occasional visiters ; and we were more especially desirous of obtaining Adams' own narrative ; for it was peculiarly
interesting to learn from one who had been impli-
cated in the mutiny, the facts of that transaction,

* Since the MS. of this narrative was sent to press, intelligence
of Adams' death has been communicated to me by our Consul at
the Sandwich Islands.

R. Beechey del.ᵗ

E. Finden sculp.ᵗ

Ætat. 65

Pub.ᵈ by H. Colburn & R. Bentley, 1831.

now that he considered himself exempt from the
penalties of his crime.

I trust that, in renewing the discussion of this
affair, I shall not be considered as unnecessarily
wounding the feelings of the friends of any of the
parties concerned; but it is satisfactory to show, that
those who suffered by the sentence of the court-mar-
tial were convicted upon evidence which is now corroborated by the statement of an accomplice who
has no motive for concealing the truth. The following account is compiled almost entirely from
Adams' narrative, signed with his own hand, of
which the following is a fac-simile.

But to render the narrative more complete, I have
added such additional facts as were derived from the
inhabitants, who are perfectly acquainted with every
incident connected with the transaction. In presenting it to the public, I vouch, only, for its being
a correct statement of the abovementioned authorities.

His Majesty's ship Bounty was purchased into
the service, and placed under the command of Lieutenant Bligh in 1787. She left England in December of that year, with orders to proceed to Otaheite,*

* This word has since been spelled *Tahiti*, but as I have a veneration for the name as it is written in the celebrated Voyages of
Captain Cook—a feeling in which I am sure I am not singular—I
shall adhere to his orthography.

and transport the bread fruit of that country to the
British settlements in the West Indies, and to bring
also some specimens of it to England.   Her crew
consisted of forty-four persons, and a gardener. She
was ordered to make the passage round Cape Horn,
but after contending a long time with adverse gales,
in extremely cold weather, she was obliged to bear
away for the Cape of Good Hope, where she underwent a refit, and arrived at her destination in October, 1788.   Six months were spent at Otaheite, collecting and stowing away the fruit, during which
time the officers and seamen had free access to the
shore, and made many friends, though only one of
the seamen formed any alliance there.

In April, 1789, they took leave of their friends at
Otaheite, and proceeded to Anamooka, where Lieutenant Bligh replenished his stock of water, and took
on board hogs, fruit, vegetables, &c., and put to sea
again on the 26th of the same month.   Throughout
the voyage, Mr. Bligh had repeated misunderstand-
ings with his officers, and had on several occasions
given them and the ship's company just reasons for
complaint.   Still, whatever might have been the
feelings of the officers, Adams declares there was no
real discontent among the crew; much less was there
any idea of offering violence to their commander.
The officers, it must be admitted, had much more
cause for dissatisfaction than the seamen, especially
the master and Mr. Christian.   The latter was a pro-
tegé of Lieutenant Bligh, and unfortunately was
under some obligations to him of a pecuniary nature,
of which Bligh frequently reminded him when any
difference arose.   Christian, excessively annoyed at
the share of blame which repeatedly fell to his lot,
in common with the rest of the officers, could ill

endure the additional taunt of private obligations; and in a moment of excitation told his commander that sooner or later a day of reckoning would arrive.

The day previous to the mutiny a serious quarrel occurred between Bligh and his officers, about some cocoa-nuts which were missed from his private stock; and Christian again fell under his commander's displeasure. The same evening he was invited to supper in the cabin, but he had not so soon forgotten his injuries as to accept of this ill-timed civility, and returned an excuse.

Matters were in this state on the 28th of April, 1789, when the Bounty, on her homeward voyage, was passing to the southward of Tofoa, one of the Friendly Islands. It was one of those beautiful nights which characterize the tropical regions, when the mildness of the air and the stillness of nature dispose the mind to reflection. Christian, pondering over his grievances, considered them so intolerable, that any thing appeared preferable to enduring them, and he determined, as he could not redress them, that he would at least escape from the possibility of their being increased. Absence from England, and a long residence at Otaheite, where new connexions were formed, weakened the recollection of his native country, and prepared his mind for the reception of ideas which the situation of the ship and the serenity of the moment particularly favoured. His plan, strange as it must appear for a young officer to adopt, who was fairly advanced in an honourable profession, was to set himself adrift upon a raft, and make his way to the island then in sight. As quick in the execution as in the design, the raft was soon constructed, various useful articles were got together, and he was on the point of launching it, when a young officer, who afterwards perished in the Pandora, to whom Christian communicated his intention, recommended him, rather than risk his life on so hazardous an expedition, to endeavour to take possession of the ship, which he thought would not be very difficult, as many of the ship's company were not well disposed towards the commander, and would all be very glad to return to Otaheite, and reside among their friends in that island. This daring proposition is even more extraordinary than the premeditated scheme of his companion, and, if true, certainly relieves Christian from part of the odium which has hitherto attached to him as the sole instigator of the mutiny.*

It however accorded too well with the disposition of Christian's mind, and, hazardous as it was, he determined to co-operate with his friend in effecting it, resolving, if he failed, to throw himself into the sea; and that there might be no chance of being saved, he tied a deep sea lead about his neck, and concealed it within his clothes.

Christian happened to have the morning watch, and as soon as he had relieved the officer of the deck, he entered into conversation with Quintal, the only one of the seamen who, Adams said, had formed any serious attachment at Otaheite; and after expatiating on the happy hours they had passed there, disclosed his intentions. Quintal, after some consideration, said he thought it a dangerous attempt, and declined taking a part. Vexed at a re-

---

* This account, however, differs materially from a note in Marshall's Naval Biography, Vol. ii. Part ii. p. 778: unfortunately this volume was not published when the Blossom left England, or more satisfactory evidence on this, and other points, might have been obtained. However, this is the statement of Adams.

pulse in a quarter where he was most sanguine of success, and particularly at having revealed sentiments which if made known would bring him to an ignominious death, Christian became desperate, exhibited the lead about his neck in testimony of his own resolution, and taxed Quintal with cowardice, declaring it was fear alone that restrained him. Quintal denied this accusation; and in reply to Christian's further argument that success would restore them all to the happy island, and the connexions they had left behind, the strongest persuasion he could have used to a mind somewhat prepared to acquiesce, he recommended that some one else should be tried—Isaac Martin for instance, who was standing by. Martin, more ready than his shipmate, emphatically declared, " He was for it; it was the very thing." Successful in one instance, Christian went to every man of his watch, many of whom he found disposed to join him, and before daylight the greater portion of the ship's company were brought over.

Adams was sleeping in his hammock, when Sumner, one of the seamen, came to him, and whispered that Christian was going to take the ship from her commander, and set him and the master on shore. On hearing this, Adams went upon deck, and found every thing in great confusion; but not then liking to take any part in the transaction, he returned to his hammock, and remained there until he saw Christian at the arm-chest, distributing arms to all who came for them; and then seeing measures had proceeded so far, and apprehensive of being on the weaker side, he turned out again and went for a cutlass.

All those who proposed to assist Christian being

armed, Adams, with others, were ordered to secure the officers, while Christian and the master-at-arms proceeded to the cabin to make a prisoner of Lieutenant Bligh. They seized him in his cot, bound his hands behind him, and brought him upon deck. He remonstrated with them on their conduct, but received only abuse in return, and a blow from the master-at-arms with the flat side of a cutlass. He was placed near the binnacle, and detained there, with his arms pinioned, by Christian, who held him with one hand, and a bayonet with the other. As soon as the lieutenant was secured, the sentinels that had been placed over the doors of the officers' cabins were taken off; the master then jumped upon the forecastle, and endeavoured to form a party to retake the ship; but he was quickly secured, and sent below in confinement.

This conduct of the master, who was the only officer that tried to bring the mutineers to a sense of their duty, was the more highly creditable to him, as he had the greatest cause for discontent, Mr. Bligh having been more severe to him than to any of the other officers.

About this time a dispute arose, whether the lieutenant and his party, whom the mutineers resolved to set adrift, should have the launch or the cutter; and it being decided in favour of the launch, Christian ordered her to be hoisted out. Martin, who, it may be remembered, was the first convert to Christian's plan, foreseeing that with the aid of so large a boat the party would find their way to England, and that their information would in all probability lead to the detection of the offenders, relinquished his first intention, and exclaimed, " If you give him the launch, I will go with him; you may as well give him the ship." He really appears to have been in earnest in making this declaration, as he was afterwards ordered to the gangway from his post of command over the lieutenant, in consequence of having fed him with a shaddock, and exchanged looks with him indicative of his friendly intentions. It also fell to the lot of Adams to guard the lieutenant, who observing him stationed by his side, exclaimed, " And you, Smith,* are you against me ?" To which Adams replied that he only acted as the others did —he must be like the rest. Lieutenant Bligh, while thus secured, reproached Christian with ingratitude, reminded him of his obligations to him, and begged he would recollect he had a wife and family. To which Christian replied, that he should have thought of that before.

The launch was by this time hoisted out ; and the officers and seamen of Lieutenant Bligh's party having collected what was necessary for their voyage,† were ordered into her. Among those who took their seat in the boat was Martin, which being noticed by Quintal, he pointed a musket at him, and declared he would shoot him unless he instantly returned to the ship, which he did. The armourer and carpenter's mates were also forcibly detained, as they might be required hereafter. Lieutenant Bligh was then conducted to the gangway, and ordered to descend into the boat, where his hands were unbound, and he and his party were veered astern,

---

* Adams went by the name of Alexander Smith in the Bounty.
† Consisting of a small cask of water, 150lbs. of bread, a small quantity of rum and wine, a quadrant, compass, some lines, rope, canvas, twine, &c.

and kept there while the ship stood towards the island. During this time Lieutenant Bligh requested some muskets, to protect his party against the natives; but they were refused, and four cutlasses thrown to them instead. When they were about ten leagues from Tofoa, at Lieutenant Bligh's request, the launch was cast off, and immediately " Huzza for Otaheite!" echoed throughout the Bounty.

There now remained in the ship, Christian, who was the mate, Heywood, Young, and Stewart, midshipmen, the master-at-arms, and sixteen seamen, besides the three artificers, and the gardener; forming in all twenty-five.

In the launch were the lieutenant, master, surgeon, a master's mate, two midshipmen, botanist, three warrant-officers, clerk, and eight seamen, making in all nineteen; and had not the three persons above mentioned been forcibly detained, the captain would have had exactly half the ship's company. It may perhaps appear strange to many, that with so large a party in his favour, Lieutenant Bligh made no attempt to retake the vessel; but the mutiny was so ably conducted that no opportunity was afforded him of doing so; and the strength of the crew was decidedly in favour of Christian. Lieutenant Bligh's adventures and sufferings, until he reached Timor, are well known to the public, and need no repetition.

The ship, having stood some time to the W. N. W., with a view to deceive the party in the launch, was afterwards put about, and her course directed as near to Otaheite as the wind would permit. In a few days they found some difficulty in reaching that island, and bore away for Tobouai, a small island about 300 miles to the southward of it, where they agreed to establish themselves, provided the natives, who were numerous, were not hostile to their purpose. Of this they had very early intimation, an attack being made upon a boat which they sent to sound the harbour. She, however, effected her purpose; and the next morning the Bounty was warped inside the reef that formed the port, and stationed close to the beach. An attempt to land was next made; but the natives disputed every foot of ground with spears, clubs, and stones, until they were dispersed by a discharge of cannon and musketry. On this they fled to the interior, and refused to hold any further intercourse with their visiters.

The determined hostility of the natives put an end to the mutineers' design of settling among them at that time; and, after two days' fruitless attempt at reconciliation, they left the island and proceeded to Otaheite. Tobouai was, however, a favourite spot with them, and they determined to make another effort to settle there, which they thought would yet be feasible, provided the islanders could be made acquainted with their friendly intentions. The only way to do this was through interpreters, who might be procured at Otaheite; and in order not to be dependent upon the natives of Tobouai for wives, they determined to engage several Otaheitan women to accompany them. They reached Otaheite in eight days, and were received with the greatest kindness by their former friends, who immediately inquired for the captain and his officers. Christian and his party having anticipated inquiries of this nature, invented a story to account for their absence, and told them that Lieutenant Bligh having found an island suitable for a settlement, had landed there with some of his officers, and sent them in the ship to procure live stock and whatever else would be useful to the colony, and to bring besides such of the natives as were willing to accompany them.* Satisfied with this plausible account, the chiefs supplied them with every thing they wanted, and even gave them a bull and cow which had been confided to their care, the only ones, I believe, that were on the island. They were equally fortunate in finding several persons, both male and female, willing to accompany them; and thus furnished, they again sailed for Tobouai, where, as they expected, they were better received than before, in consequence of being able to communicate with the natives through their interpreters.

Experience had taught them the necessity of making self-defence their first consideration, and a fort was consequently commenced, eighty yards square, surrounded by a wide ditch. It was nearly completed, when the natives, imagining they were going to destroy them, and that the ditch was intended for their place of interment, planned a general attack when the party should proceed to work in the morning. It fortunately happened that one of the

* In the Memoir of Captain Peter Heywood, in Marshall's Naval Biography, it is related that the mutineers availing themselves of a fiction which had been created by Lieutenant Bligh respecting Captain Cook, stated that they had fallen in with him, and that he had sent the ship back for all the live stock that could be spared, in order to form a settlement at a place called Wytootacke, which Bligh had discovered in his course to the Friendly Islands.

natives who accompanied them from Otaheite over-
heard this conspiracy, and instantly swam off to the
ship and apprised the crew of their danger. Instead,
therefore, of proceeding to their work at the fort, as
usual, the following morning, they made an attack
upon the natives, killed and wounded several, and
obliged the others to retire inland.

Great dissatisfaction and difference of opinion now
arose among the crew: some were for abandoning
the fort and returning to Otaheite; while others
were for proceeding to the Marquesas; but the
majority were at that time for completing what they
had begun, and remaining at Tobouai. At length
the continued state of suspense in which they were
kept by the natives made them decide to return to
Otaheite, though much against the inclination of
Christian, who in vain expostulated with them on
the folly of such a resolution, and the certain detection that must ensue.

The implements being embarked, they proceeded
therefore a second time to Otaheite, and were again
well received by their friends, who replenished their
stock of provision. During the passage Christian
formed his intention of proceeding in the ship to
some distant uninhabited island, for the purpose of
permanently settling, as the most likely means of
escaping the punishment which he well knew awaited
him in the event of being discovered. On communicating this plan to his shipmates he found only a
few inclined to assent to it; but no objections were
offered by those who dissented, to his taking the
ship; all they required was an equal distribution
of such provisions and stores as might be useful.
Young, Brown, Mills, Williams, Quintal, M'Coy,

Martin, Adams, and six natives (four of Otaheite and two of Tobouai) determined to follow the fate of Christian. Remaining, therefore, only twenty-four hours at Otaheite, they took leave of their comrades, and having invited on board several of the women with the feigned purpose of taking leave, the cables were cut and they were carried off to sea.*

The mutineers now bade adieu to all the world, save the few individuals associated with them in exile. But where that exile should be passed, was yet undecided: the Marquesas Islands were first mentioned, but Christian, on reading Captain Carteret's account of Pitcairn Island, thought it better adapted to the purpose, and accordingly shaped a course thither. They reached it not many days afterwards; and Christian, with one of the seamen, landed in a little nook, which we afterwards found very convenient for disembarkation. They soon traversed the island sufficiently to be satisfied that it was exactly suited to their wishes. It possessed water, wood, a good soil, and some fruits. The anchorage in the offing was very bad, and landing for boats extremely hazardous. The mountains were so difficult of access, and the passes so narrow, that they might be maintained by a few persons against an army; and there were several caves, to which, in case of necessity, they could retreat, and where, as long as their provision lasted, they might bid defiance to their pursuers. With this intelligence they returned on board, and brought the ship to an

---

* The greater part of the mutineers who remained at Otaheite were taken by his Majesty's ship Pandora, which was purposely sent out from England after Lieutenant Bligh's return.

anchor in a small bay on the northern side of the island, which I have in consequence named " Bounty Bay," where every thing that could be of utility was landed, and where it was agreed to destroy the ship, either by running her on shore, or burning her. Christian, Adams, and the majority, were for the former expedient; but while they went to the fore-part of the ship, to execute this business, Mathew Quintal set fire to the carpenter's store-room. The vessel burnt to the water's edge, and then drifted upon the rocks, where the remainder of the wreck was burnt for fear of discovery. This occurred on the 23d January, 1790.

Upon their first landing they perceived, by the remains of several habitations, morais, and three or four rudely sculptured images, which stood upon the eminence overlooking the bay where the ship was destroyed, that the island had been previously inhabited. Some apprehensions were, in conse-quence, entertained lest the natives should have secreted themselves, and in some unguarded mo-ment make an attack upon them; but by degrees these fears subsided, and their avocations proceeded without interruption.

A suitable spot of ground for a village was fixed upon, with the exception of which the island was divided into equal portions, but to the exclusion of the poor blacks, who being only friends of the sea-men, were not considered as entitled to the same privileges. Obliged to lend their assistance to the others in order to procure a subsistence, they thus, from being their friends, in the course of time became their slaves. No discontent, however, was manifested, and they willingly assisted in the cultivation of the soil. In clearing the space that was allotted to the village, a row of trees was left between it and the sea, for the purpose of concealing the houses from the observation of any vessels that might be passing, and nothing was allowed to be erected that might in any way attract attention. Until these houses were finished, the sails of the Bounty were converted into tents, and when no longer required for that purpose, became very acceptable as clothing. Thus supplied with all the necessaries of life, and some of its luxuries, they felt their condition comfortable even beyond their most sanguine expectation, and every thing went on peaceably and prosperously for about two years, at the expiration of which Williams, who had the misfortune to lose his wife about a month after his arrival, by a fall from a precipice while collecting birds' eggs, became dissatisfied, and threatened to leave the island in one of the boats of the Bounty, unless he had another wife; an unreasonable request, as it could not be complied with, except at the expense of the happiness of one of his companions : but Williams, actuated by selfish considerations alone, persisted in his threat, and the Europeans not willing to part with him, on account of his usefulness as an armourer, constrained one of the blacks to bestow his wife upon the applicant. The blacks, outrageous at this second act of flagrant injustice, made common cause with their companion, and matured a plan of revenge upon their aggressors, which, had it succeeded, would have proved fatal to all the Europeans. Fortunately, the secret was imparted to the women, who ingeniously communicated it to the white men in a song, of which the words were,

"Why does black man sharpen axe? to kill white man." The instant Christian became aware of the plot, he seized his gun and went in search of the blacks, but with a view only of showing them that their scheme was discovered, and thus by timely interference endeavouring to prevent the execution of it. He met one of them (Ohoo) at a little distance from the village, taxed him with the conspiracy, and, in order to intimidate him, discharged his gun, which he had humanely loaded with powder only. Ohoo, however, imagining otherwise, and that the bullet had missed its object, derided his unskilfulness, and fled into the woods, followed by his accomplice Talaloo, who had been deprived of his wife. The remaining blacks, finding their plot discovered, purchased pardon by promising to murder their accomplices, who had fled, which they afterwards performed by an act of the most odious treachery. Ohoo was betrayed and murdered by his own nephew; and Talaloo, after an ineffectual attempt made upon him by poison, fell by the hands of his friend and his wife, the very woman on whose account all the disturbance began, and whose injuries Talaloo felt he was revenging in common with his own.

Tranquillity was by these means restored, and preserved for about two years; at the expiration of which, dissatisfaction was again manifested by the blacks, in consequence of oppression and ill treatment, principally by Quintal and M'Coy Meeting with no compassion or redress from their masters, a second plan to destroy their oppressors was matured, and, unfortunately, too successfully executed.

It was agreed that two of the blacks, Timoa and

Nehow, should desert from their masters, provide themselves with arms, and hide in the woods, but maintain a frequent communication with the other two, Tetaheite and Menalee; and that on a certain day they should attack and put to death all the Englishmen, when at work in their plantations. Tetaheite, to strengthen the party of the blacks on this day, borrowed a gun and ammunition of his master, under the pretence of shooting hogs, which had become wild and very numerous; but instead of using it in this way, he joined his accomplices, and with them fell upon Williams and shot him. Martin, who was at no great distance, heard the report of the musket, and exclaimed, " Well done! we shall have a glorious feast to-day !" supposing that a hog had been shot. The party proceeded from Williams' toward Christian's plantation, where Menalee, the other black, was at work with Mills and M'Coy; and, in order that the suspicions of the whites might not be excited by the report they had heard, requested Mills to allow him (Menalee) to assist them in bringing home the hog they pretended to have killed. Mills agreed; and the four, being united, proceeded to Christian, who was working at his yam-plot, and shot him. Thus fell a man, who, from being the reputed ringleader of the mutiny, has obtained an unenviable celebrity, and whose crime, if any thing can excuse mutiny, may perhaps be considered as in some degree palliated, by the tyranny which led to its commission. M'Coy, hearing his groans, observed to Mills, "there was surely some person dying;" but Mills replied, " It is only Mainmast (Christian's wife) calling her children to dinner." The white men being yet too strong

for the blacks to risk a conflict with them, it was necessary to concert a plan, in order to separate Mills
and M'Coy. Two of them accordingly secreted
themselves in M'Coy's house, and Tetaheite ran and
told him that the two blacks who had deserted were
stealing things out of his house. M'Coy instantly
hastened to detect them, and on entering was fired
at ; but the ball passed him. M'Coy immediately
communicated the alarm to Mills, and advised him
to seek shelter in the woods ; but Mills, being quite
satisfied that one of the blacks whom he had made
his friend would not suffer him to be killed, determined to remain. M'Coy, less confident, ran in
search of Christian, but finding him dead, joined
Quintal (who was already apprised of the work of
destruction, and had sent his wife to give the alarm
to the others), and fled with him to the woods.

Mills had scarcely been left alone, when the two
blacks fell upon him, and he became a victim to his
misplaced confidence in the fidelity of his friend.
Martin and Brown were next separately murdered
by Menalee and Tenina ; Menalee effecting with a
maul what the musket had left unfinished. Tenina,
it is said, wished to save the life of Brown, and fired
at him with powder only, desiring him, at the same
time, to fall as if killed ; but, unfortunately rising
too soon, the other black, Menalee, shot him.

Adams was first apprised of his danger by Quintal's wife, who, in hurrying through his plantation,
asked why he was working at such a time ? Not
understanding the question, but seeing her alarmed,
he followed her, and was almost immediately met
by the blacks, whose appearance exciting suspicion,
he made his escape into the woods. After remain-

ing there three or four hours, Adams, thinking all
was quiet, stole to his yam-plot for a supply of provisions; his movements however did not escape the
vigilance of the blacks, who attacked and shot him
through the body, the ball entering at his right
shoulder, and passing out through his throat. He
fell upon his side, and was instantly assailed by
one of them with the butt end of the gun; but he
parried the blows at the expense of a broken finger.
Tetaheite then placed his gun to his side, but it
fortunately missed fire twice. Adams, recovering a
little from the shock of the wound, sprang on his
legs, and ran off with as much speed as he was able,
and fortunately outstripped his pursuers, who seeing
him likely to escape, offered him protection if he
would stop. Adams, much exhausted by his wound,
readily accepted their terms, and was conducted to
Christian's house, where he was kindly treated.
Here this day of bloodshed ended, leaving only four
Englishmen alive out of nine. It was a day of
emancipation to the blacks, who were now masters
of the island, and of humiliation and retribution to
the whites.

Young, who was a great favourite with the
women, and had, during this attack, been secreted
by them, was now also taken to Christian's house.
The other two, M'Coy and Quintal, who had always
been the great oppressors of the blacks, escaped to
the mountains, where they supported themselves
upon the produce of the ground about them.

The party in the village lived in tolerable tranquillity for about a week; at the expiration of
which, the men of colour began to quarrel about the
right of choosing the women whose husbands had

been killed; which ended in Menalee's shooting Timoa as he sat by the side of Young's wife, accompanying her song with his flute. Timoa not dying immediately, Menalee reloaded, and deliberately despatched him by a second discharge. He afterwards attacked Tetaheite, who was condoling with Young's wife for the loss of her favourite black, and would have murdered him also, but for the interference of the women. Afraid to remain longer in the village, he escaped to the mountains and joined Quintal and M'Coy, who, though glad of his services, at first received him with suspicion. This great acquisition to their force enabled them to bid defiance to the opposite party; and to show their strength, and that they were provided with muskets, they appeared on a ridge of mountains, within sight of the village, and fired a volley which so alarmed the others that they sent Adams to say, if they would kill the black man, Menalee, and return to the village, they would all be friends again. The terms were so far complied with that Menalee was shot; but, apprehensive of the sincerity of the remaining blacks, they refused to return while they were alive.

Adams says it was not long before the widows of the white men so deeply deplored their loss, that they determined to revenge their death, and concerted a plan to murder the only two remaining men of colour. Another account, communicated by the islanders, is, that it was only part of a plot formed at the same time that Menalee was murdered, which could not be put in execution before. However this may be, it was equally fatal to the poor blacks. The arrangement was, that Susan should murder one of them, Tetaheite, while he was sleeping by the side of his favourite; and that Young should at the same instant, upon a signal being given, shoot the other, Nehow. The unsuspecting Tetaheite retired as usual, and fell by the blow of an axe; the other was looking at Young loading his gun, which he supposed was for the purpose of shooting hogs, and requested him to put in a good charge, when he received the deadly contents.

In this manner the existence of the last of the men of colour terminated, who, though treacherous and revengeful, had, it is feared, too much cause for complaint. The accomplishment of this fatal scheme was immediately communicated to the two absentees, and their return solicited. But so many instances of treachery had occurred, that they would not believe the report, though delivered by Adams himself, until the hands and heads of the deceased were produced, which being done, they returned to the village. This eventful day was the 3d October, 1793. There were now left upon the island, Adams, Young, M'Coy, and Quintal, ten women, and some children. Two months after this period, Young commenced a manuscript journal, which affords a good insight into the state of the island, and the occupations of the settlers. From it we learn, that they lived peaceably together, building their houses, fencing in and cultivating their grounds, fishing, and catching birds, and constructing pits for the purpose of entrapping hogs, which had become very numerous and wild, as well as injurious to the yam-crops. The only discontent appears to have been among the women, who lived promiscuously with the men, frequently changing their abode.

Young says, March 12, 1794, "Going over to borrow a rake, to rake the dust off my ground, I saw Jenny having a skull in her hand: I asked her whose it was? and was told it was Jack Williams's. I desired it might be buried: the women who were with Jenny gave me for answer, it should not. I said it should; and demanded it accordingly. I was asked the reason why I, in particular, should insist on such a thing, when the rest of the white men did not? I said, if they gave them leave to keep the skulls above ground, I did not. Accordingly when I saw M'Coy, Smith, and Mat. Quintal, I acquainted them with it, and said, I thought that if the girls did not agree to give up the heads of the five white men in a peaceable manner, they ought to be taken by force, and buried." About this time the women appear to have been much dissatisfied; and Young's journal declares that, "since the massacre, it has been the desire of the greater part of them to get some conveyance, to enable them to leave the island." This feeling continued, and on the 14th April, 1794, was so strongly urged, that the men began to build them a boat; but wanting planks and nails, Jenny, who now resides at Otaheite, in her zeal tore up the boards of her house, and endeavoured, though without success, to persuade some others to follow her example.

On the 13th August following, the vessel was finished, and on the 15th she was launched: but, as Young says, "according to expectation she upset," and it was most fortunate for them that she did so; for had they launched out upon the ocean, where could they have gone? or what could a few ignorant women have done by themselves, drifting upon the waves, but ultimately have fallen a sacrifice to their folly? However, the fate of the vessel was a great disappointment, and they continued much dissatisfied with their condition; probably not without some reason, as they were kept in great subordination, and were frequently beaten by M'Coy and Quintal, who appear to have been of very quarrelsome dispositions; Quintal in particular, who proposed "not to laugh, joke, or give any thing to any of the girls."

On the 16th August they dug a grave, and buried the bones of the murdered people; and on October 3d, 1794, they celebrated the murder of the black men at Quintal's house. On the 11th November, a conspiracy of the women to kill the white men in their sleep was discovered; upon which they were all seized, and a disclosure ensued; but no punishment appears to have been inflicted upon them, in consequence of their promising to conduct themselves properly, and never again to give any cause "even to suspect their behaviour." However, though they were pardoned, Young observes, "We did not forget their conduct; and it was agreed among us, that the first female who misbehaved should be put to death; and this punishment was to be repeated on each offence until we could discover the real intentions of the women." Young appears to have suffered much from mental perturbation in consequence of these disturbances; and observes of himself on the two following days, that "he was bothered and idle."

The suspicions of the men induced them, on the 15th, to conceal two muskets in the bush, for the use of any person who might be so fortunate as to escape, in the event of an attack being made. On the 30th November, the women again collected and attacked them; but no lives were lost, and they returned on being once more pardoned, but were again threatened with death the next time they misbehaved. Threats thus repeatedly made, and as often unexecuted, as might be expected, soon lost their effect, and the women formed a party whenever their displeasure was excited, and hid themselves in the unfrequented parts of the island, carefully providing themselves with fire-arms. In this manner the men were kept in continual suspense, dreading the result of each disturbance, as the numerical strength of the women was much greater than their own.

On the 4th of May, 1795, two canoes were begun, and in two days completed. These were used for fishing, in which employment the people were frequently successful, supplying themselves with rock-fish and large mackarel. On the 27th of December following, they were greatly alarmed by the appearance of a ship close in with the island. Fortunately for them there was a tremendous surf upon the rocks, the weather wore a very threatening aspect, and the ship stood to the S.E., and at noon was out of sight. Young appears to have thought this a providential escape, as the sea for a week after was " smoother than they had ever recollected it since their arrival on the island."

So little occurred in the year 1796, that one page records the whole of the events; and throughout the following year there are but three incidents worthy of notice. The first, their endeavour to procure a quantity of meat for salting; the next, their attempt to make syrup from the tee-plant (*dracæna terminalis*) and sugar-cane; and the third, a serious accident that happened to M'Coy, who fell from a cocoa-nut tree and hurt his right thigh, sprained both his ancles and wounded his side. The occupations of the men continued similar to those already related, occasionally enlivened by visits to the opposite side of the island. They appear to have been more sociable; dining frequently at each other's houses, and contributing more to the comfort of the women, who, on their part, gave no ground for un-easiness. There was also a mutual accommodation amongst them in regard to provisions, of which a regular account was taken. If one person was suc-cessful in hunting, he lent the others as much meat as they required, to be repaid at leisure; and the same occurred with yams, taros, &c., so that they lived in a very domestic and tranquil state.

It unfortunately happened that M'Coy had been employed in a distillery in Scotland; and being very much addicted to liquor, he tried an experi-ment with the tee-root, and on the 20th April, 1798, succeeded in producing a bottle of ardent spirit. This success induced his companion, Mathew Quin-tal, to " alter his kettle into a still," a contrivance which unfortunately succeeded too well, as frequent intoxication was the consequence, with M'Coy in particular, upon whom at length it produced fits of delirium, in one of which, he threw himself from a cliff and was killed. The melancholy fate of this man created so forcible an impression on the re-maining few, that they resolved never again to touch spirits; and Adams, I have every reason to believe, to the day of his death kept his vow.

The journal finishes nearly at the period of M'Coy's death, which is not related in it : but we learned from Adams, that about 1799 Quintal lost his wife by a fall from the cliff while in search of birds' eggs; that he grew discontented, and, though there were several disposable women on the island, and he had already experienced the fatal effects of a similar demand, nothing would satisfy him but the wife of one of his companions. Of course neither of them felt inclined to accede to this unreasonable indulgence ; and he sought an opportunity of putting them both to death. He was fortunately foiled in his first attempt, but swore he would repeat it. Adams and Young having no doubt he would follow up his resolution, and fearing he might be more successful in the next attempt, came to the conclusion, that their own lives were not safe while he was in existence, and that they were justified in putting him to death, which they did with an axe.

Such was the melancholy fate of seven of the leading mutineers, who escaped from justice only to add murder to their former crimes ; for though some of them may not have actually imbrued their hands in the blood of their fellow-creatures, yet all were accessary to the deed.

As Christian and Young were descended from respectable parents, and had received educations suitable to their birth, it might be supposed that they felt their altered and degraded situation much more than the seamen who were comparatively well off : but if so, Adams says, they had the good sense to conceal it, as not a single murmur or regret escaped them ; on the contrary, Christian was always cheerful, and his example was of the greatest service

in exciting his companions to labour. He was naturally of a happy, ingenuous disposition, and won
the good opinion and respect of all who served under him ; which cannot be better exemplified than
by his maintaining, under circumstances of great
perplexity, the respect and regard of all who were
associated with him up to the hour of his death ;
and even at the period of our visit, Adams, in speaking of him, never omitted to say, " *Mr.* Christian."

Adams and Young were now the sole survivors
out of the fifteen males that landed upon the island.
They were both, and more particularly Young, of a
serious turn of mind ; and it would have been won-
derful, after the many dreadful scenes at which they
had assisted, if the solitude and tranquillity that ensued had not disposed them to repentance. During Christian's lifetime they had only once read the
church service, but since his decease this had been
regularly done on every Sunday. They now, however, resolved to have morning and evening family
prayers, to add afternoon service to the duty of the
Sabbath, and to train up their own children, and
those of their late unfortunate companions in piety
and virtue.

In the execution of this resolution Young's education enabled him to be of the greatest assistance ;
but he was not long suffered to survive his repentance. An asthmatic complaint, under which he had
for some time laboured, terminated his existence
about a year after the death of Quintal, and Adams
was left the sole survivor of the misguided and unfortunate mutineers of the Bounty. The loss of his
last companion was a great affliction to him, and
was for some time most severely felt. It was a

catastrophe, however, that more than ever disposed him to repentance, and determined him to execute the pious resolution he had made, in the hope of expiating his offences.

His reformation could not, perhaps, have taken place at a more propitious moment. Out of nineteen children upon the island, there were several between the ages of seven and nine years; who, had they been longer suffered to follow their own inclinations, might have acquired habits which it would have been difficult if not impossible for Adams to eradicate. The moment was therefore most favourable for his design, and his laudable exertions were attended by advantages both to the objects of his care and to his own mind, which surpassed his most sanguine expectations. He, nevertheless, had an arduous task to perform. Besides the children to be educated, the Otaheitan women were to be converted; and as the example of the parents had a powerful influence over their children, he resolved to make them his first care. Here also his labours succeeded; the Otaheitans were naturally of a tractable disposition, and gave him less trouble than he anticipated: the children also acquired such a thirst after scriptural knowledge, that Adams in a short time had little else to do than to answer their inquiries and put them in the right way. As they grew up, they acquired fixed habits of morality and piety; their colony improved; intermarriages occurred: and they now form a happy and well-regulated society, the merit of which in a great degree belongs to Adams, and tends to redeem the former errors of his life.

96   VOYAGE TO THE

CHAPTER IV.

Bounty Bay—Observatory landed—Manners, Customs, Occupations, Amusements, &c. of the Natives—Village—Description
of the Island—Its produce—Marriage of Adams—Barge hoisted
out—Departure—General Description.

CHAP.
IV.

Dec.
1825.

HAVING detailed the particulars of the mutiny in
the Bounty, and the fate of the most notorious of
the ring-leaders, and having brought the history of
Pitcairn Island down to the present period, I shall
return to the party who had assembled on board the
ship to greet us on our arrival.

The Blossom was so different, or to use the expression of our visiters, " so rich," compared with
the other ships they had seen,* that they were constantly afraid of giving offence or committing some
injury, and would not even move without first asking permission. This diffidence gave us full occupation for some time, as our restless visiters, anxious
to see every thing, seldom directed their attention
long to any particular object, or remained in one
position or place. Having no latches to their doors,
they were ignorant of the manner of opening ours;
and we were consequently attacked on all sides with

* It was so long since the visit of the Briton and Tagus that they
had forgotten their appearance.

F.W.Beechey delt. E.Finden sc.

Pub.by H.Colburn & R.Bentley. 1831.

LANDING IN BOUNTY BAY.

" Please may I sit down or get up, or go out of the
cabin?" or " Please to open or shut the door." Their
applications were, however, made with such good
nature and simplicity that it was impossible not to
feel the greatest pleasure in paying attention to
them. They very soon learnt the christian name of
every officer in the ship, which they always used in
conversation instead of the surname, and wherever a
similarity to their own occurred, they attached themselves to that person as a matter of course.

It was many hours after they came on board before the ship could get near the island, during which
time they so ingratiated themselves with us that we
felt the greatest desire to visit their houses; and
rather than pass another night at sea we put off in
the boats, though at a considerable distance from the
land, and accompanied them to the shore. We followed our guides past a rugged point surmounted
by tall spiral rocks, known to the islanders as St.
Paul's rocks, into a spacious iron-bound bay, where
the Bounty found her last anchorage. In this bay,
which is bounded by lofty cliffs almost inaccessible,
it was proposed to land. Thickly branched evergreens skirt the base of these hills, and in summer
afford a welcome retreat from the rays of an almost
vertical sun. In the distance are seen several high
pointed rocks which the pious highlanders have
named after the most zealous of the Apostles, and
outside of them is a square basaltic islet. Formidable breakers fringe the coast, and seem to present an
insurmountable barrier to all access.

We here brought our boats to an anchor, in consequence of the passage between the sunken rocks being much too intricate, and we trusted ourselves to

the natives, who landed us, two at a time, in their whale-boat. The difficulty of landing was more than repaid by the friendly reception we met with on the beach from Hannah Young, a very interesting young woman, the daughter of Adams. In her eagerness to greet her father, she had outrun her female companions, for whose delay she thought it necessary in the first place to apologize, by saying they had all been over the hill in company with John Buffet to look at the ship, and were not yet returned. It appeared that John Buffet, who was a seafaring man, ascertained that the ship was a man of war, and without knowing exactly why, became so alarmed for the safety of Adams that he either could not or would not answer any of the interrogations which were put to him. This mysterious silence set all the party in tears, as they feared he had discovered something adverse to their patriarch. At length his obduracy yielded to their entreaties; but before he explained the cause of his conduct, the boats were seen to put off from the ship, and Hannah immediately hurried to the beach to kiss the old man's cheek, which she did with a fervency demonstrative of the warmest affection. Her apology for her companions was rendered unnecessary by their appearance on the steep and circuitous path down the mountain, who, as they arrived on the beach, successively welcomed us to their island, with a simplicity and sincerity which left no doubt of the truth of their professions.

They almost all wore the cloth of the island: their dress consisted of a petticoat, and a mantle loosely thrown over the shoulders, and reaching to the ancles. Their stature was rather above the common height; and their limbs, from being accustomed to work and climb the hills, had acquired unusual muscularity; but their features and manners were perfectly feminine. Their complexion, though fairer than that of the men, was of a dark gipsy hue, but its deep colour was less conspicuous, by being contrasted with dark glossy hair, which hung down over their shoulders in long waving tresses, nicely oiled: in front it was tastefully turned back from the forehead and temples, and was retained in that position by a chaplet of small red or white aromatic blossoms, newly gathered from the flower-tree (*morinda citrifolia*), or from the tobacco plant; their countenances were lively and good-natured, their eyes dark and animated, and each possessed an enviable row of teeth. Such was the agreeable impression of their first appearance, which was heightened by the wish expressed simultaneously by the whole group, that we were come to stay several days with them. As the sun was going down, we signified our desire to get to the village and to pitch the observatory before dark, and this was no sooner made known, than every instrument and article found a carrier.

We took the only pathway which leads from the landing-place to the village, and soon experienced the difficulties of the ascent, which the distant appearance of the ground led us to anticipate. To the natives, however, there appeared to be no obstacles: women as well as men bore their burthens over the most difficult parts without inconvenience; while we, obliged at times to have recourse to tufts of shrubs or grass for assistance, experienced serious delay, being also incommoded by the heat of the

weather, and by swarms of house-flies which infest
the island, and are said to have been imported there
by H. M. S. Briton.

As soon as we had gained the first level, our
party rested on some large stones that lay half buried in long grass on one side of a ravine, from which
the blue sky was nearly concealed by the overlap-
ping branches of palm-trees Here, through the
medium of our female guides, who, furnished with
the spreading leaves of the tee-plant, drove away
our troublesome persecutors, we obtained a respite
from their attacks.

Having refreshed ourselves, we resumed our journey over a more easy path ; and after crossing two
valleys, shaded by cocoa-nut trees, we arrived at the
village. It consisted of five houses, built upon a
cleared piece of ground sloping to the sea, and commanding a distant view of the horizon, through a
break in an extensive wood of palms. While the
men assisted to pitch our tent, the women employed
themselves in preparing our dinner, or more properly supper, as it was eight o'clock at night.

The manner of cooking in Pitcairn's Island is
similar to that of Otaheite, which, as some of my
readers may not recollect, I shall briefly describe.
An oven is made in the ground, sufficiently large to
contain a good-sized pig, and is lined throughout
with stones nearly equal in size, which have been
previously made as hot as possible. These are
covered with some broad leaves, generally of the
tee-plant, and on them is placed the meat. If it be
a pig, its inside is lined with heated stones, as well
as the oven; such vegetables as are to be cooked are
then placed round the animal : the whole is care-

fully covered with leaves of the tee, and buried beneath a heap of earth, straw, or rushes and boughs, which, by a little use, becomes matted into one mass. In about an hour and a quarter the animal is sufficiently cooked, and is certainly more thoroughly done than it would be by a fire.

By the time the tent was up and the instruments secured, we were summoned to a meal cooked in this manner, than which a less sumptuous fare would have satisfied appetites rendered keen by long abstinence and a tiresome journey. Our party divided themselves that they might not crowd one house in particular: Adams did not entertain; but at Christian's I found a table spread with plates, knives, and forks; which, in so remote a part of the world, was an unexpected sight. They were, it is true, far from uniform; but by one article being appropriated for another, we all found something to put our portion upon; and but few of the natives were obliged to substitute their fingers for articles which are indispensable to the comfort of more polished life. The smoking pig, by a skilful dissection, was soon portioned to every guest, but no one ventured to put its excellent qualities to the test until a lengthened *Amen*, pronounced by all the party, had succeeded an emphatic grace delivered by the village parson. " *Turn to*" was then the signal for attack, and as it is convenient that all the party should finish their meal about the same time, in order that one grace might serve for all, each made the most of his time. In Pitcairn's Island it is not deemed proper to touch even a bit of bread without a grace before and after it, and a person is accused of inconsistency if he leaves off and begins again. So strict is their observance of this form, that we do not know of any instance in which it has been forgotten. On one occasion I had engaged Adams in conversation, and he incautiously took the first mouthful without having said his grace; but before he had swallowed it, he recollected himself, and feeling as if he had committed a crime, immediately put away what he had in his mouth, and commenced his prayer.

Welcome cheer, hospitality, and good-humour, were the characteristics of the feast; and never was their beneficial influence more practically exemplified than on this occasion, by the demolition of nearly all that was placed before us. With the exception of some wine we had brought with us, water was the only beverage. This was placed in a large jug at one end of the board, and when necessary, was passed round the table—a ceremony at which, in Pitcairn's Island in particular, it is desirable to be the first partaker, as the gravy of the dish is invariably mingled with the contents of the pitcher: the natives, who prefer using their fingers to forks, being quite indifferent whether they hold the vessel by the handle or by the spout. Three or four torches made with doodoe nuts (*aleurites triloba*), strung upon the fibres of a palm-leaf, were stuck in tin pots at the end of the table, and formed an excellent substitute for candles, except that they gave a considerable heat, and cracked, and fired, somewhat to the discomfiture of the person whose face was near them.

Notwithstanding these deficiencies, we made a very comfortable and hearty supper, heard many little anecdotes of the place, and derived much amusement from the singularity of the inquiries of our hosts. One regret only intruded itself upon the general conviviality, which we did not fail to mention, namely, that there was so wide a distinction between the sexes. This was the remains of a custom very common among the South-sea Islands, which in some places is carried to such an extent, that it imposes death upon the woman who shall eat in the presence of her husband; and though the distinction between man and wife is not here carried to that extent, it is still sufficiently observed to exclude all the women from table, if there happens to be a deficiency of seats. In Pitcairn's Island, they have settled ideas of right and wrong, to which they obstinately adhere; and, fortunately, they have imbibed them generally from the best source.

In the instance in question, they have, however, certainly erred; but of this they could not be persuaded, nor did they, I believe, thank us for our interference. Their argument was, that man was made first, and ought, consequently, on all occasions, to be served first—a conclusion which deprived us of the company of the women at table, during the whole of our stay at the island. Far from considering themselves neglected, they very good-naturedly chatted with us behind our seats, and flapped away the flies, and by a gentle tap, accidentally or playfully delivered, reminded us occasionally of the honour that was done us. The conclusion of our meal was the signal for the women and children to prepare their own, to whom we resigned our seats, and strolled out to enjoy the freshness of the night. It was late by the time the women had finished, and we were not sorry when we were shown to the beds prepared for us. The mattress was composed of palm-leaves, covered with native cloth; the sheets were of the same material; and we knew, by the crackling of them, that they were quite new from the loom, or beater. The whole arrangement was extremely comfortable, and highly inviting to repose, which the freshness of the apartment, rendered cool by a free circulation of air through its sides, enabled us to enjoy without any annoyance from heat or insects. One interruption only disturbed our first sleep; it was the pleasing melody of the evening hymn, which, after the lights were put out, was chaunted by the whole family in the middle of the room. In the morning also we were awoke by their morning hymn, and family devotion. As we were much tired, and the sun's rays had not yet found their way through the broad opening of the apartment, we composed ourselves to rest again; and on awaking found that all the natives were gone to their several occupations,—the men to offer what assistance they could to our boats in landing, carrying burthens for the seamen, or to gather what fruits were in season. Some of the women had taken our linen to wash; those whose turn it was to cook for the day were preparing the oven, the pig, and the yams; and we could hear, by the distant reiterated strokes of the beater,* that others were engaged in the manufacture of cloth. By our bedside had already been placed some ripe fruits; and our hats were crowned with chaplets of the fresh blossom of the nono, or flower-tree (*morinda citrifolia*), which the women had gathered in the freshness of the morning dew. On looking round the apartment, though it contained several beds, we found no par-

* This is an instrument used for the manufacture of their cloth.

tition, curtain, or screens; they had not yet been
considered necessary. So far, indeed, from concealment being thought of, when we were about to get
up, the women, anxious to show their attention, assembled to wish us a good morning, and to inquire
in what way they could best contribute to our comforts, and to present us with some little gift, which
the produce of the island afforded. Many persons
would have felt awkward at rising and dressing before so many pretty black-eyed damsels assembled
in the centre of a spacious room; but by a little
habit we overcame this embarrassment; and found
the benefit of their services in fetching water as we
required it, and substituting clean linen for such as
we pulled off.

It must be remembered, that with these people,
as with the other islanders of the South Seas, the
custom has generally been to go naked, the maro
with the men excepted, and with the women the
petticoat, or kilt, with a loose covering over the
bust, which, indeed, in Pitcairn's Island, they are
always careful to conceal; consequently, an exposure to that extent carried with it no feeling whatever of indelicacy; or, I may safely add, that the
Pitcairn's Islanders would have been the last persons
to incur the charge.

We assembled at breakfast about noon, the usual
eating hour of the natives, though they do not confine themselves to that period exactly, but take
their meal whenever it is sufficiently cooked; and
afterwards availed ourselves of their proffered services to show us the island, and under their guidance
first inspected the village, and what lay in its immediate vicinity. In an adjoining house we found two

young girls seated upon the ground, employed in the laborious exercise of beating out the bark of the cloth-tree, which they intended to present to us, on our departure, as a keepsake. The hamlet consisted of five cottages, built more substantially than neatly, upon a cleared patch of ground, sloping to the northward, from the high land of the interior to the cliffs which overhang the sea, of which the houses command a distant view in a northern direction. In the N. E. quarter, the horizon may also be seen peeping between the stems of the lofty palms, whose graceful branches nod like ostrich plumes to the refreshing trade-wind. To the northward, and north-westward, thicker groves of palm-trees rise in an impenetrable wood, from two ravines which traverse the hills in various directions to their summit. Above the one, to the westward, a lofty mountain rears its head, and toward the sea terminates in a fearful precipice filled with caverns, in which the different sea-fowl find an undisturbed retreat. Immediately round the village are the small enclosures for fattening pigs, goats, and poultry ; and beyond them, the cultivated grounds producing the banana, plantain, melon, yam, taro, sweet potatoes, appai, tee, and cloth plant, with other useful roots, fruits, and shrubs, which extend far up the mountain and to the southward ; but in this particular direction they are excluded from the view by an immense banyan tree, two hundred paces in circumference, whose foliage and branches form of themselves a canopy impervious to the rays of the sun. Every cottage has its out-house for making cloth, its baking-place, its sty, and its poultry-house.

Within the enclosure of palm-trees is the cemetery

F.W. Beechey del.ᵗ — E. Finden SC.

INTERIOR OF PITCAIRN [ISLAND]

Pub.ᵈ by H. Colburn & R. Bentley. 1831.

where the few persons who had died on the island, together with those who met with violent deaths, are deposited. Besides the houses above mentioned, there are three or four others built upon the plantations beyond the palm groves. One of these, situated higher up the hill than the village, belonged to Adams, who had retired from the bustle of the hamlet to a more quiet and sequestered spot, to enjoy the advantages of an elevated situation, so desirable in warm countries; and in addition to these again there are four other cottages to the eastward which belong to the Youngs and Quintals.

All these cottages are strongly built of wood in an oblong form, and thatched with the leaves of the palm-tree bent round the stem of the same branch, and laced horizontally to rafters, so placed as to give a proper pitch to the roof. The greater part have an upper story, which is appropriated to sleeping, and contain four beds built in the angles of the room, each sufficiently large for three or four persons to lie on. They are made of wood of the cloth-tree, and are raised eighteen inches above the floor; a mattress of palm-leaves is laid upon the planks, and above it three sheets of the cloth-plant, which form an excellent substitute for linen. The lower room generally contains one or more beds, but is always used as their eating-room, and has a broad table in one part, with several stools placed round it. The floor is elevated above a foot from the ground, and, as well as the sides of the house, is made of stout plank, and not of bamboo, or stone, as stated by Captain Folger; indeed they have not a piece of bamboo on the island; nor have they any mats. The floor is a fixture, but the sideboards are

let into a groove in the supporters, and can be removed at pleasure, according to the state of the
weather, and the whole side may, if required, be
laid open. The lower room communicates with the
upper by a stout ladder in the centre, and leads up
through a trap-door into the bedroom.

From the village several pathways (for roads
there are none) diverge, and generally lead into the
valleys, which afford a less difficult ascent to the
upper part of the island than the natural slope of
the hills; still they are very rugged and steep, and
in the rainy season so slippery that it is almost impossible for any person, excepting the natives, to
traverse them with safety. We selected one which
led over the mountain to the landing-place, on the
opposite side of the island, and visited the several
plantations upon the higher grounds, which extend
towards the mountain with a gentle slope. Here
the mutineers originally built their summer-houses,
for the purpose of enjoying the breeze and overlooking the yam grounds, which are more productive than those lower down. Near these plantations
are the remains of some ancient morais; and a spot
is pointed out as the place where Christian was
first buried. By a circuitous and, to us, difficult
path, we reached the ridge of the mountain, the
height of which is 1109 feet above the sea; this is
the highest part of the island. The ridge extends
in a north and south direction, and unites two small
peaks: it is so narrow as to be in many parts scarcely
three feet wide, and forms a dangerous pass between
two fearful precipices. The natives were so accustomed to climb these crags that they unconcernedly
skipped from point to point like the hunters of cha-

mois; and young Christian actually jumped upon the very peak of a cliff, which was so small as to be scarcely sufficient for his feet to rest upon, and from which any other person would have shuddered even to look down upon the beach, lying many hundred feet at its base. At the northern extremity of this ridge is a cave of some interest, as being the intended retreat of Christian, in the event of a landing being effected by any ship sent in pursuit of him, and where he resolved to sell his life as dearly as he could. In this recess he always kept a store of provisions, and near it erected a small hut, well concealed by trees, which served the purpose of a watchhouse. So difficult was the approach to this cave, that even if the party were successful in crossing the ridge, as long as his ammunition lasted, he might have bid defiance to any force. An unfrequented and dangerous path leads from this place to a peak which commands a view of the western and southern coasts: at this height, on a clear day, a perfect map of the bottom is exhibited by the different coloured waters. On all points the island is terminated by cliffs, or rocky projections: off which lie scattered numerous fragments of rock, rising like so many black pinnacles amid the surf, which on all sides rolls in upon the shore.

We descended by a less abrupt slope than that by which we advanced, and took our way through yam grounds to a ravine which brought us to the village. The path leading down this ravine is, in many places, so precipitous, that we were constantly in danger of slipping and rolling into the depths below, which the assistance of the natives alone prevented.

While we were thus borrowing help from others,

and grasping every tuft of grass and bough that
offered its friendly support, we were overtaken by a
group of chubby little children, trudging unconcernedly on, munching a water melon, and balancing
on their heads calabashes of water, which they had
brought from the opposite side of the island. They
smiled at our helplessness as they passed, and we felt
their innocent reproof; but we were still unpractised
in such feats, while they, from being trained to
them, had acquired a footing and a firmness which
habit alone can produce.

It was dark when we reached the houses, but we
found by a whoop which echoed through the woods,
that we were not the last from home. This whoop,
peculiar to the place, is so shrill, that it may be heard
half over the island, and the ear of the natives is so
quick, that they will catch it when we could distinguish nothing of the kind. By the tone in which
it is delivered, they also know the wants of the person, and who it is. These shrill sounds, which we
had just heard, informed us, and those who were at
the village, that a party had lost their way in the
woods. A blazing beacon was immediately made,
which, together with a few more whoops to direct
the party, soon brought the absentees home. Their
perfection in these signals will be manifest from the
following anecdote: I was one day crossing the
mountain which intersects the island with Christian;
we had not long parted with their whale-boat on the
western side of the island, and were descending a
ravine amidst a thicket of trees, when he turned
round and said, "The whale-boat is come round to
Bounty Bay;" at which I was not a little surprised,
as I had heard nothing, and we could not see through

the wood; but he said he heard the signal; and when we got down it proved to be the case.

In this little retreat there is not much variety, and the description of one day's occupation serves equally for its successor. The dance is a recreation very rarely indulged in; but as we particularly requested it, they would not refuse to gratify us. A large room in Quintal's house was prepared for the occasion, and the company were ranged on one side of the apartment, glowing beneath a blazing string of doodoe nuts; the musicians were on the other, under the direction of Arthur Quintal. He was seated upon the ground, as head musician, and had before him a large gourd, and a piece of musical wood (porou), which he balanced nicely upon his toes, that there might be the less interruption to its vibrations. He struck the instrument alternately with two sticks, and was accompanied by Dolly, who performed very skilfully with both hands upon a gourd, which had a longitudinal hole cut in one end of it; rapidly beating the orifice with the palms of her hands, and releasing it again with uncommon dexterity, so as to produce a tattoo, but in perfect time with the other instrument. A third performed upon the Bounty's old copper fish-kettle, which formed a sort of bass. To this exhilarating music, three *grown-up* females stood up to dance, but with a reluctance which showed it was done only to oblige us, as they consider such performances an inroad upon their usual innocent pastimes. The figure consisted of such parts of the Otaheitan dance as were thought most decorous, and was little more than a shuffling of the feet, sliding past each other, and snapping their fingers; but even this produced,

at times, considerable laughter from the female spectators, perhaps from some association of ridiculous
ideas, which we, as strangers, did not feel; and no
doubt had our opinion of the performance been consulted, it would have essentially differed from theirs.
They did not long continue these diversions, from
an idea that it was too great a levity to be continued
long; and only the three beforementioned ladies
could be prevailed upon to exhibit their skill.   One
of the officers, with a view of contributing to the
mirth of the colonists, had obligingly brought his
violin on shore, and, as an inducement for them to
dance again, offered to play some country dances
and reels, if they would proceed; but they could
not be tempted to do so.   They, however, solicited
a specimen of the capabilities of the instrument,
which was granted, and, though very well executed,
did not give the satisfaction which we anticipated.
They had not yet arrived at a state of refinement to
appreciate harmony, but were highly delighted with
the rapid motion of the fingers, and always liked
to be within sight of the instrument when it was
played.   They were afterwards heard to say, that
they preferred their own simple musical contrivance
to the violin.   They did not appear to have the least
ear for music: one of the officers took considerable
pains to teach them the hundredth psalm, that they
might not chaunt all the psalms and hymns to the
same air; but they did not evince the least aptitude
or desire to learn it.

   The following day was devoted to the completion
of our view of the island, of which the natives were
anxious we should see every part.   We accordingly
set out with the same guides by a road which

brought us to " the Rope," a steep cliff, so called
from its being necessary to descend it by a rope. It
is situated at the eastern end of the island, and over-
looks a small sandy bay lined with rocks, which render
it dangerous for a boat to attempt to land there.

At the foot of " the Rope" were found some
stone axes, and a hone, the manufacture of the aborigines, and upon the face of a large rock were some
characters very rudely engraved, which we copied ;
they appeared to have been executed by the Bounty's people, though Adams did not recollect it. To
the left of " the Rope" is a peak of considerable
height, overlooking Bounty Bay. Upon this eminence the mutineers, on their arrival, found four
images, about six feet in height, placed upon a platform ; and, according to Adams's description, not
unlike the morais at Easter Island, excepting that
they were upon a much smaller scale. One of these
images, which had been preserved, was a rude representation of the human figure to the hips, and was
hewn out of a piece of red lava.

Near this supposed morai, we were told that human bones and stone hatchets were occasionally dug
up, but we could find only two bones, by which we
might judge of the stature of these aborigines.
These were an os femoris and a part of a cranium
of an unusual size and thickness. The hatchets, of
which we obtained several specimens, were made of
a compact basaltic lava, not unlike clinkstone, very
hard and capable of a fine polish. In shape they
resembled those used at Otaheite, and by all the
islanders of these seas that I have seen. A large
stone bowl was also found, similar to those used at
Otaheite, and two stone huts. That this island

should have been inhabited is not extraordinary, when it is remembered that Easter Island, which is much more distant from the eastern world, was so, though nothing is known of the fate of the people.

From these images, and the large piles of stones on heights to which they must have been dragged with great labour, it may be concluded that the island was inhabited a considerable time ; and from bones being found always buried under these piles, and never upon the surface, we may presume that those who survived quitted the island in their canoes to seek an asylum elsewhere.

Having this day seen every part of the island, we had no further desire to ramble ; and as the weather did not promise to be very fair, I left the observatory in the charge of Mr. Wolfe, and embarked, accompanied by old Adams. Soon after he came on board it began to blow, and for several days afterwards the wind prevented any communication with the shore. The natives during this period were in great apprehension : they went to the top of the island every morning to look for the ship ; and once, when she was not to be seen, began to entertain the most serious doubts whether Adams would be returned to them; but he, knowing we should close the island as soon as the weather would permit, was rather glad of the opportunity of remaining on board, and of again associating with his countrymen. And although he had passed his sixty-fifth year, joined in the dances and songs of the forecastle, and was always cheerful.

On the 16th the weather permitted a boat to be sent on shore, and Adams was restored to his anxious friends. Previous to quitting the ship, he said it

would add much to his happiness if I would read
the marriage ceremony to him and his wife, as he
could not bear the idea of living with her without
its being done. He had long wished for the arrival
of a ship of war to set his conscience at rest on that
point. Though Adams was aged, and the old woman had been blind and bed-ridden for several years,
he made such a point of it, that it would have been
cruel to refuse him. They were accordingly the
next day duly united, and the event noted in a
register by John Buffet.

The islanders were delighted at having us again
among them, and expressed themselves in the warmest terms. We soon found, through our intercourse
with these excellent people, that they had no wants
excepting such as had been created by an intercourse
with vessels, which have from time to time supplied
them with European articles. Nature has been extremely bountiful to them; and necessity has taught
them how to apply her gifts to their own particular
uses. Still they have before them the prospect of
an increasing population, with limited means of supporting it. Almost every part of the island capable
of cultivation has been turned to account; but what
would have been the consequences of this increase,
had not an accident discovered their situation, it is
not difficult to foresee; and a reflecting mind will
naturally trace in that disclosure the benign interference of the same hand which has raised such a
virtuous colony from so guilty a stock. Adams
having contemplated the situation which the islanders would have been reduced to, begged, at our first
interview, that I would communicate with the government upon the subject, which was done; and I

am happy to say that, through the interference of the Admiralty and Colonial Office, means have been taken for removing them to any place they may choose for themselves; and a liberal supply of useful articles has recently been sent to them.*

Some books of travels which were left from time to time on the island, and the accounts they had heard of foreign countries from their visiters, has created in the islanders a strong desire to travel, so much so that they one day undertook a voyage in their whale-boat to an island which they learnt was not very far distant from their own; but fortunately for them, as the compass on which they relied, one of the old Bounty's, was so rusty as to be quite useless, their curiosity yielded to discretion, and they returned before they lost sight of their native soil.

The idea of passing all their days upon an island only two miles long, without seeing any thing of the world, or, what was a stronger argument, without doing any good in it, had with several of them been deeply considered. But family ties, and an ardent affection for each other, and for their native soil, had always interposed to prevent their going away singly. George Adams, however, having no wife to detain him, but, on the contrary, reasons for wishing to employ his thoughts on subjects foreign to his home, was very anxious to embark in the Blossom; and I would have acceded to his wishes, had not his mother wept bitterly at the idea of parting from him, and imposed terms touching his return to the island to which I could not accede. It was a sore disappointment to poor George, whose

* I have been informed since that they have changed their mind, and are at present contented with their situation.

case forms a striking instance of the rigid manner
in which these islanders observe their word.

Wives upon Pitcairn Island, it may be imagined, are very scarce, as the same restrictions with regard to relationship exist as in England. George,
in his early days, had fallen in love with Polly
Young, a girl a little older than himself; but
Polly, probably at that time liking some one else,
and being at the age when young ladies' expectations are at the highest, had incautiously said, she
never would give her hand to George Adams. He,
nevertheless, indulged a hope that she would one
day relent; and to this end was unremitting in his
endeavours to please her. In this expectation he
was not mistaken; his constancy and attentions,
and, as he grew into manhood, his handsome form,
which George took every opportunity of throwing
into the most becoming attitudes before her, softened
Polly's heart into a regard for him, and, had nothing
passed before, she would willingly have given him
her hand. But the vow of her youth was not to be
got over, and the love-sick couple languished on from
day to day, victims to the folly of early resolutions.

The weighty case was referred for our consideration; and the fears of the party were in some measure relieved by the result, which was, that it would
be much better to marry than to continue unhappy,
in consequence of a hasty determination made before the judgment was matured; they could not,
however, be prevailed on to yield to our decision,
and we left them unmarried.*

Another instance of a rigid performance of promise was exemplified in old Adams, who is anxious

* They have since been united, and have two children.

that his own conduct should form an example to the
rising generation.

In the course of conversation, he one day said he
would accompany me up the mountain, if there was
nobody else near; and it so happened, that on the
only day I had leisure to go the young men were
all out of the way. Adams, therefore, insisted upon
performing his engagement, though the day was
extremely hot, and the journey was much too laborious, in any weather, for his advanced period of
life. He nevertheless set out, adding, " I said I
would go, and so I will; besides, without example
precept will have but little effect." At the first
valley he threw off his hat, handkerchief, and jacket,
and left them by the side of the path; at the second his trousers were cast aside into a bush; and
had he been alone, or provided with a maro, his
shirt would certainly have followed: thus disen-
cumbered, he boldly led the way, which was well
known to him in earlier days; but it was so long
since he had trodden it, that we met with many
difficulties. At length we reached the top of the
ridge, which we were informed was the place where
M'Coy and Quintal had appeared in defiance of
the blacks. Adams felt so fatigued that he was
now glad to lie down. The breeze here blew so
hard and cold, that a shirt alone was of little use,
and had he not been inured to all the changes of
atmosphere, the sudden transition upon his aged
frame must have been fatal.

During the period we remained upon the island
we were entertained at the board of the natives,
sometimes dining with one person, and sometimes
with another: their meals, as I have before stated,

were not confined to hours, and always consisted of
baked pig, yams, and taro, and more rarely of sweet
potatoes.

The productions of the island being very limited,
and intercourse with the rest of the world much restricted, it may be readily supposed their meals cannot be greatly varied. However, they do their best
with what they have, and cook it in different ways,
the pig excepted, which is always baked. There are
several goats upon the island, but they dislike their
flesh, as well as their milk. Yams constitute their
principal food; these are boiled, baked, or made into
pillihey (cakes), by being mixed with cocoa nuts;
or bruised and formed into a soup. Bananas are
mashed, and made into pancakes, or, like the yam,
united with the milk of the cocoa-nut, into pillihey,
and eaten with molasses, extracted from the tee-
root. The taro-root, by being rubbed, makes a
very good substitute for bread, as well as the bananas, plantains, and appai. Their common beverage
is pure water, but they made for us a tea, extracted from the tee-plant, flavoured with ginger, and
sweetened with the juice of the sugar-cane. When
alone, this beverage and fowl soup are used only
for such as are ill. They seldom kill a pig, but live
mostly upon fruit and vegetables. The duty of
saying grace was performed by John Buffet, a recent settler among them, and their clergyman; but
if he was not present, it fell upon the eldest of the
company. They have all a great dislike to spirits,
in consequence of M'Coy having killed himself by
too free an indulgence in it; but wine in moderation is never refused. With this simple diet, and
being in the daily habit of rising early, and taking a

great deal of exercise in the cultivation of their grounds, it was not surprising that we found them so athletic and free from complaints. When illness does occur, their remedies are as simple as their manner of living, and are limited to salt water, hot ginger tea, or abstinence, according to the nature of the complaint. They have no medicines, nor do they appear to require any, as these remedies have hitherto been found sufficient.

After their noontide meal, if their grounds do not require their attention, and the weather be fine, they go a little way out to sea in their canoes, and catch fish, of which they have several kinds, large and sometimes in abundance; but it seldom happens that they have this time to spare; for the cultivation of the ground, repairing their boats, houses, and making fishing-lines, with other employments, generally occupy the whole of each day. At sunset they assemble at prayers as before, first offering their orison and thanksgiving, and then chaunting hymns. After this follows their evening meal, and at an early hour, having again said their prayers, and chaunted the evening hymn, they retire to rest; but before they sleep, each person again offers up a short prayer upon his bed.

Such is the distribution of time among the grown people; the younger part attend at school at regular hours, and are instructed in reading, writing, and arithmetic. They have very fortunately found an able and willing master in John Buffet, who belonged to a ship which visited the island, and was so infatuated with their behaviour, being himself naturally of a devout and serious turn of mind, that he resolved to remain among them; and in addition to the instruction of the children, has taken upon himself the duty of clergyman, and is the oracle of the community.* During the whole time I was with them, I never heard them indulge in a joke, or other levity, and the practice of it is apt to give offence: they are so accustomed to take what is said in its literal meaning, that irony was always considered a falsehood, in spite of explanation. They could not see the propriety of uttering what was not strictly true, for any purpose whatever.

The Sabbath-day is devoted entirely to prayer, reading, and serious meditation. No boat is allowed to quit the shore, nor any work whatever to be done, cooking excepted, for which preparation is made the preceding evening. I attended their church on this day, and found the service well conducted; the prayers were read by Adams, and the lessons by Buffet, the service being preceded by hymns. The greatest devotion was apparent in every individual, and in the children there was a seriousness unknown in the younger part of our communities at home. In the course of the Litany they prayed for their sovereign and all the royal family with much apparent loyalty and sincerity. Some family prayers, which were thought appropriate to their particular case, were added to the usual service; and Adams, fearful of leaving out any essential part, read in addition all those prayers which are intended only as substitutes for others. A sermon followed, which was very well delivered by Buffet; and lest any part of it should be forgotten or escape attention, it was read three times. The whole concluded with

* Another seaman has settled amongst them, and is married to one of Adams' daughters; but he is not liked.

hymns, which were first sung by the grown people, and afterwards by the children. The service thus performed was very long; but the neat and cleanly appearance of the congregation, the devotion that animated every countenance, and the innocence and simplicity of the little children, prevented the attendance from becoming wearisome. In about half an hour afterwards we again assembled to prayers, and at sunset service was repeated; so that, with their morning and evening prayers, they may be said to have church five times on a Sunday.

Marriages and christenings are duly performed by Adams. A ring which has united every person on the island is used for the occasion, and given according to the prescribed form. The age at which this is allowed to take place, with the men, is after they have reached their twentieth, and with the women, their eighteenth year.

All which remains to be said of these excellent people is, that they appear to live together in perfect harmony and contentment; to be virtuous, religious, cheerful, and hospitable, beyond the limits of prudence; to be patterns of conjugal and parental affection; and to have very few vices. We remained with them many days, and their unreserved manners gave us the fullest opportunity of becoming acquainted with any faults they might have possessed.

In the equipment of the Blossom a boat was built purposely for her by Mr. Peake of Woolwich dockyard, upon a model highly creditable to his professional ability, and finished in the most complete manner. As we were now about to enter a sea crowded with islands which rise abruptly to the surface, without

any soundings to give warning of their vicinity, this
little vessel was likely to be of the greatest service,
not only in a minute examination of the shore, but,
by being kept a-head of the ship during the night,
to give notice of any danger that might lie in her
route. She was accordingly hoisted out while we
were off this island, and stowed and provisioned for
six weeks. I gave the command of her to Mr. Elson, the master, an officer, well qualified to perform
the service I had in view; having with him Mr. R.
Beechey, midshipman, and a crew of eight seamen
and marines. Instructions were given to Mr. Elson
for his guidance, and proper rendezvous appointed
in case of separation. We first experienced the utility of this excellent sea-boat, in bringing off water
from the shore through seas which in ordinary cases
would have proved serious obstacles ; and had there
not been so much surf upon the rocks, that the casks
could only be got through it by the natives swimming out with them, we should in a short time have
completed our stock of water. This process, however, was very harassing to them, who, besides this
arduous task, had to bring the water from a distance
in calabashes ; so, that with the utmost despatch,
our daily supply scarcely equalled the consumption,
and we were compelled to trust to the hope of being
more fortunate at some other island.

During the period of our stay in the vicinity of
the island, we scarcely saw the sun, and I began to
despair of being able to fix our position with sufficient accuracy. On the 20th, however, the clouds
cleared away, and the night was passed in obtaining
lunar distances with stars east and west of the moon,
several meridional altitudes, and transits which, com-

pared with those taken the first night the instrument was put up, gave good rates to the chronometers. Our labours having thus terminated more
successfully than we expected, we hastened our
embarkation, which took place on the 21st. In return for the kindness we experienced from the
islanders, we made them presents of articles the
most useful to them which we could spare, and they
were furnished with a blue cloth suit each from the
extra clothing put on board for the ship's company,
and the women with several pieces of gowns and
handkerchiefs, &c.

When we were about to take leave, our friends
assembled to express their regret at our departure.
All brought some little present for our acceptance,
which they wished us to keep in remembrance of
them; after which they accompanied us to the
beach, where we took our leave of the female part
of the inhabitants. Adams and the young men
pushed off in their own boat to the ship, determined
to accompany us to sea as far as they could with
safety. They continued on board, unwilling to
leave us, until we were a considerable distance from
land, when they shook each of us feelingly by the
hand, and, amidst expressions of the deepest concern
at our departure, wished us a prosperous voyage,
and hoped that we might one day meet again. As
soon as they were clear of the ship they all stood up
in their boat, and gave us three hearty cheers, which
were as heartily returned. As the weather became
foggy, the barge towed them towards the shore, and
we took a final leave of them, unconscious until the
moment of separation of the warm interest their
situation and good conduct had created in us.

The Pitcairn islanders are tall, robust, and healthy. Their average height is five feet ten inches; the tallest person is six feet and one quarter of an inch; and the shortest of the adults is five feet nine inches and one-eighth. Their limbs are well-proportioned, round, and straight; their feet turning a little inwards. The boys promise to be equally as tall as their fathers; one of them whom we measured was, at eight years of age, four feet one inch; and another, at nine years, four feet three inches. Their simple food and early habits of exercise give them a muscular power and activity not often surpassed. It is recorded among the feats of strength which these people occasionally evince, that two of the strongest on the island, George Young and Edward Quintal, have each carried, at one time, without inconvenience, a kedge anchor, two sledge hammers, and an armourer's anvil, amounting to upwards of six hundred weight; and that Quintal, at another time, carried a boat twenty-eight feet in length. Their activity on land has been already mentioned. I shall merely give another instance which has been supplied by Lieutenant Belcher, who was admitted to be the most active among the officers on board, and who did not consider himself behindhand in such exploits. He offered to accompany one of the natives down a difficult descent, in spite of the warnings of his friend that he was unequal to the task. They, however, commenced the perilous descent, but Mr. Belcher was obliged to confess his inability to proceed, while his companion, perfectly assured of his own footing, offered him his hand, and undertook to conduct him to the bottom, if he would depend on him for safety. In the water they are almost as much at home as on land, and can remain nearly a whole day in the sea. They frequently swam round their little island, the circuit of which is at the least seven miles. When the sea beat heavily on the island they have plunged into the breakers, and swam to sea beyond them. This they sometimes did pushing a barrel of water before them, when it could be got off in no other way, and in this manner we procured several tons of water without a single cask being stove.

Their features are regular and well-looking, without being handsome. Their eyes are bright and generally hazel, though in one or two instances they are blue, and some have white speckles on the iris; the eyebrows being thin, and rarely meeting. The nose, somewhat flat, and rather extended at the nostrils, partakes of the Otaheitan form, as do the lips, which are broad, and strongly sulcated. Their ears are moderately large, and the lobes are invariably united to the cheek; they are generally perforated when young, for the reception of flowers, a very common custom among the natives of the South Sea Islands. The hair, in the first generation, is, with one exception only, deep black, sometimes curly, but generally straight; they allow it to grow long, keep it very clean, and always well supplied with cocoa-nut oil. Whiskers are not common, and the beards are thin. The teeth are regular and white; but are often, in the males, disfigured by a deficiency in enamel, and by being deeply furrowed across. They have generally large heads, elevated in the line of the occiput. A line passed above the eyebrows, over the ears, and round the back of the head, in a line with the occipital spine, including the hair, measured twenty-two inches; another, twenty-one inches and three-quarters; and in Polly Young, surnamed Big-head, twenty-three inches,—the hair would make a difference of about three-quarters of an inch. The coronal region is full; the forehead of good height and breadth, giving an agreeable openness to the countenance; the middle of the coronal suture is rather raised above the surrounding parts. Their complexion, in the first generation, is, in general, a dark gipsy hue: there are, however, exceptions to this; some are fairer, and others, Joseph Christian in particular, much darker.*

The skin of these people, though in such robust health, compared with our own, always felt cold; and their pulses were considerably lower than ours. Mr. Collie examined several of them: in the fore-noon he found George Young's only sixty; three others, in the afternoon, after dinner, were sixty-eight, seventy-two, and seventy-six; while those of the officers who stood the heat of the climate best were above eighty. Constant exposure to the sun, and early training to labour, make these islanders look at least eight years older than they really are.

The women are nearly as muscular as the men, and taller than the generality of their sex. Polly Young, who is not the tallest upon the island, mea-sured five feet nine inches and a half. Accustomed to perform all domestic duties, to provide wood for cooking, which is there a work of some labour, as it

* This man was idiotic, and differed so materially from the others in colour that he is in all probability the offspring of the men of colour who accompanied the mutineers to the island, and who, un-less he be one, have left no progeny.

must be brought from the hills, and sometimes to
till the ground, their strength is in proportion to
their muscularity; and they are no less at home in
the water than the men.

The food of the islanders consists almost entirely
of vegetable substances. On particular occasions,
such as marriages or christenings, or when visited by
a ship, they indulge in pork, fowls, and fish. Although, as has already been mentioned, they discovered a method of distilling a spirit from the tee-
root, the miseries it entailed on them have taught
them to discontinue the use of it, and to confine
themselves strictly to water, of which, during meals,
they partake freely, but they seldom use it at other
times. The spirit, which was first distilled by
M'Coy, and led to such fatal consequences, bears
some affinity to peat-reeked whisky.

The treatment of their children differs from that
of our own country, as the infant is bathed three
times a day in cold water, and is sometimes not
weaned for three or four years; but as soon as that
takes place it is fed upon " popoe," made with ripe
plantains and boiled taro rubbed into a paste. Upon
this simple nourishment children are reared to a more
healthy state than in other countries, and are free
from fevers and other complaints peculiar to the
greater portion of the world. Mr. Collie remarks in
his journal, that nothing is more extraordinary in the
history of the island than the uniform good health of
the children; the teething is easily got over, they
have no bowel complaints, and are exempt from
those contagious diseases which affect children in
large communities. He offered to vaccinate the
children as well as all the grown persons; but they

deemed the risk of infection to be too small to render that operation necessary.

In rainy weather, and after the occasional visits of vessels, the islanders are more affected with plethora and boils than at other periods; to the former the whole population appear to be inclined, but they are usually relieved from its effects by bleeding at the nose; and, without searching for the real cause, they have imbibed a belief that these diseases are contagious, and derived from a communication with their visiters, although there may not be a single case of the kind on board the ship. The result naturally leads to such a conclusion; but a little reflection ought to have satisfied them, that a deviation from their established habits, an unusual indulgence in animal food, and additional clothing, were of themselves sufficient to account for the maladies. They are, however, unaccustomed to trace effects to latent causes. Hence they assert, that the Briton left them headaches and flies; a whaler infected with the scurvy (for which several of her crew pursued the old remedy of burying the people up to the necks in the earth) left them a legacy of boils and other sores; and though we had no diseases on board the Blossom, they fully expected to be affected by some cutaneous disorder after our departure; and even attributed some giddiness and headaches that were felt during our stay to infection from the ship's company.

The women have all learned the art of midwifery: parturition generally takes place during the night-time; the duration of labour is seldom longer than five hours, and has not yet in any case proved fatal. There is no instance of twins, nor of a single miscarriage, except from accident.

We found upon Pitcairn Island, cocoa-nuts, bread-fruit (*artocarpus incisa*), plantains (*musa paradaisa-ca*), bananas (*musa sapientum*), water-melons (*cucur-bita citrullus*), pumpkins (*cucurbita pépo*), potatoes (*solanum esculentum*), sweet potatoes (*convolvulus ba-tatas*), yams (*dioscoria sativum*), taro (*caladium escu-lentum*), peas, yappai* (*arum costatum*), sugar-cane, ginger, turmeric, tobacco, tee-plant* (*dracæna ter-minalis*), doodoe* (*aleurites triloba*), nono* (*morinda citrofolia*), another species of morinda, parau* (*hibis-cus tiliaceus*), fowtoo* (*hibiscus tricuspis*), the cloth-tree (*broussonetia papyrifera*), pawalla* (*pandanus odoratissimus?*), toonena* (?), and banyan-tree. A species of metrosideros, and several species of ferns.

The first twelve of these form the principal food of the inhabitants. The sugar-cane is sparingly cul-tivated; they extract from it a juice which is used to flavour the tea of such as are ill, by pounding the cane, and boiling it with a little ginger and cocoa-nut grated into a pulp, as a substitute for milk. In this manner a pleasant beverage is produced.

The tee-plant is very extensively cultivated. Its leaves, which are broad and oblong, are the common food of hogs and goats, and serve the natives for wrappers in their cooking. The root affords a very saccharine liquor, resembling molasses, which is ob-tained by baking in the ground; it requires two or three years after it is planted to arrive at the proper size for use, being then about two inches and a half in diameter; it is long, fusiform, and beset with fibres: from this root they also make a tea, which when flavoured with ginger is not unpleasant. The

---

* Native names.—A more correct account of the botany will be published by Dr. Hooker, Professor of Botany, &c. of Glasgow.

doodoe is a large tree, with a handsome blossom, and
supplies ornaments for the ears and hair, and nuts
containing a considerable quantity of oil, which, by
being strung upon sticks, serve the purpose of candles.   The porou and fowtoo are trees which supply
them with fishing-lines, rope, and cord of all sorts.
The tree is stripped of the bark while the sap is in
full circulation, and dried; a fibrous substance is
then procured from it, which is twisted for use; but
it is not strong, and is very perishable.

The cloth-tree is pre-eminently useful; and here,
as in all places in the South-Seas where it grows,
supplies the natives with clothing.   The manner
in which the cloth is manufactured has been frequently described, and needs no repetition.   There
is, however, a fashion in *the beater*, some preferring
a broad, others a very closely ribbed garment ; for
which purpose they have several of these instruments with large and small grooves.   If the cloth is
required to be brown, the inner bark of which the
cloth is made is wrapped in banana leaves, and put
aside for about four days; it is then beaten into a
thick doughy substance, and again left till ferment-
ation is about to take place, when it is taken out,
and finally beat into a garment both lengthwise and
across.   The colour thus produced is of a deep reddish brown hue.   The pieces are generally sufficiently large to wrap round the whole body, but
they are sometimes divided.

The toonena is a large tree, from which their
houses and canoes are made.   It is a hard, heavy,
red-coloured wood, and grows on the upper parts of
the island.   There was formerly a great abundance
of this wood, but it is now become so scarce as to

require considerable search and labour to find sufficient to construct a house. The young trees have
thriven but partially, arriving at a certain growth,
and then stopping. A tree of this kind, which was
the largest in the island, measured, at the time of
our visit, twelve feet in circumference ; another was
nine feet seven and a half inches in girth, at five
feet from the root; its trunk grew to the height
of thirty feet, perfectly straight, and without
branching.

The banyan is one of those large spreading trees
common in India. Nature has been so provident to
this island that there are very few trees in it which
cannot be turned to account in some way, and this
tree, though it yields no fruit, and produces wood
so hard and heavy as to be unserviceable, still con-
tributes to the assistance of the islanders, by supplying them with a resin for the seams of their
boats, &c. This useful substance is procured by
perforating the bark of the tree, and extracting the
liquor which exudes through the aperture.

We saw dyes of three colours only in Pitcairn
Island, yellow, red, and brown. The yellow is procured from the inner bark of the root of the nono
tree (*morinda citrifolia*), and also from the root of a
species of ginger. We did not see this plant growing, but it was described as having leaves broader
and longer than the common ginger, a thicker root
in proportion to its length, a darker hue, and not so
tubercular. The red dye is procured from the inner bark of the doodoe tree, and may have its intensity varied by more or less exposure to the rays of
the sun while drying. These dyes are well coloured,
but for want of proper mordants the natives cannot

fix them, and they must be renewed every time the linen is washed. The method of producing the brown dye has already been described.

The temperate climate of Pitcairn Island is extremely favourable to vegetation, and agriculture is attended with comparatively light labour. But as the population is increasing, and wants are generated which were before unthought of, the natives find it necessary to improve their mode of culture; and for this purpose they make use of sea-weed as manure. They grow but one crop in a year of each kind. The time of taking up the yams, &c. is about April. The land is not allowed time to recover itself, but is planted again immediately. Experience has enabled them to estimate, with tolerable precision, the quantity that will be required for the annual consumption of the island; this they reckon at 1000 yams to each person. The other roots, being considered more as luxuries, are cultivated in irregular quantities. The failure of a crop, so exactly estimated, must of course prove of serious consequence to the colony, and much anxiety is occasionally felt as the season approaches for gathering it. At times cold south-westerly winds nip the young plants, and turn such as are exposed to them quite black : during our visit several plantations near the sea-coast were affected in this manner. At other times caterpillars prove a great source of annoyance.

The yam is reproduced in the same manner as potatoes in England. The taro (*caladium esculentum*) requires either a young shoot to be broken off and planted, or the stem to be removed from the root, and planted after the manner of raising pine-apples. The yappe is a root very similar to the taro, and is treated in the same manner. All the above-mentioned farinaceous roots thrive extremely well in Pitcairn Island; but this is not the case with English potatoes, which cannot be brought even to a moderate growth. Peas and beans yield but very scanty crops, the soil being probably too dry for them, and are rarely seen at the repasts of the natives. Onions, so universally dispersed over the globe, cannot be made to thrive here. Pumpkins and water-melons bear exceedingly well, but the bread-fruit, from some recent cause, is beginning to give very scanty crops. This failure Adams attributes to some trees being cut down, that protected them from the cold winds, which is not improbable; for at Otaheite, where the trees are exposed to the south-west winds, the crops are very indifferent.

Having given this short sketch of the soil and vegetation of the island, I shall add a few words on the climate and winds.

The island is situated just without the regular limit of the trade-winds, which, however, sometimes reach it. When this is the case, the weather is generally fine and settled. The south-west and north-west winds, which blow strong and bring heavy rains, are the chief interruptions to this serenity. Though they have a rainy season, it is not so limited or decided as in places more within the influence of the trade-winds. During the period of our visit, from the 5th of December to the 21st, we had strong breezes from N. E. to S. E., with the sky overcast. The wind then shifted to N. W., and brought a great deal of rain: though in the height of summer, we had scarcely a fine day during our stay.

The temperature of the island during the above period was $70\frac{1}{2}$°. On shore the range from nine A. M. to three P. M. was 76° to 80°: on board at the same time from 74° to 76°. Taking the difference between these comparisons, we may place the mean temperature on shore for the above-mentioned period at $76\frac{1}{2}$°. In the winter the south-westerly winds blow very cold, and even snow has been known to fall.

The number of persons on Pitcairn Island in December, 1825, amounted to sixty-six, and for the information of such as may be disposed to give their particular attention to such an inquiry, I subjoin a notice of the population from the period of its first establishment on the island.

		Males.	Females.
The first settlers consisted of . . { white . .		9	0
coloured .		6	12
27 Total.		15	12
Of these were killed in the quarrel { white . .		6	0
coloured .		6	0
. . . . . by accident . . . white . .		1	3
. . . died a natural death		1	3
1 went away. Total deaths		14	6
The original settlers therefore whom we found on the island were . . . . }		1	5
The children of the white settlers (the men of colour having left none) . . . . . . }		10	10
Their grandchildren		22	15
Recent settlers		2	0
Child of one of them		1	0
66 present population.		36	30

The total number of children left by the white settlers was fourteen, of whom two died a natural death; one was seized with fits, to which he was subject, while in the water, and was drowned; and one was killed by accident, leaving ten, as above. Of the grand-children, or second generation, there was also another male who died an accidental death. There have, therefore, been sixty-two births in the period of thirty-five years, from the 23d January, 1790, to the 23d December, 1825, and only two natural deaths.

In a climate so temperate, with but few probabilities of infection, with simple diet, cleanly habits, moderate exercise, and a cheerful disposition, it was to be expected that early mortality would be of rare occurrence; and accordingly we find in this small community that the difference in the proportion of deaths to births is more striking than even in the most healthy European nations.

CHAPTER V.

Visit Oeno Island—Description of it—Loss of a Boat and one
Seaman—Narrow escape of the Crew—Crescent Island—Gambier
Groupe—Visited by Natives on Rafts—Discover a Passage into
the Lagoon—Ship enters—Interview with the Natives—Anchor
off two Streams of Water—Visited by the Natives—Theft—
Communication with them suspended—Morai—Manner of preserving the Dead—Idols and Places of Worship.

As soon as Adams and his party left us we spread
every sail in the prosecution of our voyage, and to
increase our distance from a climate in which we
had scarcely had the decks dry for sixteen days;
but the winds were so light and unfavourable, that
on the following morning Pitcairn Island was still
in sight. The weather was hazy and moist, and the
island was overhung with dense clouds, which the
high lands seemed to attract, leaving no doubt with
us of a continuation of the weather we had experienced while there. At night there was continued
lightning in this direction. Several birds of the
pelican tribe (*pelicanus leucocephalus*) settled upon
the masts and allowed themselves to be taken by
the seamen.

About ninety miles to the northward of Pitcairn
Island there is a coral formation, which has been

named Oeno Island, after a whale-ship, whose master supposed it had not before been seen; but the discovery belongs to Mr. Henderson of the Hercules. It is so low that it can be discerned at only a very few miles distance, and is highly dangerous to a night navigation. As this was the next island I intended to visit, every effort was made to get up to it; and at one o'clock in the afternoon of the 23d December it was seen a little to leeward of us. We had not time to examine it that evening, but on the following morning we passed close to the reefs in the ship, in order to overlook the lagoon that was formed within them, and to search narrowly for an opening into it. While the ship took one side of the island, the barge closely examined the other, and we soon found that the lagoon was completely surrounded by the reef. Near the centre of it there was a small island covered with shrubs; and towards the northern extremity, two sandy islets a few feet above the water. The lagoon was in places fordable as far as the wooded island; but, in other parts, it appeared to be two or three fathoms deep. The reef is entirely of coral formation, similar to Ducie's Island, and has deep water all round it. Just clear of the breakers there are three or four fathoms water; the next cast finds thirteen fathoms; then follow rapidly thirty fathoms, sixty fathoms, and no bottom at a hundred fathoms. We found the south-western part of the reef the highest, and the lagoon in that direction nearly filled up as far as the island with growing coral. There were, of course, no inhabitants upon so small a spot; nor should we have been able to communicate had there been any, in consequence of a surf rolling heavily over all parts of the reef, and with such unequal violence that the treacherous smoothness would one moment tempt a landing, while the next wave, as we unfortunately experienced, would prove fatal to any boat that should hazard it.

Lieutenant Belcher was sent to ascertain the depth of water round the island, with permission to land if unattended with danger; and Mr. Collie accompanied him, Mr. Barlow being midshipman of the boat. Pulling round the island, they came to a place where the sea appeared tolerably smooth, and where in the opinion of the officers a landing might be effected. The boat was accordingly anchored, and Messrs. Belcher and Collie prepared to land, by veering the boat into the surf, and jumping upon the reef. They had half filled two life-preservers, with which they were provided, when Mr. Belcher observed a heavy roller rising outside the boat, and desired the crew to pull and meet it, which was done, and successfully passed; but a second rose still higher, and came with such violence that the sitters in the stern of the boat were thrown into the sea; a third of still greater force carried all before it, upset the boat, and rolled her over upon the reef, where she was ultimately broken to pieces. Mr. Belcher had a narrow escape, the boat being thrown upon him, the gunwale resting upon his neck and keeping him down; but the next sea extricated him, and he went to the assistance of his companions; all of whom were fortunately got upon the reef, except one young lad, who probably became entangled with the coral, and was drowned. The accident was immediately perceived from the ship, and all the boats sent to the assistance of the survivors. But the

surf rolled so furiously upon the shore as to occasion much anxiety about rescuing them. At last a
small raft was constructed, and Lieutenant Wainwright finding no other means of getting a line to
them, boldly jumped overboard, with a lead-line in
his hand, and suffered himself to be thrown upon
the reef. By this contrivance all the people were
got off, one by one, though severely bruised and
wounded by the coral and spines of the echini.

Mr. Belcher had here another escape, by being
washed off the raft, his trousers getting entangled in
the coral at the bottom of a deep chasm. Fortunately they gave way, and he rose to the surface,
and by great effort swam through the breakers.
Lieutenant Wainwright was the last that was hauled
off. To this young officer the greatest praise is due
for his bravery and exertions throughout. But for
his resolution, it is very doubtful whether the party
would have been relieved from their perilous situation, as the tide was rising, and the surf upon the
reef momentarily increasing. In the evening we
made sail to the westward, and on the 27th saw
Crescent Island; and shortly afterwards the high
land of Gambier's groupe.

Both these islands were discovered by Mr. Wilson
during a missionary voyage, but he had no communication with the natives. The first was so
named in consequence of its supposed form; but in
fact it more nearly resembles an oblong. It is exactly
three miles and a half in length, and one and a half
in width, and of similar formation to Oeno and
Ducie's Islands. It consists of a strip of coral about
a hundred yards or less in width, having the sea on
one side and a lagoon on the other. Its general

height is two feet above the water. Upon this strip several small islands, covered with trees, have their foundation. The soil, where highest, reaches just six feet above the sea ; and the tops of the trees are twenty feet higher. We saw about forty naked inhabitants upon this small spot ; but from the mast-head of the boat, which overlooked the land, could perceive no cultivation ; and there were no fruit-trees upon the island but the pandanus, which has not been mentioned in any voyage that I am acquainted with as constituting a food for the natives of these seas : indeed, from the fibrous nature of the nut it bears, it did not appear to us possible that it could be serviceable as food. We were, consequently, curious to know upon what the natives subsisted, independently of the shell-fish which the reefs supplied ; but nothing occurred to satisfy us on that head. The surf was too high for the boats to land, and our only communication was by signs and an exchange of sentences unintelligible on both sides.

Upon the angles of the island there were three square stone huts, about six feet high, with a door only to each ; they did not appear to be dwelling-houses, and were probably places of interment or of worship. Several sheds thatched with the boughs of trees, some open on one side only and others on both, which were seen on different parts of the island, were more appropriate residences in such a climate.

The natives were tall and well made, with thick black hair and beards, and were very much tattooed. Their signs intimated a disposition to be friendly, and an invitation to land, which we could not do ;

but none of them ventured to swim off to the boats,
probably on account of the sharks, which were very
numerous.

We quitted Crescent Island at day-light on the
29th, and about noon the same day were close off
Gambier's groupe.   Several of these islands had a
fertile appearance, especially the largest, on which is
situated the peak we had seen the day before, and
which Mr. Wilson, in passing to the northward
of the groupe, named Mount Duff.   It was probable,
that among these islands we should find a stream of
water from which our stock might be replenished,
provided an opening through the reef which surrounds the volcanic islands could be found; and as it
was of the highest importance that our wants in this
respect should be supplied, I determined closely to
examine every part of the groupe for an entrance;
for in the event of not being so fortunate as to succeed here, it would be necessary to alter the plan of
operations, and proceed direct to Otaheite, the only
place where a supply of that indispensable article
could be depended upon. On approaching the island,
with the ship, we were gratified by perceiving that
the coral chain, which to the northward was above
water, and covered with trees, to the southward
dipped beneath it; and though the reef could be
traced by the light blue-coloured sea, still it might
be sufficiently covered to admit of the ship passing
over it, and finding an anchorage in the lagoon. As
we were putting off from the ship in the boats to
make this interesting inquiry, several small vessels
under sail were observed bearing down to us. When
they approached we found they were large katamarans or rafts, carrying from sixteen to twenty men

Drawn by William Smyth.

A RAFT OF GAMBIER ISLAND.

Pub.d by H.Colburn.&R.Bentley, 1831.

each. At first several of them were fastened together, and constituted a large platform, capable of
holding nearly a hundred persons; but before they
came near enough to communicate they separated,
furled their sails, and took to their paddles, of which
there were about twelve to each raft. We were
much pleased with the manner of lowering their
matting sail, diverging on different courses, and
working their paddles, in the use of which they had
great power, and were well skilled, plying them
together, or, to use a nautical phrase, keeping stroke.
They had no other weapons but long poles; and
were quite naked, with the exception of a banana
leaf cut into strips, and tied about their loins, and
one or two persons who wore white turbans. Their
timidity in approaching both the ship and the barge
was immediately apparent; but they had no objection to any of the small boats; which they were
probably aware they could, if necessary, easily upset
when within their reach; and, indeed, it required
considerable caution to prevent such an occurrence,
not from any malicious intention on the part of the
natives, but from their thoughtlessness and inquisi-
tiveness. I approached them in the gig, and gave
them several presents, for which, they in return,
threw us some bundles of paste tied up in large
leaves. Not knowing at first what it was, I caught
it in my arms, and was overpowered with an odour
that made me drop it instantly. They made signs
that it was to be eaten, and we afterwards found it
was the common food of the natives. It was what
is called mahie at the Marquesas, but with a higher
gout than I ever heard that article possessed in those
islands, and very much resembled the first opening of

a cask of sour krout, though considerably more over-powering. We soon perceived they had a previous knowledge of iron, but they had no idea of the use of a musket. When one was presented to induce them to desist from their riotous conduct, instead of evad-ing the direction of the fatal charge, they approached it; and imagining the gun was offered to them, they innocently held out their hands to accept it. Before we came close to them, they tempted us with cocoa-nuts and roots, performed ludicrous dances, and in-vited our approach; but as soon as we were within reach, the scene was changed to noise and confusion. They seized the boat by the gunwale, endeavoured to steal every thing that was loose, and demanded whatever we held in our hand, without seeming in the least disposed to give any thing of their own in return. At length some of them grasped the boat's yoke, which was made of copper, and others the rudder, which produced a scuffle, and obliged me to fire my gun over their heads. Upon the discharge, all but four instantly plunged into the sea; but these, though for a moment motionless with astonish-ment, held firmly by the rudder, until they were rejoined by their companions, and then forcibly made it their prize. We could only have prevented this by the use of fire-arms, but I did not choose to resort to such a measure for so trifling an end, espe-cially as the barge was approaching, and afforded the most likely means of recovering our loss without the sacrifice of life on their part, or the risk of being upset on our own. As I intended to remain some days at these islands, I wished by all means to avoid a conflict; at the same time it was essential to our future tranquillity to show a resolution to resist such unwarrantable conduct, and to convince them of our determination to enforce a respect of property. As soon, therefore, as we were joined by the barge, we grappled the raft that contained our rudder; on which the greater part of the natives again threw themselves into the sea; but those who remained appeared determined to resist our attack, and endeavoured to push the boat off. Finding, however, they could not readily do this, a man whose long beard was white with age, offered us the disputed article, and we were on the point of receiving it, which would have put an end to all strife, when one of the natives disengaged the raft, and she went astern. Again free, the rudder was replaced on the raft, and the swimmers regained their station. They were followed by the gig and jolly boat, and a short skirmish ensued, in which Mr. Elson fell. The boat's crew imagining him hurt, and seeing the man he had been engaged with aiming another blow at him, fired and wounded his assailant in the shoulder. The man fell upon the raft, and his companions, alarmed, threw the rudder into the sea and jumped overboard. As this man took a very leading part, he was probably a chief. No other wound was inflicted, nor did this happen before it was merited; for our forbearance had extended even beyond the bounds of prudence; and had less been evinced, we should sooner have gained our point, and probably have stood higher in the estimation of our antagonists. After this rencontre, some of the rafts again paddled towards us, and waved pieces of white cloth; but the evening being far spent, and anxious to find anchorage for the ship, I proceeded to examine the islands. We passed the bar, formed by the chain before mentioned dipping under water, in five, seven, and eight fathoms over a rugged coral bottom, and entered the lagoon, gradually deepening the water to twenty-five fathoms. There was a considerable swell upon the shallow part of the reef, but within it the water was quite smooth. The first island we approached had a bay formed at its eastern angle, where the ship might ride in safety with almost all winds. Night coming on, we anchored the boat upon the bar, and caught a large quantity of fish, consisting of several sorts of perca (*vittata, maculata*), a labrus, and many small sharks. After daylight we returned to the ship, and in the evening anchored in the spot we had selected the day before. As we entered the bay, the natives were observed collected upon a low point, at one extremity of it, hallooing, and waving pieces of white cloth. Almost all of them had long poles, either pointed or tipped with bone. Some had mats thrown over their shoulders, and their heads and loins covered with banana leaves cut into strips. They were much startled at the noise occasioned by letting go the anchor, and at the chain-cable running out, and gazed intently at the different evolutions necessary to be gone through in bringing the ship to an anchor, in furling sails, &c.

No person came on board that night; but daylight had scarcely dawned when one of the natives paddled off to the ship upon a small katamaran: he was quite naked, and had only a pole and a paddle on the raft. For a considerable time he hesitated to come alongside; but on our assuring him, in the Otaheitan language, we were his friends, he was persuaded to make the attempt. After a little further conciliation he made his raft fast by a rope that was thrown to him, and ascended the side of the ship, striking her several times with his fist, and examining her at every step.   His surprise on reaching the deck was beyond all description ; he danced, capered, and threw himself into a variety of attitudes, accompanying them with vehement exclamations ; and entered into conversation with every person, not suspecting that his language was unintelligible ; and was so astonished at all he saw that his attention wandered from object to object without intermission. He very willingly accepted every present that was offered him ; and having satisfied himself of our friendly disposition, hastened on shore to his companions, who were collected in great numbers upon the low point, anxiously awaiting his return.   The report which he gave was undoubtedly of a favourable nature, as several katamarans, laden with visiters, immediately pushed off, and came fearlessly alongside.

The decks were soon crowded with delighted spectators, wondering at every thing they beheld, and expressing their feelings by ludicrous gestures. The largest objects, such as the guns and spars, greatly attracted their attention : they endeavoured to lift them, with a view, no doubt, of bundling them overboard ; but finding they could not be moved, the smaller articles became the more immediate subjects of curiosity and desire, and it required a vigilant watch to prevent their being carried off. They were pleased with many articles that were shown them ; but nothing made them so completely happy as the sight of two dogs that we had on board. The largest of these, of the Newfoundland breed, was big and surly enough to take care of himself ; but the other, a terrier, was snatched up by one of the natives, and was so much the object of his solicitation that it was only by force he was prevented carrying him away. To people who had never seen any quadruped before but a rat, so large an animal as a Newfoundland dog, and that perfectly domesticated and obedient to his master, naturally excited intense curiosity, and the great desire of these people to possess themselves of it is not to be wondered at. Had there been a female dog on board, they certainly should have had them both; but one would have been of no use, except, probably, to furnish a meal, which is the fate of all the rats they can catch.

One of the rafts that came off to the ship, a smaller one than any of the others, brought a person of superior appearance; his complexion was much fairer than that of his countrymen, and his skin beautifully tattooed; his features were of the true Asiatic character : he had long black mustaches and hair, and wore a light turban, which gave him altogether the appearance of a descendant of Ishmael. It was natural to infer that this was a person of some authority; for as yet we had seen no distinction whatever between our visiters, except that some were more unruly than others; but we found we were mistaken : he mingled indiscriminately with his companions, and was deficient in those little points which are inseparable from a person accustomed to command. Indeed, by the total disregàrd they paid to each other, as also to every person in the ship, we might have concluded that our visiters were ignorant of any distinctions in society.

Among the many katamarans that came off, not one of them brought any articles to give or sell,

which did not argue much in favour of the supplies
of the place, or the good will of the islanders. A
green banana, lying upon one of the rafts, was the
only eatable thing among them, excepting some
boiled tee-root, and bundles of that execrable paste,
which they had provided for their own breakfast.
Almost all our visiters were naked, with the exception of a girdle made of a banana leaf, cut into strips,
which by no means answered our idea of the intended purpose. Maros were worn only by the
aged, and instead of them ligatures of straw were
applied in the manner described at St. Christina and
Nukahiwa.* The average height of the islanders
was five feet nine; they were, generally speaking,
well made, their limbs round, without being muscular, and their figure upright and flexible. Tattooing was very extensively practised, in which respect,
as also in the arrangement of the lines, they again
reminded us of the Marquesans. This general practice in the South Seas, when judiciously executed,
besides having its useful effects, is highly ornamental.
In the Gambier Islanders there is a greater display of taste than I have seen or heard of anywhere
else, not excepting the Marquesans: but the Nuka-
hiwers, as well as the Otaheitans and others, attend
principally to device; whereas the Gambier Islanders dispose the lines so as materially to improve the
figure, particularly about the waist, which, at a little
distance, has the appearance of being much smaller
than it really is. Whether this has been accidental
or designed we had no opportunity of learning.

The number of visiters on board was considera-

* Krusenstern's Embassy to Japan, 4to.

ble; yet there was very little to interest us beyond
the first gratification of our curiosity. They were
so engrossed by their own efforts to purloin some of
the many things which they saw, that it was impossible to engage their attention in other matters.
It was besides necessary to keep so strict a watch
over the stores of the ship, and their conduct was so
noisy and importunate, that our desire for their
company was hourly lessened, and we were not
sorry when, on preparing the boats to land, we saw
the rafts put off from the ship, and every man upon
our decks throw himself into the sea and swim
ashore.

On approaching the beach, we found the coral
animals had reared their structure all round the
island, and had brought it so near to the surface
that the large boats could not come within two hundred yards of the landing-place, and the smaller
ones could approach only by intricate windings between the rocks.

The natives were very numerous upon the shore,
the usual population being greatly increased by
parties which curiosity had brought from the other
islands. The women and children at first formed
part of the noisy multitude, all of whom were clamorous for us to effect a landing; but the females
shortly retired out of sight, and the men formed
themselves into two lines, and ceremoniously proceeded to a place where their katamarans usually
disembarked, humming in chorus a sullen tune not
devoid of harmony. Some of them seeing we were
greatly impeded by the coral rocks, waded out and
laid hold of the boats, while others pushed off upon
rafts, and attempted to drag us in, by fixing their

poles under the seats of the boat, and pressing upon the gunwale as a fulcrum; an ingenious contrivance, from which we found it difficult to free ourselves, especially as the poles were very large. Others, again, prepared cords to fasten the boats to their raft, unconscious of our possessing any instrument sufficiently sharp to disengage them. In short, they were determined we should land; but as I did not like the place, and as their conduct appeared to be a repetition of what we had experienced outside the harbour, we disappointed their expectations, and went to the next island.

We were there joined by some of our visiters who had been on board the ship, who reminded us of our former acquaintance, and greeted us with a hearty rub of their noses against ours. This salutation, it was thought by some of us, sealed a friendship between the parties; but we had not sufficient opportunity of ascertaining whether it was considered inviolable. The manner of effecting this friendly compact is worthy of description. The lips are drawn inward between the teeth, the nostrils are distended, and the lungs are widely inflated; with this preparation, the face is pushed forward, the noses brought into contact, and the ceremony concludes with a hearty rub, and a vehement exclamation or grunt: and in proportion to the warmth of feeling, the more ardent and disagreeable is the salutation.

Finding, from communication with our friends, that water was to be had at Mount Duff, we quitted them and crossed to that point, where we had the satisfaction to see two streams trickling down the sides of the hill, either of them sufficiently ample for our purpose, and so situated that the ship could, if necessary, be placed near enough to cover the parties sent to procure it. This gratifying discovery was of the greatest importance, and the ship was immediately removed to a convenient spot opposite the place.

We were late getting across the lagoon from our first anchorage, in consequence of the necessity of proceeding with the utmost caution to prevent striking upon rocks of coral, which were numerous, and in some instances rose from twenty-eight fathoms to within twelve feet of the surface; so that it was dark before the sails were furled, and we had no communication with the natives that night. One man only, probably by way of ascertaining whether we kept watch, paddled silently off upon a small katamaran; but on being hailed, went quietly away. At daylight, the shore opposite the ship was lined with the natives, and katamarans commenced coming off to her laden with visiters, who, encouraged by their former reception, fearlessly ascended the side, and in a short time so crowded the decks, that the necessary duties of the crew were suspended. Their surprise was, if possible, greater than that of the other islanders; but it did not appear to be excited by any particular object.

It is said that as a people become civilized, their curiosity increases. Here, however, it was excited more from a desire to ascertain what was capable of being pilfered than from any thirst for knowledge. Through this propensity, every thing underwent a rigid examination. We had taken the precaution to put all the moveable articles that could be spared below, and nothing was stolen from the upper decks; but in the midshipmen's berth, things had not been so carefully secreted, and a soup-tureen, a spyglass, and some crockery were soon missing; the former was detected going over the side, and one of the tea-cups was observed in the possession of a person swimming away from the ship. This afforded a favourable opportunity of showing our determination to resist all such depredations; and indeed it was absolutely necessary to do so, as every person appeared to consider he had a right to whatever he could carry away with him; and the number of our visiters amounted to double that of our own crew, so that it was impossible to watch every one of them. Besides, this conduct, if not checked in time, might lead to serious consequences, which I wished by every means to avoid. One of our small boats was consequently sent in pursuit of the thief, who was swimming at a considerable rate towards a raft with his prize in his hand. His countrymen, observing that he was pursued, would not permit him to mingle with them, lest they should participate in the blame; but he eluded detection by diving underneath their rafts, until he became exhausted, when he threw the cup to the bowman of the boat, and made his escape. Immediately the boat was sent off, all the rafts left the ship, and every man upon the decks jumped overboard as if by instinct; but when tranquillity was restored, they returned for fresh plunder. The rapidity with which the news of a theft spreads among such a community has been noticed by Captain Cook, and here it was no less remarkable.

I determined, since the main deck was cleared,

that it should be kept so, and placed a marine at
each of the ladders; but as the natives tried every
method to elude their vigilance, the sentinels had an
arduous task to perform, and disturbances must inevitably have arisen in the execution of their orders
had it not been for our Newfoundland dog. It fortunately happened that this animal had taken a dislike to our visiters, and the deck being cleared, he
instinctively placed himself at the foot of the ladder, and in conjunction with the little terrier, who
did not forget his perilous hug of the day before,
most effectually accomplished our wishes. The natives, who had never seen a dog before, were in the
greatest terror of them; and Neptune's bark was
soon found to be more efficacious than the point of
a sentry's bayonet, and much less likely to lead to
serious disturbances. Besides, his activity cleared
the whole of the main deck at once, and supplied
the place of *all* the sentinels. The natives applied
the name of *boa* to him, a word which in the Otaheitan language properly signifies a hog. But it
may be observed that *boa* is applied equally to a
bull, or to a horse, which they call *boa-afae-taata*,
(literally, man-carrying pig), or to all foreign quadrupeds.

Upon one of the rafts which came alongside there
was an elderly man with a grey beard, dressed in
white cloth. The paddles of his raft were of superior workmanship to the others, and had the extremity of the handle ornamented with a neatly
carved human hand. He carried a long staff of
hard black wood, finely polished, widened at one
end like a chisel. But though he was thus distinguished, he exercised no authority over his unruly

countrymen. Several of the people upon the rafts
had provided themselves with food, which consisted
of boiled root of the tee-plant, of pearl oysters, and
the sour pudding before mentioned. We endeavoured to tempt them to taste some of our food;
which they willingly accepted, but declined to partake of it, and placed it upon the raft, with nails,
rags, and whatever else they had collected. A piece
of corned beef that was given them passed from
hand to hand with repeated looks of inquiry, until
it was at last deposited in the general heap. I
took some pains to explain to them it was not
human flesh, which they in all probability at first
imagined it to be; and from their behaviour on
the occasion I think it quite certain they are not
cannibals.

As the curiosity of one party of our visiters became satisfied, they quitted the ship, and others supplied their place. One of these favoured us with a
song, which commenced with a droning noise, the
words of which we could not distinguish; they
then gave three shouts, to which succeeded a short
recitation, followed by the droning chorus and
shouts as before. In this manner the song proceeded, each recitation differing from the former, until
three shouts, louder than the others, announced the
finale. The singers arranged themselves in a semicircle round the hatchway, and during the performance pointed to the different parts of the ship,
to which their song was undoubtedly applicable;
but it was impossible to say in what way, though I
have every reason to believe it was of a friendly
nature.

While the decks were so crowded with visiters,

the duty of watering the ship could not be carried on, and it was of the greatest consequence that it should be got through speedily, as the boats were required to survey the group, upon which I could not bestow many days. My hope was, that the natives would quit us as their curiosity became satisfied, especially as they had nothing to barter, except some sour paste, which, being extremely unpalatable to every one on board, was not marketable. After breakfast, two small boats, the only ones we had in repair, were equipped for landing, and the barge was ordered to be in attendance; for though there was every reason to expect a friendly reception, yet in a country where the language is not understood, and among a barbarous people, whose principal aim is plunder, it is extremely difficult to avoid disputes, especially when the force to which they are opposed is greatly inferior to their own. We felt the loss of the cutter at this moment, as she was a boat so much better calculated for the service we had to perform than the gig or whale-boat.

As we had anticipated, the boats had no sooner put off from the ship, than all the natives quitted her as before, and joined their companions on shore, who were assembled in a wood skirting the beach. At the approach of the boats, there was much bustle among the trees; every one appeared to be arming himself, and many who had long poles broke them in halves to supply those who had none. These preparations made it necessary to be cautious how the boats were placed in their power, as they were small, and easily upset, and the natives very numerous. We found the shore, as at the other island, surrounded by coral rocks, upon which the boats grounded about two hundred yards from the beach, and they could not advance without imminent danger of being stove. The natives, whose rafts drew so little water that they could be floated over these impediments, could not understand our motives in delaying, or searching for any other place than that to which they had been accustomed, and kept continually vociferating " Ho-my! Ho-my!" It was natural that they, ignorant of the cause, should suppose we had other things in view than that of landing ; and one of them who had received a bottle as a present from some of our people, imagining we were come in search of it, ran into the water as far as he could, holding it up at arm's length, and when he could advance no farther, threw it towards the boat, and, in spite of our signs for him to keep it, he followed the boats, and kept throwing the bottle towards us, until he found it was of no use.

A short distance below the place where the multitude were assembled, the rocks admitted a freer access to the shore than above, and we effected a landing.

Directly the boats touched the beach, one of the natives who was near them took off his turban and waved it to his countrymen, who instantly answered the signal with a shout, and rushed towards the spot. The foremost of their party stopped within a short distance of us until the crowd came up to him, and then advanced and saluted Mr. Belcher, who was unarmed, by rubbing noses. Observing there was some distrust of a fowling-piece which I held in my hand, I placed it against a rock for an instant among our own party, while I advanced a step to salute a person who appeared to be the leading man of the islanders. The opportunity this afforded the

natives of indulging their favourite propensity was
not overlooked ; and one of them, regardless of all
risk, thrust himself between our people, snatched
up the gun, and, mingling with the mob with the
greatest adroitness, succeeded in making his escape.
The crowd instantly fled into the wood, and along
the beach, but shortly rallied, and with loud shouts
advanced upon us, until the discharge of a carronade
from the barge, which was fortunately near, put
them to flight. The man who had sealed the compact of friendship, if so indeed it be, by rubbing
noses, sat quietly upon a large stone close to us during this affair, as if he relied upon the pledge that
had been given for his security. It would have
been treacherous, and perhaps pregnant with serious evils to them and to ourselves, had any violence
been offered, or any thing done that might appear
like an infringement of this understanding, or I
should certainly have detained his person, in the
hope of the gun being returned. As it was, I allowed him to go quietly away.

The boats were at this time unavoidably very
awkwardly situated, by being aground upon the
rocks, and in a situation from which it would have
been extremely difficult to extricate them, had a determined attack been made by the natives. The
consequences in that case would have been very
serious ; though their weapons consisted only of
long poles and bone-headed spears, yet they were
sufficiently powerful, from their numbers alone, to
have rendered the most determined defence on our
part doubtful.

As soon as we were free, we followed the natives
along the beach, approaching them, whenever the

rocks would allow, to offer terms of reconciliation; but our overtures were answered only by showers of stones. This conduct, which we now began to think was only a part of their general character, rendered it extremely difficult, nay almost impossible, to have any dealings with them without getting into disputes. No time, place, or example, made any difference in the indulgence of their insatiable propensity to theft. Explanations and threats, which in some instances will prevent the necessity of acting, were unfortunately not at our command, in consequence of our ignorance of their language, and the only option left us was to yield up our goods unresistingly, or to inflict a more severe chastisement than the case might deserve. Captain Cook, who managed the natives of these seas better than any other navigator, pursued a system which generally succeeded, though in the end it cost him his life. It was rigid, but I am certain it was better adapted to preserve peace than the opposite plan adopted by Perouse, at Easter Island, who, though one of the most enlightened navigators, was, of all, the most unfortunate.

To seize one of the natives, or upon something that was of more value to them than the goods they had stolen, was the most effectual way of recovering what was lost, and by at once adopting this mode of proceeding might prevent a recurrence of such a circumstance; I consequently took away a net and some rafts that were lying upon the shore. The net was about forty feet in length, made with the bark of the porou tree (*hibiscus tiliaceus*), precisely in the same manner as our seins are, but weighted with stones and rounded pieces of coral instead of lead.

To obtain possession of these articles without strife, it was necessary to drive away a party that was seated upon a large tree near them, and a carronade was fired over their heads: but of this they took no notice, probably considering themselves safe at so great a distance, and having had no experience to the contrary, supposed that such weapons were calculated only to intimidate by noise. The next gun dispersing the sand amongst them, they speedily resigned their seats, and with all the inhabitants went to the upper village. After this our communication was for a time suspended, as the natives kept aloof, and the boats were required to proceed with the watering.

At daylight on the 2d of January, we commenced filling our casks from two good streams, which supplied water much faster than it could be got off.

We perceived the natives collected in a large body at the village, and soon afterwards some men stole along the beach to reconnoitre the watering party; but they were prevented offering any molestation by a gun being fired from the ship. On this day I observed the old custom of taking possession of the groupe, and hoisted the English ensign upon the shore, turned a turf, and sowed several useful seeds, which it is to be hoped will spring up to the benefit of the natives. I named the island on which Mount Duff stood, after my first Lieutenant, Mr. Peard, and the others in succession, Belcher, Wainwright, Elson, Collie, and Marsh, after the other officers, and the lagoon in which the ship was anchored after herself.

Before our party reached the shore the next morning, one of the natives was perceived carrying off a small cask that had been left there the preceding night. We watched him through our telescopes, and observed him conceal it with a large mat which he carried with him. He had doubtless no suspicion that his actions could be observed at so great a distance, as he began to retrace his steps along the open beach; but seeing he was not sufficiently quick to escape the boats that were going on shore, he quitted his prize, and hid himself in the wood. The watering had not long been renewed before a large party collected upon the height above, headed by two men, who appeared to be chiefs, clad in loose white turbans and cloaks: the eldest led the party cautiously down the hill, and made a stand at a large stone, which one of his party ascended, and there waved a banana leaf. We answered this friendly signal by waving in return a white flag from the ship: but here our amity ended; for while this was going forward the other chief stood upon the ridge, and beckoned to the natives on the other side of the hill to join him, which greatly augmented his numbers; and some of them loosened large stones, apparently with a view of annoying our watering party, who were so situated under the hill that a few such fragments precipitated upon them would cause very serious mischief. As there was every appearance of treachery, the boats were put upon their guard by signal; but the barge mistaking its purport, fired two shot to dislodge the islanders, both of which, to their great astonishment, fell very close to them, and induced them to retire to the other side of the ridge. Some, however, had the curiosity to return and examine the place, and, after a little digging, found one of the shot, which they carried to their comrades, many of whom assembled round the prize, never probably having had so large a piece of iron in their possession before.

At noon on the 5th the watering was completed, and without any accident or sickness, which, considering the difficulty of getting the casks off, and the constant exposure of the seamen to a vertical sun while in the water, there was every reason to apprehend. It was further satisfactory to find that this service had been effected without any harm to the natives, except in one instance, when a marine inconsiderately fired at a party who were lurking in the wood, and wounded one of them in the foot. From the disposition of the inhabitants, and the superiority of their numbers, there was reason to apprehend a different result; and the quietness with which it was conducted must be attributed to their being kept at a distance during its performance.

The boats were now sent to survey the groupe, and were kept constantly employed upon it from daylight to dusk. In the course of this examination every part was visited, and we had frequent communication with the natives, who on such occasions were always civil, and brought such supplies of fruit and food as their scanty means afforded, and generally abstained from the indulgence of their propensity for thieving, which when numerous they so fully indulged. Their behaviour was indeed so different from what it had been, that we must attribute it to the operation of fear, as their numbers were then very small, in consequence of our visits being unexpected and the population of each village very limited. The net we had taken off the shore was carried round to the principal village and offered in return for the articles that had been stolen, but

whether our meaning was understood or not, they
were never produced.

This village is situated in a bay, at the eastern
foot of Mount Duff, and is rendered conspicuous by
a hut of very large dimensions, which we shall describe hereafter, and by a quadrangular building of
large blocks of coral erected in the water, at a few
yards distance from the shore, which appeared to us
to be a morai. Upon its northern extreme stood a
small hut, planted round with trees, which it was
conjectured contained images and offerings ; but, as
the door was closed, and the natives were watching
us, we would not examine it. Contiguous to it there
was a body placed upon boards, wrapped in thick
folds of paper cloth ; and, not far from it, another
enveloped in a smaller quantity of the same material.
There was no offensive smell whatever from either
of these corpses, though the one last mentioned did
not appear to have been long exposed. The heads
of both were lying to the N.E. ; both bodies were
more abundantly surrounded by cloth than any we
had seen here ; and from the nature of the platform
on which they were placed, which must have required considerable labour to construct, we concluded
they were the bodies of chiefs ; and we were, on that
account, more tenacious of subjecting them to the
scrutiny our curiosity prompted, lest the natives
should suppose we were offering them some indig-
nity. An old man whom we interrogated as to the
nature of the building gave us no information : but
looked very serious whenever he was referred to the
place, and seemed disposed to believe we were inclined
to place his body there to keep the others company.

Though we were prevented from examining these mummies by the watchfulness of the natives, we were more successful at the island to the eastward, off which we first anchored. We there found six bodies under a projecting part of a cliff, which overhung them sufficiently to protect them from the inclemency of the weather. Above them we noticed a child suspended by a string round its waist tied to a projecting crag. The bodies of the adults were placed parallel, with their heads to the N. E., as in the other instance. They were wrapped first in cloth, then in matting, and again covered over with thick folds of cloth secured by a small cord lashing. Mr. Collie, the surgeon, made an incision into the stomach of one of the newest mummies, which appeared the most hardened, and found the membraneous part of the abdomen dried and shrivelled up, enclosing an indurated earthy substance, which at first induced him to believe it had undergone the process of embalming; but finding afterwards membranes and earthy matter within a cranium similarly dried, and knowing that there was no way in which any extraneous substance could have been introduced there, except by the vertebral canal, he was induced to alter his opinion, which, he says, had nothing to support it, but the idea that putrefaction must have taken place without some counteracting agent. This complete desiccation of the human frame is not unfrequent in these seas, nor indeed in other places; but it requires considerable care and attention to do it effectually. The method formerly pursued at Otaheite, was to keep the corpse constantly wiped dry, and well lubricated with cocoa-nut oil. Our intercourse with the Gambier Islanders did not

afford us the opportunity of ascertaining if this were
their practice also, but we noticed the precaution of
exposing the bodies upon frames three or four feet
above the ground, that the air might freely cir-
culate about them, and of keeping them well covered
with folds of cloth. It is remarkable that none of
these bodies had any offensive smell, not even those
that had been recently exposed upon the drying-
board. Lieutenant Belcher, whose duty carried him
a great deal about the islands, saw some bodies that
were exposed to dry, covered with a matted shed to
protect them from the rain; and in one he found
the head and right arm separated from the trunk,
wrapped in separate pieces of cloth, and secured by
a lashing to the body. On no part of the shore did
we see skulls or bones exposed and heaped together,
as about the morais common to Polynesia; and although Mr. Belcher found some human bones partly
burned lying loose upon a rock, together with a
body deposited in a grave with a wicker-work frame
over it, there is every reason to believe that these
exposures are very rare indeed, and that almost all
the bodies are wrapped in cloth, and deposited as
first described. This custom furnishes a satisfactory
reason for their cloth being so scarce; and though
we cannot commend their policy in clothing the
dead at the expence of the living, yet they must
be allowed the merit due to their generosity and
respect for their departed friends.

On the 7th I visited a village at the south extremity of Belcher Island. It was situated in a little
bay, at the foot of a ridge of hills which intersected
the island. We were received by about a dozen
men and women, who behaved in a very friendly

manner, and brought down cocoa-nuts (some of which, by the by, had been previously emptied of their contents), sugar-cane, tee-roots, one bunch of bananas, and several clusters of pandanus nuts; these they threw into the boat without soliciting any return; and, what is more extraordinary, without evincing any desire to steal. All the men then quitted us, excepting one, who was as anxious that we should depart as the women were that we should land. Two of these females behaved in a manner which attracted our attention, although we could not account for their conduct; they waded out to the boats, crying most piteously, striking their breasts, and pulling their hair, which hung loose over their shoulders, with every demonstration of the deepest distress; and, to our surprise, threw their arms round our necks, and hugged us so close that we could not disengage ourselves from their embrace without violence. As we were quite unconscious of the nature of their grief, we could offer them no consolation beyond that of kindness, and giving them some beads and trinkets. After a few minutes they disengaged their arms, began dancing, laughing, and saluting us occasionally with a rub of the nose: in the midst of this mirth they would suddenly relapse into grief, and throw their arms about in a frantic way, until I began to fear they might injure themselves; but this paroxysm was as short as that of the mirth by which it was succeeded; they again began to dance, and were afterwards quite cheerful. The only cause to which we could attribute this extraordinary conduct, or at least for the melancholy part of it, was that they might in some way be connected with the man who had been wounded upon the raft. And if this were the case, it affords a presumption that the custom of self-mutilation on such occasions, so common to many of the islands in the Pacific, does not exist here.

As the sun went down the natives pointed to it, and signified to us to be gone, exclaiming " Bobo mai." We got from them a few articles of manufacture, very similar to those of Pitcairn Islands. In return for these we made them useful presents, and took our leave with the promise of " Bobo mai," which we understood to mean " come to-morrow." We rowed round the rest of the island, and soon satisfied ourselves of its extreme poverty. There were two villages upon its western side, situated in deep sandy bays, which would form excellent harbours for shipping, if they could be entered ; but this is impracticable from the many coral knolls on the outside.

Lieutenant Belcher describes a morai, which he visited, in the following manner. A hut, about twenty feet in length by ten wide, and seven high, with a thatched roof, of which the eaves were three feet from the ground, contained the deity. There were only two apertures, about two feet six inches square, furnished with thatched shutters. In front of the building, a space about twenty feet square was paved with hewn coral slabs, with curbstones at the edges, as neatly fitted as the pavements in England. Along the whole length of the interior of the hut was a trough elevated about three feet from the ground ; in the centre of which was an idol three feet high, neatly carved and polished ; the eyebrows were sculptured, but not the eyes ; and from the manner in which the muscles were defined it was evident that these people were not regardless of the anatomy of the figure. It was placed in an upright position on the trough or manger, and fastened by the extremities to the side of the hut: the head was bound with a piece of white cloth, as were also the loins, and those parts which the natives themselves never conceal, the aged excepted. In the trough beneath the image were several paddles, mats, coils of line, and cloth, offerings which had been made to the deity; and at his feet was placed a calabash, which the natives said contained water " *avy*." On each side of the image was a stand, having three carved arms, to the hands of which several articles were suspended, such as carved cocoa-nut shells, and pieces of bamboo, perhaps musical instruments; but Mr. Belcher abstained from trespassing on this sacred ground, for fear of giving offence to the natives, who did not much like this exercise of curiosity. Indeed, the whole time he was there, the women were anxious to get him away, and the men looked serious, and were very glad when he left the place. The females accompanied him to the threshold of the morai; but the men studiously avoided treading upon the sacred pavement, and knelt down the whole time he was there, without, however, any apparent devotion. Mr. Belcher endeavoured to purchase this idol; but valuable as his offers must have been to these poor people, the temptation did not prove sufficient. Another image about the same size was found upon one of the coral islands of the groupe, clothed in the same way, but more rudely carved, and deficient in the offerings above-mentioned.

CHAPTER VI.

Second interview with the Natives—Visit to the principal Village —Bodies exposed to dry—Areghe or Chief—Lieutenant Wainwright attacked by Natives—Advantage of the Port—Further Description of the Island, its Soil and Productions.

ON the afternoon of the 8th, we again landed under Mount Duff, to try the feeling of the natives. Our party was not large, and we carefully avoided every thing that might appear offensive, carrying with us a white flag upon a staff. One man only, at first, ventured near us, rubbed noses, and received several presents, with which he was highly delighted. His companions, who, during the interview, had been peeping from behind the trees, noticing his friendly reception, laid aside their weapons, came out of the wood, and saluted us in their usual manner, singing, as they approached, the chorus we had heard on board, which strengthened our opinion of its being a song of welcome.

The next day I landed with a party in the bay where the principal village is situated, and was met at the landing-place, which was about half a mile from the village, by two or three men who rubbed noses, and seemed glad to see us. They took us by the arm and conducted us to the village along a narrow pathway, through long grass and loose stones, overshadowed by a wood of bread-fruit and cocoa-nut trees.   In this distance we passed a few patches of cultivation, but they were rare, and indicated very little attention to agriculture.   The natives increased greatly in numbers as we advanced, and all were officious to pay us attention, and assist us to the village: they were armed, yet their manner showed it was, as with us, only a precautionary measure: nothing in appearance could exceed their amicable behaviour.   We had each two or more friends, who officiously passed their arms under ours, helped us over the stones and conducted us along the right pathway to the village; a species of escort, however, which, by depriving us of the use of our limbs, placed us entirely in their power.   We passed several huts open on the south side, and one, which was full of fishing nets, closed up; near these there were two bodies wrapped in a great many cloths, exposed upon stalls raised about a yard from the ground, and supported upon forked props, as repre-sented.

The natives were unwilling that we should touch
any of these, and we did not offend them by so
doing, but approached within a few feet to ascertain
whether there were any offensive smell from the
corpse, but none could be discerned.

Further on we came to an open area, partly paved
with blocks of coral, and divided off from the cultivated land by large slabs of the same material very
evenly cut, and resembling those at the Friendly
Islands. At one end of this area stood the large
hut which had before excited our curiosity: it was
about thirteen yards in length by six or seven in
width, and proportionably high, with a thatched
roof. On the south side it was entirely open, and
the gables nearly so, being constructed with upright
poles, crossed by smaller ones, forming an open
frame-work, through which the sea breeze circulated,
and refreshed the area within. Beneath the roof on
the open side, about four feet within the eaves, there
was a low broad wall well constructed with blocks
of coral, hewn out and put together in so workmanlike a style, and of such dimensions, as to excite our
surprise how with their rude instruments it could
have been accomplished. The blocks were five feet
long by three wide, and one foot thick; and were
placed upon their narrow edge in a manner in which
we traced a resemblance to the walls in Hapae, as
described by Captain Cook. Upon this eminence
was seated a venerable looking person about sixty
years of age, with a long beard entirely grey; he
had well proportioned features, and a commanding
aspect; his figure was rather tall, but lassitude and
corpulency greatly diminished his natural stature;
he was entirely naked except a maro, and a crown

made from the feathers of the frigate-bird, or black tern; his body was extensively tattooed, and from the loins to the ankles he was covered with small lines, which at a distance had the appearance of pantaloons. Long nails, and rolls of skin overhanging his hips, pointed out his exemption from labour, and an indulgence in luxuries which in all probability attached to him in virtue of his birthright. He was introduced to us as an areghe or chief; he did not rise from his seat, but gave the nasal salutation in his squatting posture, which in the Friendly Islands is considered a mark of respect.

An exchange of presents succeeded this meeting. Some scarlet cloth, which I had brought on shore for the purpose, was placed over his shoulders, and closed by a buckle in front, which delighted the subjects as much as the chief, who, in return, presented me with his crown, and intimated that I should wear it by placing it upon my hat. This friendly understanding I endeavoured to turn to our advantage by making him understand, as well as I could by signs and Otaheitan words, that we would barter articles we had brought with us for fruit and vegetables; and in the hope of this being acceded to, we waited longer at the village than we should otherwise have done; but the only answer we got was " Bobo mai," which from the Otaheitan vocabulary we should interpret " Here to-morrow;" but its application in the Gambier groupe was so various as to leave us much in doubt whether they were not disposed to turn our imperfect use of it to their own advantage. Our visit to the village brought a great accession to its usual inhabitants, and several hundred people had collected about us, but the greatest order prevailed; nor did their curiosity to scrutinize our persons once lead them to acts of rudeness, notwithstanding we were the first Europeans that had ever landed on their island. Indeed, throughout this visit, or at least until we were coming away, there was a marked improvement in their behaviour; not a single act of theft was attempted, while, on the contrary, one of honesty occurred, which, as it is the only instance I have to record, must not be omitted; —it consisted in restoring to one of our officers a handkerchief which he left at a place where he had been sitting. This propriety of conduct no doubt originated in the strictness of the discipline which we observed towards them. It certainly did not proceed from the example of the chief, for the only act of acquisitiveness from which we had reason to apprehend any dispute proceeded from that personage himself. To oblige him, I had consented to his looking into the bag of presents, with which he became so enamoured that he retained it in his grasp, and once or twice endeavoured to appropriate it to himself by force.

We had not remained many minutes in the hut where we were first introduced, when the areghe rose, and, taking me with him, went to a large stone, in the centre of the paved area, where we both sat down, and were immediately surrounded by some hundreds of his subjects. The exchange of place was by no means agreeable, as we quitted a cool and refreshing retreat for a spot scarcely screened from a scorching sun by a few scanty leaves of the breadfruit tree. After being seated here a few minutes, a tall good-looking young man was introduced, also as an areghe, to whom the old chief transferred the

cloth I had given him. I made him in addition a
similar present, and distributed others of smaller
value to several natives around us, in the hope of
quieting their solicitations; but I soon perceived
that this generosity had the opposite effect.

The young chief was handsomely tattooed; he
had a turban of white cloth, and a girdle of banana-
leaf as his only covering. He was more anxious to
communicate with us than the old man; pointed to
the road leading over the hill to a village on the
opposite side, and made many signs, which we in-
terpreted as promising us the restoration of the
articles that had been stolen, and also some supplies;
at the same time he intimated that a person of superior dignity resided on the other side of the hill.
But if this were true, the distance was only half a
mile, and we remained long enough in the village
for a person to have traversed it five or six times.
We were next introduced, by the chief, to several
women, who saluted us in the usual manner, and
thankfully accepted our presents. The chief wished
me particularly to notice one of them, a fine tall
woman about thirty-five years of age, with sharp
black eyes, long black hair, rather sunburnt, white
and even teeth, a complexion lighter than the generality of her country-women, and with a good-natured countenance which the coarseness of feature
only prevented being pretty. She had an armlet
tattooed on each arm, and was without any other
ornament whatever; her ears even were not pierced
for the reception of rings. Her dress consisted
of a piece of white cloth wrapped round the hips,
and another round the waist below the breast,
which was exposed. There was something com-

manding in her manner, and from her intimacy with the chief she was evidently a person of superior rank. She addressed her conversation to me with a volubility and earnestness which showed she felt confident of being understood, but I regret that our total ignorance of their language denied me the pleasure of interpreting even one word; and I could only infer from her tears and actions, that her tale was of a serious and distressing nature. She soon however dried her tears, and sat beside us with the greatest composure.

While I was engaged with the chief, the officers strolled about, each accompanied by a circle of friends, and were kindly treated. Mr. Belcher, in his researches, discovered three drums, very similar to those at Otaheite, as described by Captain Cook. The largest was about five feet six inches high, and fourteen inches in diameter. It was made of the trunk of a porou tree *(hibiscus tiliaceus)*, hollowed out, and covered with a shark's skin, which had been strained over it when wet; the edges were secured with sinnet, neatly made, and finished with pieces of cloth plaited in with fine line: it was otherwise ornamented about the trunk, and stood upon four feet. It was brought to me, and I offered the areghe some knives in exchange, which he refused until the number was increased. When the bargain was concluded, the young chief showed the manner of playing upon the instrument, and convinced us that his skill must have been the result of long practice.— The art consisted in giving rapid strokes with the palm of the right hand, and placing the left at the same time so nicely as to check the vibrations without stopping them, which produced a harmonic sound, differing from that of any instrument of the kind I had ever heard.

The other drums were about three feet and a half in height by nine inches in diameter, similar in other respects to the large one. The proficiency in execution to which the natives have attained, and the perfection in the manufacture of these instruments, leave little doubt of their taking much delight in the amusement of dancing, though, generally speaking, they do not appear to be a lively people. I used every endeavour, but in vain, to persuade the areghe to favour us with one of these exhibitions, and among others, I made the marines go through some of their manœuvres, in the hope that he would exhibit something in return; this, however, had a very different effect from what was intended; for the motions of the marines were misinterpreted, and so alarmed some of the bystanders, that several made off, while others put themselves into an attitude of defence, so that I speedily dismissed the party.

This interview was deficient in those ceremonies which threw such a lively interest over the voyages of Captain Cook, and, what was equally mortifying to us, it did not obtain those supplies of fruit and vegetables which generally attended his visits; although we waited a considerable time in the hope of inducing the chiefs to come on board the ship, and in the expectation of some supplies before we quitted them, but to no purpose. I therefore summoned our party together, and we took leave of the chiefs, both of whom retired, leaving us in the hands of the mob. On removing the drum which had been sold by the areghe, two of the

natives laid violent hands upon it, and demanded something more than had been given. To avoid disturbance I complied with their request by doubling the
original sum ; but this, so far from securing the drum,
rendered the probability of our obtaining it without
force more remote. I brought the old chief back to
explain the matter to him, but he shewed no disposition to interfere; and foreseeing the consequence of per-
sisting, I left our purchase in the hands of the islanders, disgusted with their dishonesty and cunning.

On our return, about two o'clock in the afternoon, we observed the meals of the natives laid out
upon tables, made of slabs of coral, raised about a
yard from the ground, and standing in the middle
of the paved areas in front of the huts. These
tables again resembled those in the Friendly Islands,
and the execrable sour pudding tied up in bundles
with banana-leaves, of which the fare of the natives
consisted, is the same as the mahie used there, at
Otaheite, and at the Marquesas, &c. ; but in flavour it
more immediately reminded us of the Nukahiwans.

We found fewer companions in our retreat from
the village than at our introduction to it, and were
attended by three individuals only, who had attached themselves to some of the officers, though
many followed at a distance. I was a little behind
the party, when a man whom I did not recollect
to have seen before, grasped me by the arm in
which I held my gun, with a feigned view of helping me over the rugged path, while a second, putting his arms across, stopped up the road ; several
others, at the same time, joined in the demand of
Homy ! homy !' and prepared us for what shortly
took place. I managed to get rid of my unruly

assistants without force, and joined the marines; but Lieutenant Wainwright (who, unknown to us all, was left in the village, ignorant of our having quitted it until informed by one of the natives), was not so fortunate. He had passed through the village, where the natives were assembled in circles, apparently in debate, without molestation, and in a few minutes would have been among our party; when several of the natives, seeing him alone, assailed him, and endeavoured to throw him down and rob him. Finding they could not succeed, they attacked him with their poles: but he was then fortunately within a short distance of us; and we became for the first time apprised of his danger by hearing him call for assistance. Mr. Belcher, and those who were nearest, ran to him; but the islanders assailing them with stones, and the attack on their part becoming general, I ordered the marines to fire, which put them to flight, and I am happy to say that we saw only one of them wounded.

Thus this interview with the natives terminated in a manner which their general conduct might have led us to expect, though the result is much to be deplored. It confirmed my opinion, that the natural disposition of the people is highly unfavourable to intercourse, and that they are restrained from acts of violence and aggression by the operation of fear alone. With this impression, and finding the island so extremely deficient in supplies, that the natives could not spare us any thing, I was careless about renewing our visit, and we embarked without further molestation, and proceeded to the ship.

The bay in which this village is situated lies on the N. E. side of Mount Duff; it is bordered by a

Rich.d B. Beechey del.t   E. Finden sc.

ATTACK OF THE NATIVES OF GAMBIER ISLANDS.

Pub.d by H. Colburn & R. Bentley, 1831.

sandy beach, behind which there is a thick wood of
bread-fruit and cocoa-nut trees; above it, to the
left, there is a second or upper village, upon a level
piece of ground, where the natives retreat in case of
necessity. The bay would be very desirable for an
anchorage, were it not for the coral knolls at its
entrance, which make the navigation difficult even
for a boat. After this visit, the boats were again
sent surveying ; and on the 12th we had completed
all that our time would admit of, by fixing the
position of a number of coral knolls which are dispersed over the navigable part ·' the lagoon, the
greater part of which may be seen from a ship's
mast-head before she comes upon them. Our only
want afterwards was a little fire-wood ; and having
noticed several logs lying upon the shore abreast of
the ship, Mr. Belcher was sent to purchase them.
The natives readily disposed of their property, and
were very friendly as long as they were receiving presents; but directly he attempted to take
away the trees, the islanders collected in the wood,
and pelted the boat's crew with stones. Three logs
were however got off, and Mr. Belcher was putting
in for more, when, the natives again beginning to
throw stones, he desisted.

It is to be regretted that the disposition of the
natives obstructed the friendly intercourse we were
anxious to establish. The task of correcting their
evil propensities unfortunately devolved upon us, as
the first visiters to the islands; and we could not
prolong our stay, or devote the time that was necessary while we did remain, to conciliate their friendship. But though unsuccessful in this respect, it is
to be hoped that our visit will prove beneficial to

others, by directing them to a port in which ships may be refitted or repaired, and where they may procure a supply of good water, than which nothing is more important to the navigation of these seas; as that indispensable article is not found to exist in a pure state anywhere between Otaheite and the coast of Chili, a distance of 4000 miles, Pitcairn Island excepted, where the difficulty of getting it off has already been mentioned. It is also presumed, that the position of the islands having been ascertained, the peaks of Mount Duff, which are high and distinguishable at a great distance, will serve as a guide to the labyrinth of coral islands which the navigator, after passing this groupe, has to thread on his way to the westward.

This groupe was discovered by the ship Duff, on a missionary voyage, in 1797, and named by Mr. Wilson, her commander, after Admiral Lord Gambier. It consists of five large islands and several small ones, all situated in a lagoon formed by a reef of coral. The largest is about six miles in length, and rises into two peaks, elevated 1248 feet above the level of the sea. These peaks, which were called after the Duff, are in the form of wedges, very conspicuous at a distance, and may be seen fourteen or fifteen leagues. All the islands are steep and rugged, particularly Marsh Island, which at a distance resembles a ship. The external form of these islands at once conveys an impression of their volcanic origin; and, on examination, they all appeared to have been subjected to the action of great heat.

" The general basis of the rocks is a porous basaltic lava, in one place passing into a tuffacious slate; in another, into the solid and angular column

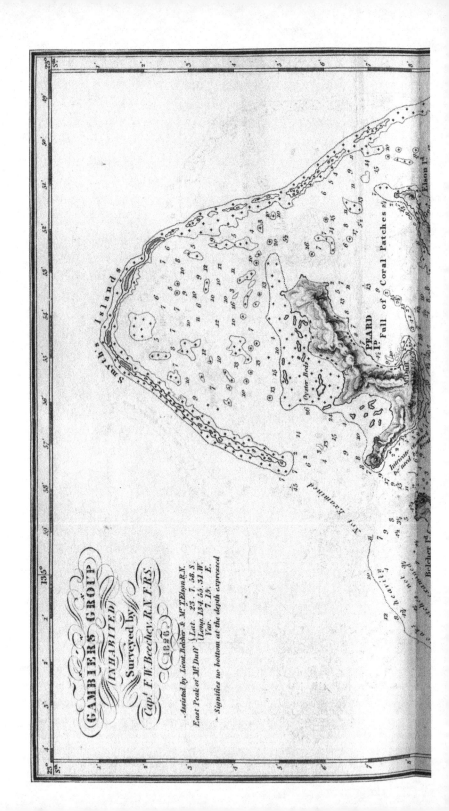

GAMBIER'S GROUP
(INHABITED)
Surveyed by
Capt. F. W. Beechey. R.N. F.R.S.
1826

Assisted by Lieut. Belcher & M.r I. Elson. R.N.
East Peak or M.r Duff { Lat. 23 . 7 . 58. S.
{ Long. 134. 55 . 31. W.
Var. 7. 13. E.

.. Signifies no bottom at the depth expressed

Smith's Islands

PEARD
P.t Full of Coral Patches

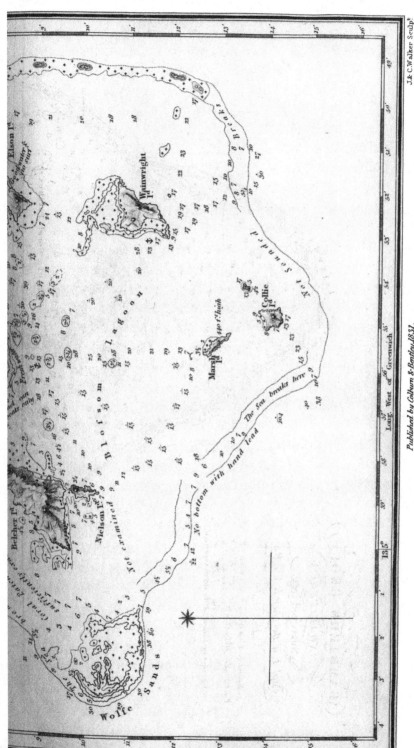

J. & C. Walker Sculp.t

Published by Colburn & Bentley, 1831.

of compact basalt, containing the imbedded minerals which characterise that formation, and bearing a close resemblance in this particular to the basaltic formation of the county of Antrim in Ireland. There is, however, less of the basalt and more of the porous. The zealites, soapstone, chalcedony, olivine, and calcareous spar, are formed in, and connect the relationship of these distant formations; whilst the different-coloured jaspers are peculiar to these islands. There is also another obvious distinctive feature produced by the numerous dykes of a formation differing in composition and texture, and marked by a defined line. They are generally more prominent than the common rock; traversing a great many, if not all the islands, in a direction nearly east and west; generally about eighteen inches wide, nearly perpendicular to the horizon, or dipping to the southward. Their texture is sometimes compact, sometimes vesicular, with few if any imbedded minerals, excepting one on Marsh Island, which contained great quantities of olivine. Upon a small island contiguous to this, the harder dyke crosses the highest ridge, and divides on the eastern side into two parts which continue down to the water's edge."*

Lieutenant Belcher, whose scientific attainments also enabled him to appreciate what fell under his observation, noticed every where the trap formation abounding in basaltic dykes also lying N. E. and S. W., and seldom deviating from the perpendicular; or if they did, it was to the eastward. We are indebted to him for specimens of zealite, carbonate

* Mr. Collie's Journal.

of lime, calcareous spar, crystals, an alcime, olivine, jasper, and chalcedony; and had our stay, and his other duties admitted, we should, no doubt, have received from him a more detailed account of this interesting groupe.

There are no appearances of pseudo-craters on any of the islands, nor do they seem to have been very recently subjected to fire, being clothed with verdure, and for the most part with trees. Conspicuously opposed to these lofty rugged formations, raised by the agency of fire, is a series of low islands, derived from the opposite element, and owing their construction to myriads of minute lithophytes endowed with an instinct that enables them to separate the necessary calcareous matter from the ocean, and with such minute particles to rear a splendid structure many leagues in circumference. A great wall of this kind, if we may use the expression, already surrounds the islands, and, by the unremitting labour of these submarine animals, is fast approaching the surface of the water in all its parts. On the N. E. side, it already bears a fertile soil beyond the reach of the sea, sustains trees and other subjects of the vegetable kingdom, and affords even an habitation to man.

In the opposite direction it dips from thirty to forty feet beneath the surface, as if purposely to afford access to shipping to the lagoon within. Whether this irregularity be the consequence of unequal growth, or of the original inclination of the foundation, is a question that has excited much interest. All the islands we subsequently visited were similar to these in having their weather or eastern side more advanced than the opposite one.

The outer side of the wall springs from unfathomable depths; the inner descends with a slope to about 120 or 150 feet below the surface. This abruptness causes the sea to break and expend its fury upon the reef without disturbing the waters in the lagoon. The coral animals consequently rear their delicate structure there without apprehension of violence; and form their submarine grottoes in all the varied shapes which fancy can conceive. They have already encircled each of the islands with a barrier, which they are daily extending; and have reared knolls so closely as almost to occupy all the northern part of the lagoon. More independent tribes are in other parts bringing to the surface numerous isolated columns, tending to the same end; and all seems to be going on with such activity, that a speculative imagination might picture to itself at no very remote period, one vast plain covering the whole surface of the lagoon, yielding forests of bread-fruit, cocoa-nuts, and other trees, and ultimately sustenance to a numerous population, and a variety of animals subservient to their use.

The general steepness of the volcanic islands of this groupe is such, that the soil finds a resting-place on a comparatively small portion of them; and on the coral islands it is scarcely deep and rich enough, exposed as it is to the sea air, to contribute much to the support of man. A soil formed from the decomposition of the basaltic rocks, irrigated by streams from the mountains, requires nothing but a due proportion of care and labour on the part of the natives to render it very productive. There is, however, a sad neglect in this respect, which is the more extraordinary, as there are no quadrupeds or

poultry on the islands, and without vegetable productions the natives have only the sea to depend
upon for their subsistence. The wild productions
are a coarse grass (*Saccharum fatuum*), which covers
such parts of the mountains as are neglected or are
too steep for cultivation. Lower down we noticed
the capparidia, a procumbent pentandrous shrub, the
nasturtium, sesuvium of Pitcairn Island, the euge-
nia, and scævola kœnigii; and close down to the
shore a convolvulus covering the brown rock with
its clusters of leaves and pink blossoms. The porou
and miroe (*Thespesia popularia*) were more abundant,
the nono not common. They must also have the
auti and amai, as their weapons are made of it,
though we did not see it. The timber of which
their rafts are constructed is a red wood, somewhat
porous, and of softer grain than the amai. Some of
these trunks are so large as at first to excite a suspicion of their having been drifted from a more
extensive shore; but the quantity which they possess, several logs of which were newly shaped out,
affords every reason for believing that it is the produce of their own valleys. They are not deficient in
variety of edible fruits and roots, nor in those kinds
which are most productive and nutritious. Besides
the tee-plant, sweet potatoe, appé, sugar-cane, water-
melon, cocoa-nut, plantain, and banana, they possess
the bread-fruit, which in Otaheite is the staff of life,
and the taro, a root which in utility corresponds
with it in the Sandwich Islands. Were they to pay
but a due regard to the cultivation of the two last
of these valuable productions, an abundance of
wholesome food might be substituted for the nauseous mixture mahie, which, though it may, as in-

deed it does, support life, cannot be said to do more. Rats and lizards were the only quadrupeds we saw upon the islands. Of the feathered tribe, oceanic birds form the greater part; but even these are rare, compared with the numbers that usually frequent the islands of the Pacific, arising, no doubt, from the Gambier Islands being inhabited. The whole consist of three kinds of tern, the white, black, and slate-coloured—of which the first is most numerous, and the last very scarce; together with a species of procellaria, the white heron, and the tropic and egg birds. Those which frequent the shore are a kind of pharmatopus, curlew, charadrine, and totanus; and the woods, the wood-pigeon, and a species of turdus, somewhat resembling a thrush in plumage, but smaller, possessing a similar though less harmonious note. The insects found here were very few, the common house-fly excepted, which on almost all the inhabited islands in the Pacific is extremely numerous and annoying. Of fish there is a great variety, and many are extremely beautiful in colour; as well those of large dimensions, which we caught with lines, consisting of several sorts of perca, as the numerous family of the order of branchiostigi, which sported about the coral.

The largest portion of the natives of the Gambier Islands belong to a class which Mr. J. R. Forster would place among the first variety of the human species in the South Seas. Like the generality of uncivilized people, they are good-natured when pleased, and harmless when not irritated; obsequious when inferior in force, and overbearing when otherwise; and are carried away by an insatiable desire of appropriating to themselves every thing

which attracts their fancy—an indulgence which
brings them into many quarrels, and often costs
them their lives. If respect for the deceased be considered a mark of civilization and humanity, they
cannot be called a barbarous people; but they possess no other claims to a worthier designation. In
features, language, and customs, they resemble the
Society, Friendly, Marquesa, and Sandwich Islanders; but they differ from those tribes in one very
important point—an exemption from those sensual
habits and indecent exhibitions which there pervade
all ranks. It may be said of the Gambier Islanders
what few can assert of any people inhabiting the
same part of the globe—that during the whole of
our intercourse with them we did not witness an
indecent act or gesture. There is a great mixture
of feature and of colour among them; and we
should probably have found a difference of dialect
also, could we have made ourselves masters of their
language. It seems as if several tribes from remote
parts of the Pacific had here met and mingled their
peculiarities. In complexion and feature we could
trace a resemblance even to the widely separated
tribes of New Zealand, New Caledonia, and Ma-
lacca. Their mode of salutation is the same as that
which existed at the Friendly, Society, and Sandwich Islands: they resemble the inhabitants of the
latter almost exclusively in tattooing the face, and
the inhabitants of the former in staining their skin
from the hips to the knees. Their huts, coral tables,
and pavements, are nearly the same as at the Friendly Islands and the Marquesas; but they are more
nearly allied to the latter by a custom which otherwise, I believe, is at present confined to them, and

without a due observance of which, Krusenstern says, it is in vain to seek a matrimonial alliance at St. Christina.*   In the preservation of their dead, wrapping them in an abundance of cloth and mats, they copy the Otaheitans and Hapaeans; though in the ultimate disposal of them in caves, and keeping them above ground, they differ from all the other islanders. Their language and religion are closely allied to several, yet they differ essentially from all the above-mentioned tribes in having no huge carved images surmounting their morais, and no fiatookas or wattas. Unlike them also, they are deficient in canoes, though they might easily construct them; they have neither clubs, slings, nor bows and arrows; and are wanting in those marks of self-mutilation which some tribes deem indispensable on the death of their chiefs or esteemed friends, or in cases when they wish to appease their offended deity.

They are for the most part fairer and handsomer than the Sandwich Islanders, but less effeminate than the Otaheitans. The average height of the men is above that of Englishmen, but they are not so robust. One man who came on board measured six feet and half an inch, and one on shore six feet, two and a half inches. The former measured round the thorax, under the arms, three feet two inches and a half; and a person of less stature three feet one inch. The thickest part of the middle of this person's arm, when at rest, was eleven inches and three-eighths. These dimensions of girth will, I believe, be found less in proportion than those of the labouring class of our own countrymen, though the general appear-

* See Krusenstern's Embassy to Japan.

ance of these islanders at first leads to a different conclusion. They are upright in figure, and round, but not robust. In their muscles there is a flabbiness, and in the old men a laxity of integument, which allows their skin to hang in folds about the belly and thighs to a greater degree than those I afterwards noticed at Otaheite or Woahoo. Two causes may be assigned for this; the nature of their food, and their indolent habits.

In general the Gambier Islanders have a fine Asiatic countenance, with mustachios and beards, but no whiskers; and when their heads are covered with a roll of white cloth, a very common custom, they might pass for Moors. It is somewhat remarkable that we perceived none of the fourth class, or those more nearly allied to negroes, thus habited, but that it seemed to be confined to those of the lightest complexion. The colour of their eyes is either hazel or dark brown: they are small, deep in the head, and have generally an expression of cunning. Their eyebrows are naturally arched, and seldom meet in front; the cheek-bones are not so prominent as in the fourth class, and the lips are thinner; the ears are moderately large, and the lobes attached to the cheek, as in all the Pitcairn Islanders, but not perforated: the nose in general is aquiline; the teeth, in the fourth class especially, not remarkable for evenness or whiteness, and seem to fall out at an early period; the hair is turned back and cut straight, and would be quite black, were it less subjected to the sun, or, like that of the islanders just mentioned, well oiled; but, exposed as it is to a scorching sun, it becomes dried up and of different hues on the same head; and combs being unknown, it is bushy and impervious: the mustachios grow long, but the beards, which are kept from three to four inches in length, are sometimes brought to a point, at others divided into two; one man, however, was observed with a beard which hung down to the pit of the stomach: the hands are large, but the feet small and elegant, and the toes close together, from which it is probable that they pass a great portion of their time upon their rafts, or idly basking in the sun,—perhaps in lying upon their stone pavements like the Hapaeans. The women are below the common standard height, and in personal shape and beauty far inferior to the males. The wife of the chief, who has been already described, was the finest woman I saw among them. Her dress may be considered a fair specimen of the general covering of the women, who have no ornaments of any kind, and appeared quite indifferent to the beads and trinkets which were offered them.

Tattooing is here so universally practised that it is rare to meet a man without it; and it is carried to such an extent that the figure is sometimes covered with small checkered lines from the neck to the ankles, though the breast is generally exempt, or only ornamented with a single device. In some, generally elderly men, the face is covered below the eyes, in which case the lines or net-work are more open than on other parts of the body, probably on account of the pain of the operation, and terminate at the upper part in a straight line, from ear to ear, passing over the bridge of the nose. With these exceptions to which we may add the fashion, with some few, of blue lines, resembling stockings, from the middle of the thigh to the ankle, the effect is be-

coming, and in a great measure destroys the appearance of nakedness. The patterns which most
improve the shape, and which appear to me peculiar
to this groupe, are those which extend from the
armpits to the hips, and are drawn forward with a
curve which seems to contract the waist, and at a
short distance gives the figure an elegance and outline not unlike that of the figures seen on the
walls of the Egyptian tombs. It would be useless
to describe the various fanciful attempts to efface
the natural colour of the skin; the most common
only will be noticed: — A large cross, about eight
inches in diameter, left white on each side, on the
latissimus dorsi; and a smaller one on each shoulder,
or on the upper part of the arm: also a narrow
stripe passing from one shoulder to the other in a
curved line over the lower part of the neck, uniting
the tattooing over the fleshy part of the deltoid
muscle; and in many so joined as to leave the natural skin in the form of a cross in the middle.
Imitations of blue pantaloons and breeches are also
very common, and sleeves which divide at the wrist,
and extend along the convexity of the metatarsal
bones to the tips of the fingers and thumbs, leaving
a space between the thumb and forefinger, on which
the mark V is punctured. The chief had this mark,
the crosses, the slender waist, and the pantaloons.
The women are very little subjected to this torture.
The wife of the areghe had an armlet on each arm;
a female who came with her had a square upon her
bosom, and some few had stockings. From the circumstance of none of the boys being tattooed, it is
probable the practice commences here, as in many
other islands, after puberty.

The lines in all cases are drawn with great precision, and almost always with taste, and bespeak
great proficiency. The practice undoubtedly improves the appearance of the figure, and may perhaps, as in the Marquesas, distinguish certain classes
or tribes. At Otaheite it is supposed to harden the
skin, and render it less liable to be blistered by the
sun. Covering the face with lines is very rare
in the South Seas, being almost entirely confined,
according to Cook, to the Sandwich Islanders and
New Zealanders. In no instance did we observe
the lips or tongues tattooed, as is the practice with
the Sandwich Islanders on the death of an intimate
friend.

I have estimated the number of souls inhabiting
these islands at 1500, from the number and size of
the villages. Mr. Collie, who estimates them from
other data, says, " On the 1st January, when the
boats went to land, 200 people, for the most part in
the prime of life, were counted on the beach. On
the 9th, in the village, we enumerated 300 persons,
men and women. On both these occasions it is
highly probable that the men in the vigour of life
had come from the adjoining parts of the island,
and from the islands contiguous. We may then
assume, on the nearest approximation to the truth,
that there were between 250 and 300 males between the ages of twenty and fifty — say 275;
which, according to the most accurate census of population and bills of mortality in Sweden and Swit-
zerland, where the modifying circumstances are in
all probability not very different, would give 1285
for the total number of inhabitants."

The diseases and deformities of these people are

very few   After we quitted the islands, the surgeon favoured me with the following report:—

" Among more than three hundred men, women,
and children, who indiscriminately surrounded us
at the village on the 9th; among those who had
previously come on board, and at other times, whether upon the shore or on their rafts; we saw very
few labouring under any original deformity or annoying disease. The only case of mal-conformation was
a wide fissure in the palate of one man, whose speech
was considerably affected by it. No external mark
of cicatrization in the upper lip denoted that the
internal defect was the remains of a hare-lip or any
injury. One man had a very uneven and ragged
stump of the right arm, but without any discharge.
Another had a steatomatous tumour over one shoul-
der-blade, about the size of a billiard-ball. One disease was so common that I have no doubt it was
endemic: this was, patches of the lepra vulgaris,
which being void of any inflammatory appearance,
and confined to the back in all who were affected
with it, and in a considerable proportion of these to
a small space between the shoulders, appeared to
create no alarm, and most probably called forth no
curative application. The frequent and alternate
exposure of the men to the salt water and rays of the
sun, with a scanty supply of the anointing oil of the
cocoa-nut, would favour the breaking out of this cutaneous affection. The mats which they tied round
their necks, and frequently allowed to hang down
behind, whether through accident or design, would
tend to avert the effects of exposure. A few had
lost some of their front teeth; and we saw one man,
on the 9th, with two uncicatrized and bare but clean

wounds, one before and another behind the middle of the right deltoid muscle, where the flies were feeding without molestation, and the person seemed almost unconscious of them and of the ulcers. No preternatural tumefaction denoted any excess of inflammation. No unhealthy hue in the countenance of man or woman intimated any internal disease lurking within the body." By far the greater part of the males go entirely naked, except a girdle, which is made of a banana-leaf split into shreds, and tied round the loins, not intended to answer the purpose of concealment; and they differ from all other inhabitants of the Pacific in having no maro. Some wear a turban; others a piece of paper cloth thrown over the shoulders.

The huts of the Gambier Islanders are so small that they can only be intended as sleeping-places during bad weather: they are in length from eight or ten feet to fifteen, excepting the larger houses of the areghe; they are built of the porou wood, and covered in with a pointed roof thatched over with the leaves of the palm-tree. In some the door is scarcely three feet high, and it is necessary to creep on all-fours to enter. On the inside they are neat, and the floor is covered with mats or grass. The larger huts of the village on Mount Duff are so constructed that one side can be conveniently removed, by which means they are rendered cool and comfortable.

The large house, or that of the areghe, was about thirty-nine feet in length by eighteen or twenty in width; the pitch of the roof was about twenty-five feet in height, and that of the perpendicular sides of the house about ten feet; but these dimensions were obtained by estimation only, the natives appearing to have an objection to our pacing the ground for the purpose of measurement. The south side of the house was left open, and the ends were made of an open framework of upright poles traversed at right angles by smaller spars, so that the roof and the north side were the only parts covered in. They served as an excellent protection from the sun, while the trade-wind traversing every corner of the apartment rendered it agreeably cool. On that part of the house where the side was deficient, there was a foundation for the wall about three feet in height thrown up, composed of large blocks of coral, shaped in a very workmanlike style, similar to those mentioned by Cook at the Friendly Islands, and well put together: it stood about three feet within the outer part of the roofing, and served as a seat for the chiefs as well as for many others.

We perceived no furniture in their houses, and some of our officers thought it was purposely put out of sight. The only utensils were gourds and cocoa-nut shells. The tables were made of slabs of coral, or sometimes of wood, in which case they are carved: they are about a yard long, and are placed upon wooden or stone pedestals sufficiently high to prevent the depredations of the rats. They stand in the middle of the paved areas in front of the houses, from which we infer the practice of eating in the open air. Their food has already been described as consisting principally of sour paste (the mahie of the Friendly Islands, Otaheite, Marquesas, &c.), made with plantains, bread-fruit, and boiled tee-root. The paste or mahie, when fresh and hot, has not a disagreeable taste; a slight flavour of baked apples may be distinguished : but it soon begins to smell very offensively; so much so, that the seamen would not touch it with their hands to throw it overboard. The tee-plant *(dracœna terminalis)* is a fusiform root about two feet long, and as thick as the arm ; its flavour is not unpleasant, but from its coarseness it must, to ordinary stomachs, be very indigestible. The natives collect the fibres in their mouths, and spit them out in round balls. Fish and shell-fish, of which the large pearl oysters and chama are in the greatest abundance, must form a material part of the food of these people : they have, besides, the sweet potatoe, taro, and the before-mentioned fruits ; but these cannot be abundant, as they never brought any of them to us for sale, and frequently deceived us with empty cocoa-nuts.

Their method of procuring fish is by lines and nets, and a contrivance still resorted to in Otaheite, consisting of casting into the sea a great many branches of the cocoa-nut tree, and other boughs, tied together, and allowing them to remain some time, during which the small fish become entangled, and are dragged out with them. The nets and lines, as well as cord, sinnet, &c., are all made from the bark of the porou, as in all the islands of Polynesia. One net which we measured was ninety feet in length. In the manufacture of these, they display a greater proficiency than in their cloth, which is much inferior to that at Pitcairn Island or Otaheite. Their implements for this purpose are the same in shape as those at the above-mentioned places; but the one which we got differed in not being grooved.

Their weapons consist of spears, and a staff flattened at the end like a whale-lance : they are made of a hard wood, and highly polished. The spears were headed with bone, or the sting rays of the raia (*pastinea*); a custom which once existed at Otaheite, and now extends to many of the low islands. The antiquity of this practice is traced to very remote periods, as it is said that the head of the spear presented by Circe to Telegonus, and with which he unceremoniously slew his father Ulysses, was of this kind. At Gambier Island they remove the heads of the spears when not required, a square piece being left at the end of the staff to receive it. Besides these weapons, they always carry large sticks.

Contrary to the general custom, no canoes are seen at Gambier Islands, but rafts or katamarans are used instead. They are from forty to fifty feet in length, and will contain upwards of twenty persons. They consist of the trunks of trees fastened together by rope and cross-beams : upon this a triangular sail is hoisted, supported by two poles from each end; but it is only used when the wind is very favourable; at which time, if two or three katamarans happen to be going the same way, they fasten on and perform their voyage together. At other times they use very large paddles made of a dark hard wood, capable of a good polish, and neatly executed. Some of them had a hand or a foot, carved at the extremity of the handles, very well finished. They are above five feet and a half in length, including two feet eight inches of blade, which is about a foot in width, curved, and furnished with a small point or nail at the extremity. In shallow water they make use of long poles for punting, in preference to their paddles.

CHAPTER VII.

ON the morning of the 13th of January we
weighed from Gambier Islands, and deepened the
water so much that, after quitting our anchorage,
we could get no soundings with the hand-lines
until near the bar, which was plainly distinguished
by its colour long before we came upon it.   There
was not less than seven fathoms where we passed,
and yet the sea, which rolled in heavily from the
S. W., all but broke, notwithstanding the wind
had been blowing strong in the opposite direction
for a week before.   This effect of the prevalent
south-westerly gales in the high latitudes, which is
felt many hundred miles from the place whence it
proceeds, occasions a material obstacle to landing
upon the low islands, by rolling in upon the shore,

CHAP.
VII.

Jan.
1826.

in an opposite direction to the trade wind, and thereby making it more dangerous to land on the lee-side of the island than on the other. In the Gambier groupe there are several small sandy islands at the S. W. extremity of the chain that surrounds it, over which the sea broke so heavily that they were entirely lost amidst the foam. I named them Wolfe Islands, after Mr. James Wolfe, one of the midshipmen of the ship. We passed them tolerably close, admiring the grand scene which they presented, and then stood on a northerly course with the intention of visiting Lord Hood's Island.

In the forenoon of the following day several white tern, noddies, and black gulls came about us, and gradually increased in numbers as we proceeded on our course. A few hours afterwards Lord Hood's Island was reported from aloft. On nearing it, we found it to consist of an assemblage of small islets, rising from a chain of coral, even with, or a little above, the water's edge. Upon these grew a variety of evergreen trees thickly intertwined, among which the broad leaves and clusters of fruit of the pandanus were conspicuous, and beneath them a matted surface of moss and grass, so luxuriant and invitingly cool, that we were almost tempted to land at any risk. The sea, however, broke so heavily upon all parts of the shore that the attempt would have been highly dangerous, and we consequently collected all the information that was required, and hastened our departure. Krusenstern states in his " Mémoire sur la Pacifique," that this island is inhabited: such must undoubtedly have been the case once, as we saw a square stone hut, similar to those described at Crescent Island, on one of its angles; but there are no human beings upon it at present, which indeed we conjectured to be the case before our boats made the circuit of it, from the number of sea birds in its vicinity, and also from the shoals of sharks which followed the boats, and even bit at the oars; for these animals, like most others, seem to have learned by experience to avoid the haunts of man. The only living thing seen upon the shore was a grey heron gorging itself with black star-fish.

Lord Hood's Island was discovered by Mr. Wilson in the Missionary ship Duff; it is 11·2 miles in length, and 4·7 miles in width, in a north and south direction; and like almost all the coral islands it contains a lagoon, and is steep on all its sides.

After quitting it, we looked in vain, the next day, for an island which Mr. Wilson supposed he saw; but not finding it in or near the situation assigned, and he being himself doubtful whether it might not have been a cloud, I did not bestow longer time in the search, but steered for the island of Clermont Tonnere, which was seen on the 18th. This island bore a very close resemblance to Hood's Island, but was inhabited, and clothed with cocoa-nut trees. The sea broke so heavily upon all parts of the shore that there was no possibility of landing in our boats; yet the natives put off in their canoes and paddled to us. They were a very inferior race to those of the Gambier Islands, and seemed more nearly allied in feature to those of Mangea and New Caledonia; yet here also there was among them a great diversity of complexion. In one of the canoes there was a man nearly as dark as an African negro,

with woolly hair, tied in a knot like the Radackers;
and another with a light complexion, sandy hair,
and European features.

About forty of the natives came down to the beach
when we approached it, with bunches of feathers
and leaves fastened upon sticks, and with bludgeons
in their hands. Both sexes were naked with the
exception of their maros, and without any ornaments or tattooing. Iron, which they called " toki,"
was the most marketable article, but the surf was so
high that there was very little communication with
them. The men, who came off to us in their canoes,
would not suffer our boats to approach them. After
having made a number of presents to one of them,
we thought we might at least examine his canoe ;
but he and his comrade paddled away with the
greatest precipitation, and were so terrified at the
approach of the boat that they jumped overboard
and swam towards the shore.

The canoe was constructed with small pieces of
wood well put together and sewed with the bark of
a tree, and, like all the single canoes of Polynesia,
was provided with an outrigger. She carried two
men, but was propelled almost entirely by one, the
other being fully occupied in throwing out the
water, which came in plentifully at both sides and
over the stern. Could they have avoided this and
applied the efforts of both to the paddles, her rapidity would have surpassed that of our boat; but
as it was she was soon overtaken. We did not
keep these poor fellows longer in the water than we
could help, but quitted the canoe as soon as we had
examined its construction, and had the satisfaction
to observe them return to it, and get in, one at a time,
at the stern, and then paddled ashore.

The dialect of the people of Clermont Tonnere was quite different from that of the Gambier Islanders, though, from a few words which we distinguished, there is no doubt of the language being radically the same. According to our calculation the whole population did not exceed two hundred.

The island is ten miles in length, but very narrow, particularly at the extremities, and, when seen at a distance, does not appear to be half a mile wide. It is of the same formation as Lord Hood's Island, but more perfect. With the exception of a few breaks in the southern shore, by which the sea, when high, may at times communicate with the lagoon, it is altogether above water. At the extremities and angles the soil is more elevated than in other parts, as if the influence of the sea had been more felt upon them, and heaped up the coral higher. They are, also, better provided with shrubs, and particularly cocoa-nut trees, the soil resting upon the debris being, I suppose, deeper. The lagoon had several small islets in it, and the shores all round are steep, and abound with fish, but we did not see any sharks.

Captain Duperrey, in his voyage round the world in the Coquille, visited this island, and, supposing it to be a new discovery, named it Clermont Tonnere, after the French minister of marine. It is evident, however, from its situation agreeing very nearly with that of an island discovered by the Minerva, that it must be the same; no other being found sufficiently near to answer the description. Captain Duperrey has, no doubt, been misled by the dimensions given of the island by the Minerva; but that may be easily accounted for, by supposing the island to have been seen from the Minerva lengthwise, and at a distance.

While we were off Clermont Tonnere, we had a
narrow escape from a water-spout of more than
ordinary size. It approached us amidst heavy rain,
thunder, and lightning, and was not seen until it
was very near to the ship. As soon as we were
within its influence, a gust of wind obliged us to
take in every sail, and the topsails, which could not
be furled in time, were in danger of splitting. The
wind blew with great violence, momentarily changing its direction, as if it were sweeping round in
short spirals; the rain, which fell in torrents, was
also precipitated in curves with short intervals of
cessation. Amidst this thick shower the water-
spout was discovered, extending in a tapering form
from a dense stratum of cloud, to within thirty feet
of the water where it was hid, by the foam of the
sea being whirled upwards with a tremendous gira-
tion. It changed its direction after it was first seen,
and threatened to pass over the ship; but being
diverted from its course by a heavy gust of wind, it
gradually receded. On the dispersion of this magnificent phenomenon, we observed the column to
diminish gradually, and at length to retire to the
cloud, from whence it had descended, in an undu-
lating form.

Various causes have been assigned for these formations, which appear to be intimately connected
with electricity. On the present occasion a ball of
fire was observed to be precipitated into the sea, and
one of the boats, which was away from the ship, was
so surrounded by lightning, that Lieutenant Belcher
thought it advisable to get rid of the anchor, by
hanging it some fathoms under water, and to cover
the seamen's muskets. From the accounts of this

A — page 149.

B — page 149.

C — page 149

Appearance during an Eclipse of the Sun.—see Page 4.

officer and Mr. Smyth, who were at a distance from the ship, the column of the water-spout first descended in a spiral form, until it met the ascending column a short distance from the sea;* a second and a third were afterwards formed,† which subsequently united into one large column,‡ and this again separated into three small spirals, and then dispersed. It is not impossible that the highly rarefied air confined by the woods encircling the lagoon islands may contribute to the formation of these phenomena.

A canoe near the ship very wisely hastened on shore at the approach of the bad weather, for had it been drawn within the vortex of the whirlwind it must have perished. We had the greatest apprehension for our boats, which were absent during the storm, but fortunately they suffered no injury.

Neither the barometer nor sympeisometer were sensibly affected by this partial disturbance of the atmosphere; but the temperature underwent a change of eight degrees, falling from 82° to 74°; at midnight it rose to 78°. On the day succeeding this occurrence, several water-spouts were seen in the distance, the weather being squally and gloomy.

After examining the vicinity of Clermont Tonnere for the island of the Minerva, and seeing no other land, we steered for Serle Island, which was discovered at daylight on the 21st January, bearing west. Its first appearance was that of a low strip of land with a hillock at each extremity, but these, on a nearer approach, proved to be clumps of large trees. Admiral Krusenstern, in his valuable Memoir on the South Pacific,§ observes, that Serle

* See plate (A).   † (B.)   ‡ (C).
§ Page 276, 4to. edition.

Island is higher than any other island of the low archipelago; that it has two hills at its extremities, and a third near its centre; and on this account recommends it as a place of reconnoissance for ships entering the archipelago. In this, however, he has been misled by some navigator who mistook the trees for hills, and over-estimated the height of them, as the tallest does not exceed fifty feet.

Some columns of smoke rising from the island showed that it was inhabited, and on rounding the N. W. extreme we perceived several men and women running along the beach, dragging after them long poles or spears. The population altogether cannot exceed a hundred. The men were entirely naked, but the women had the usual covering. They were of the same dark swarthy colour with the natives of Clermont Tonnere, with the hair tied in a similar knot on the top of the head, and like them they were deficient in tattooing and ornaments. Their weapons were poles about twenty feet in length, similar to those of the Friendly Islanders, and heavy clubs. We could not perceive any canoes.

This island is seven miles and a half in length in a N. W. direction, and two and a quarter miles in width in its broadest part. It is of coral formation, and very similar to that just described; its windward side is the most perfect: the southern side of the chain, however, differs in being wider, and having a barren flat full an eighth of a mile outside the trees. On this account it is necessary for a ship to be cautious in approaching it during the night, as it is so low that the breakers would be the first warning of the danger of her situation. The lagoon is very narrow, and apparently shallow, with several islands in the

middle.   Besides the clumps of trees at the extremities, which at a distance have the appearance of
banyan trees, there are several clusters of palms; a
distinction which I would recommend to the attention of commanders of vessels; as, besides assisting
them in identifying the islands, it will enable them
to estimate their distance from them with tolerable
precision.

We left Serle Island on the morning of the 22d,
and at sunset hove to in the parallel of Whitsunday
Island. This island, discovered by Captain Wallis
in 1767, is situated forty miles to the westward of
the place he has assigned to it, and we consequently
ran to the westward all the next day, in expectation
of seeing it, but it was not reported from the mast-
head until late in the evening.   In the morning of
the 23d the boats succeeded in landing, though with
some difficulty; and found indubitable proofs of the
island having been thickly inhabited; but no natives
were seen.   Under a large clump of trees we observed
several huts, eight feet by three, thatched with dried
palm leaves, the doors of which were so low that it
was necessary to crawl upon the ground to enter the
apartments within.   Near these dwellings were some
sheds and several piles of chewed pandanus nuts.

The island was traversed in various directions by
well trodden pathways: not far from the huts were
several reservoirs of water cut about eighteen inches
into the coral, and about five feet from the general
surface of the soil; the water in them was fresh, but
from neglect the reservoirs were nearly filled with
decayed leaves, and emitted a putrid smell.   In
another direction we saw several slabs of coral placed
erect, to denote burial places; and near the opening

to the lagoon there were several rows of stakes driven into the ground for the purpose of taking fish. But what most attracted our attention was a heap of fish bones, six feet by five, neatly cleaned, and piled up very carefully with planks placed upon them to prevent their being scattered by the wind.

We found the island only a mile and a half in length, instead of four miles, as stated by Captain Wallis; steep all round; of coral formation; well wooded, and containing a lagoon. The general height of the soil was six feet above the level of the sea, of which nearly two feet were coral rock; from the trees to the surf there was a space of hard rock nearly 150 yards in length, covered with about a foot of water, beyond which it descended rapidly, and at 500 yards distance no bottom could be found with 1500 feet of line. On the inner side, from the trees to the lake, there was a gentle declivity of muddy sand filled with shells of the cardium, linedo, tridacnæ, gigas, and a species of trochus. The trees, which formed a tolerably thick wood round the lagoon, were similar to those at Clermont Tonnere, consisting principally of pandanus and cocoa-nut, interwoven with the tournefortia, scœvola, and lepidium piscidium.

On the south side of the island there was a very narrow entrance to the lagoon, too shallow for the passage of boats, even had the water been smooth. It was of this opening, I presume, that Captain Wallis observes that the surf was too high upon the rocks for his boats to attempt the passage.

The lagoon was comparatively shallow; the edges, for a considerable distance, sloped gradually toward the centre and then deepened suddenly; the edge of the bank being nearly perpendicular. This bank, as well as numerous islets in the lagoon, were formed of coral and dead and live tridacnæ shells. The space between the islets was very rugged, and full of deep holes.

In the lagoon there were several kinds of brilliantly coloured fish ; on the reef, some fistularia ; and in the surf a brown and black chætodon with a black patch at the junction of the tail with the body. Upon the land were seen a few rats and lizards, a white heron, a curlew, some sandpipers, and a species of columba resembling the columba australis.

In the evening we bore up for Queen Charlotte's Island, another coral formation also discovered by Captain Wallis, and so grown up that we could not see any lagoon in its centre, as we had done in all the others. Several huts and sheds similar to those at Whitsunday Island occur in a bay on its northern shore, but there were no inhabitants. It may be remembered that when Captain Wallis visited this island, the natives took to their canoes and fled to the next island to the westward : whether they did so on the present occasion we could not determine, but in all probability we should have seen them if they had. Queen Charlotte's Island afforded Captain Wallis a plentiful supply of cocoa-nuts, but at present not a tree of that description is to be seen. The shore is more steep than either Whitsunday or Clermont Tonnere, and the huts more numerous.

At two o'clock in the afternoon we quitted Queen Charlotte's Island, and in two hours afterwards saw Lagoon Island, which was discovered by Captain Cook ; the former bearing S. 6° W. true, the latter due north, by which an excellent opportunity occurred of comparing the longitudes of those celebrated navigators.

The next morning we coasted the north side of Lagoon Island very closely, while the barge navigated the other. It is three miles in length in a W. by S. direction, and a mile and a quarter in width. Its general figure has been accurately described by Captain Cook: the southern side is still the low reef of breakers which he saw, and the three shallow openings on the north shore still exist, though one of them has almost disappeared. Two cocoa-nut trees in the centre of the island, which Cook observes had the appearance of flags, are still waving; " the tower" at the western end is also there, but has increased to a large clump of cocoa-nut trees: a similar clump has sprung up at the eastern end. The lagoon is, in some parts, very shallow and contracted, and has many dry islets upon it. The shore is steep, as at the other coral islands, excepting on the south side, which should not be approached within a quarter of a mile.

We brought to off a small village at the N. W. extremity of the island, and sent two boats on shore. The natives seeing them approach came down to the beach armed with poles from twenty to twenty-five feet in length, with bone heads, and short clubs shaped like a bill-hook; but before they reached the surf they laid down their weapons. At first they beckoned our people to land; but seeing the breakers too high, they suffered themselves to be bribed by a few pieces of iron, and swam off to them. A brisk traffic soon began, and all the disposable articles of the natives were speedily purchased for a

few nails, broken pieces of iron, and beads: they
then brought down cocoa-nuts, and exchanged six of
them for a nail or a bit of iron, which is known here,
as at Clermont Tonnere, by the name of "toki."
The strictest integrity was observed by these people
in all their dealings. If one person had not the
number of cocoa-nuts demanded for a piece of iron,
he borrowed from his neighbour; and when any of
the fruit fell over-board in putting it into the boat,
they swam after it and restored it to the owner.
Such honesty is rare among the natives of Polynesia, and the Lagoon Islanders consequently ingratiated themselves much with us. We got from
them nearly two hundred cocoa-nuts, and several
ornamental parts of their dress, one of which consisted of thin bands of human hair, very neatly
plaited, about five feet in length, with four or five
dozen strings in each. To some of these were
attached a dried doodoe-nut (*aleurites triloba*), or a
piece of wood. We also got some of their mats and
sinnet made of the porou bark (*hibiscus tiliaceus*).

The men were a fine athletic race, with frizzled
hair, which they wore very thick. In complexion
they were much lighter than the islanders of Clermont Tonnere: one man, in particular, and the only
one who had whiskers, was so fair, and so like an
European, that the boat's crew claimed him as a
countryman. No superfluous ornaments were worn
by either sex, nor were any of them tattooed: the
dress of the males was simply a maro of straw, and
sometimes a straw sack hung over their shoulders to
prevent the sun from scorching their backs: two of
them were distinguished by crowns of white feathers. The women had a mat wrapped about their

loins as their only covering: some wore the hair tied in a bunch upon one side of the head, others had a plaited band tied round it. They were inferior to the men in personal appearance, and mostly bow-legged; but they exercised an authority not very common among uncivilized people, by taking from the men whatever articles they received in exchange for their fruit, as soon as they returned to the shore. The good-natured countenances of these people, the honesty observed in all their dealings, and the great respect they paid their women, bespeak them a more amiable race than the avaricious Gambier Islanders.

We quitted them about three o'clock in the afternoon, and in a few hours after saw Thrum Cap Island, bearing N. 56° 54′ W.; the clump on Lagoon Island at the same time bearing S. 58° 14′ E., thirteen miles distant. This island, discovered and so named by Captain Cook, is also of coral, three-quarters of a mile in length, well wooded, and steep all round. At a mile distance from it we could get no bottom with 400 fathoms. We could perceive no lagoon; and the surf ran too high to admit of landing. Some slabs placed erect, and a hut, showed it had once been inhabited; but the only living things we saw were birds and turtle. M. Bougainville gave this island the name of Les Lanciers, in consequence of the men whom he saw on it, being armed with long spears, and who probably were visiters from the island we had just left.

From Thrum Cap we steered for Egmont Island, the second discovery of Captain Wallis, which we shortly saw from the mast-head, and by sunset were close to it. The next morning the shore was very carefully examined, and we found the reef so low toward the centre that in high tides there can be no communication with the extremities. The island is steep, like all the other coral islands, and well wooded with cocoa-nut and pandanus-trees, and has one of the large clumps at its N. W. extremity.

Upon the windward island we perceived about fifty inhabitants collected upon the beach; the men in one groupe, armed in the same manner as the Lagoon Islanders, and the women in another place more inland. No boat could land on this or on any other part of the island: to leeward the S. W. swell rolled even more heavily upon the shore than that occasioned by the trade-wind on the opposite side: we were in consequence obliged to trade with the natives in the manner pursued at Lagoon Island. Two of the islanders, when they thought we were going to land, advanced with slow strides, and went through a number of pantomimic gestures, which we could not understand, except that they were of a friendly nature. This lasted until the boats anchored outside the reef, and they were invited to accept some pieces of "toki." Gold and silver are not more valued in European countries, than iron, even in its rudest form, is by the islanders of Polynesia. At the sound of the word, the two spokesmen, and all the natives, who had before been seated under the shade of the trees, ran off to their huts, and brought down whatever they thought likely to obtain a piece of the precious substance,—mats, bands, nets, oyster-shells, hooks, and a variety of small articles similar to those before described were offered for sale. The only article they would not part with, though we offered a higher price than it seemed to deserve, was a stick with a bunch of black tern feathers suspended to it. At Lagoon, and other islands which we visited both before and afterwards, the natives carried one or more of these sticks: they are mentioned as being seen by the earliest voyagers, and are probably marks of distinction or of amity.

These people so much resemble the Lagoon Islanders in person, manners, language, and dress, as to need no description: the island is also of the same formation, and has apparently the same productions. We noticed only one canoe; but no doubt they have others, as a constant communication is kept up with the islands to windward. It may be recollected that it was upon this island Captain Wallis found all the natives collected who had deserted Queen Charlotte's Island on his approach. Though these two places are many miles out of sight of each other, yet their canoes took the exact direction which, being afterwards followed by Captain Wallis, led to the discovery of the island.

Next morning we saw land to the S. by E., which proved to be a small coral island, answering in situation nearly to that of Carysfort Island, discovered by Captain Edwards, but so small as to render it very unlikely that it should be the same. Though we ranged the shore very closely, we did not perceive any inhabitants. It was well wooded, and had several clusters of cocoa-nut trees. The next morning parties were sent to cut down some of the trees for fire-wood. The surf ran high upon the shore; but, with the assistance of a small raft, a disembarkation was effected without any serious accident. Several of the officers, anxious to land upon this our first discovery in these seas, joined the party in spite of a sound ducking, which was the smallest penalty attached to the undertaking. In one of these attempts the Naturalist was unfortunately drawn into a deep hole in the coral by the recoil of the sea, and, but for prompt assistance, would in all probability have lost his life.

The island proved to be only a mile and three-quarters in length, from north to south, and a mile and three-tenths in width. It consisted of a narrow strip of land of an oval form, not more than two hundred yards wide in any part, with a lagoon in its centre, which the colour of the water indicated to be of no great depth. In places this lake washed the trunks of the trees; in others it was separated from them by a whitish beach, formed principally of cardium and venus-shells. Shoals of small fish of the chætodon genus, highly curious and beautiful in colour, sported along the clear margin of the lake, and with them two or three species of fistularia; several moluscous animals and shell-fish occupied the hollows of the coral (principally madrepora cervi-cornis); and the chama giganteus was found so completely overgrown by the coral that just sufficient space was left for it to open its shell; a fact which tends to show the rapidity with which coral increases.

Upon the shores of the lagoon, the pandanus, cocoa-nut, toufano, scœvola kœnigii, the suriana (whose aroma may be perceived at the distance of several miles,) the large clump-tree, pemphis acidula, tournefortia sericea, and other evergreens common to these formations, constituted a thick wood, and afforded a cool retreat from the scorching rays of a vertical sun, and the still greater annoyance arising from the reflection of the bright white sand; a luxury which until our arrival was enjoyed only by a few black and white tern, tropic and frigate birds, and some soldier-crabs which had taken up their abode in the vacated turbo-shells.

Under these trees were three large pits containing several tons of fresh water, and not far from them some low huts similar to those described at the other islands, and a tomb-stone shaped like that at Whitsunday Island. We judged that the huts had been long deserted, from the circumstance of the tern and other aquatic birds occupying some cala-bashes which were left in them. Among several things found in this deserted village were part of a scraper used by merchant-ships, and a large fish-hook, which we preserved, without suspecting that they would at a future day clear up the doubt that these articles were calculated to throw upon the merit of discovering this island, to which we other-wise felt an indisputable claim. Our suspicions on this head were also strengthened by noticing that a cocoa-nut tree had been cut down with an instru-ment sharper than the stone axes of the Indians. We had, however, no direct proof that the island had been before visited by any ship; and we con-soled ourselves with the possibility of the instru-ments having been brought from a distance by the natives, who might be absent on a temporary visit, and several of whose canoes we found in the lagoon: the largest of these was eighteen feet in length by fifteen inches in breadth, hollowed out of the large tree (which we at first mistook for a banyan-tree,) and furnished with outriggers similar to the canoes of Clermont Tonnere.

This island, the north end of which is situated in latitude 20° 45′ 07″ S., and longitude 4° 07′ 48″ West, of Gambier Island, I named Barrow Island, in compliment to the Secretary of the Admiralty, whose literary talents and zeal for the promotion of geographical science have been long known to the world.

The party on shore succeeded in the course of a few hours in collecting a tolerable supply of hard wood, very well adapted for fuel, and some brooms, after which we beat to windward in search of Carysfort Island; and at four o'clock in the afternoon had the satisfaction to see land in that direction; but, in consequence of a strong current setting to the southward, we did not get near it until the afternoon of the following day. It answered in every respect to Captain Edwards's description of Carysfort Island. The strip of land is so low, that the sea, in several places, washes into the lagoon. Like all the other islands of this formation we had visited, the weather side and the points of the island were most wooded, but the vegetation was on the whole scanty. There is no danger near this island. The outer part of the bank descends abruptly as follows: at sixty yards from the breakers, 5 fathoms water

Eighty yards . .	13	ditto
One hundred and twenty do.	18	ditto
Two hundred yards	24	ditto

On the edge of the bank immediately after, no bottom with 35 fathoms.

During the night we stood quietly to the southward in search of Matilda Rocks and Osnaburgh Island. At daylight we saw large flocks of tern, and at eleven o'clock land was reported bearing

W. by S.   The barge and the ship circumnavigated
this island before dark, and then kept under easy
sail during the night.   I learnt from Mr. Belcher,
who passed round the eastern side of the island,
that he had found an opening into the lagoon in
that direction, and had discovered near it two
anchors lying high up on the reef.

At daylight next morning land was seen to the
southward, which on examination proved to be
another small coral island, three miles and three-
quarters in length, by three in width: its form is
nearly an oblong with the southern side much
curved.   The lagoon in the centre was deep, its
boundary very low and narrow, and in places it
overflowed.   Several ripplings were observed about
these islands, but we passed through them without
obtaining soundings.

As soon as the plan of this island was completed,
we returned to that upon which the anchors were
observed, and spent the whole day in its examination.   The lagoon was entered in the boats by a
channel sufficiently wide and deep for a vessel of the
class of the Blossom, and proved in every respect an
excellent harbour: in entering, however, it is necessary to look out carefully for rocks, which rise suddenly to the surface, or within a very short distance
of it.

On landing at the back of the reef, we perceived
unequivocal signs of a shipwreck—part of a vessel's
keel and fore-foot, broken casks, a number of staves,
hoops, a ship's hatch marked VIII., some copper,
lead, &c., and the beach strewed with broken iron
hoops, and in their vicinity the anchors which were
discovered the preceding day : there were also broken

harpoons, lances, a small cannon, cast metal boilers, &c. &c., and a leaden pump which had a crown and the date 1790 raised upon it. All the iron-work was much corroded, and must have been a considerable time exposed to the action of the sea and air, but it was not overgrown in the least by the coral. Two of these anchors weighed about a ton each; the other was a stream anchor, and with one of the bowers, was at the break of the sea; the other bower, together with the boiler, and all the before-mentioned materials, were lying about two hundred yards from it. The situation in which they were found, the size of the anchors, the harpoons, staves, &c. and the date of the pump, render it highly probable that they belonged to the Matilda, a whaler which was wrecked in 1792, in the night-time, upon a reef of coral rocks, in latitude 22° S., and longitude 138° 34′ W. But whether they had been washed up there by some extraordinarily high tide and sea, or the reef had since grown upward, and raised them beyond the present reach of the waves, we could not decide: the former is most probable; though it is evident, if the above-mentioned remains be those of the Matilda, of which there can be very little doubt, that a considerable alteration has taken place in the island, as the crew of that vessel describe themselves to have been lost on *a reef of rocks*, whereas the island on which these anchors are lying extends fourteen miles in length, and has one of its sides covered nearly the whole of the way with high trees, which, from the spot where the vessel was wrecked, are very conspicuous, and could not fail to be seen by persons in the situation of her crew.

The island differs from the other coral formations

before described, in having a greater disproportion in the growth of its sides. The one to windward is covered with tall trees as before mentioned, while that to leeward is nearly all under water. The dry part of the chain enclosing the lagoon is about a sixth of a mile in width, but varies considerably in its dimensions: the broad parts are furnished with low mounds of sand, which have been raised by the action of the waves, but are now out of their reach, and mostly covered with vegetation. The violence of the waves upon the shore, except at low water, forces the sea into the lake at many points, and occasions a constant outset through the channel to leeward.

On both sides of the chain the coral descends rapidly: on the outer part there is from six to ten fathoms close to the breakers, the next cast is thirty to forty, and at a little distance there is no bottom with two hundred and fifty fathoms. On the lagoon side, there are two ledges: the first is covered about three feet at high water: at its edge the lead descends to three fathoms to the next ledge, which is about forty yards in width; it then slopes to about five fathoms at its extremity, and again descends perpendicularly to ten; after which there is a gradual descent to twenty fathoms, which is the general depth of the centre of the lagoon. The lake is dotted with knolls or columns of coral, which rise to all intermediate heights between the bottom and the surface, and are dangerous even to boats sailing in the lagoon with a fresh breeze, particularly in cloudy weather, as at that time it is difficult to distinguish even those which are close to the surface.

No cocoa-nut or other fruit-trees have yet been

planted on this isolated shore, nor are there any ves-
tiges of its ever having been inhabited, excepting by
the feathered tribe, a few lizards, soldier-crabs, and
occasionally by turtle. The birds, unaccustomed to
molestation, were so ignorant of their danger that
we lifted them off their nests ; and the fish suffered
as much by our sticks and boat-hooks, as by our
fishing-lines. The sharks, as in almost all uninhabited islands within the tropics, were so numerous
and daring, that they took the fish off our lines as
we were hauling them in, and the next minute were
themselves taken by a bait thrown over for them ; a
happy thought of our fishermen, who by that means
not only recovered many of their hooks, but got
back the stolen fish in a tolerably perfect state.

In several small lakes, occasioned by the sea at
times overflowing the land, we saw an abundance of
fish of the chætodon and sparus genera, of the same
beautiful colours as those at Barrow Island, and in
one of them caught a species of gymnothorax about
two feet in length. There were but few echini upon
the reef, but an abundance of shell-fish, consisting of
the arca, ostrea, cardium, turbo, helix, conus, cyprea,
voluta, harpa, haliotis, patella, &c. ; also several
aphroditæ holuthuriæ *(biche la mer)* and asteriæ, &c.

The position of this island differed so considerably
from that of Osnaburgh Island, discovered by Captain Carteret, that I beat two days to the eastward
in the parallel of 22° S. in the expectation of finding
another ; but when the view from the mast-head
extended half a degree beyond the longitude he had
assigned to his discovery, and we had not even any
indication of land, I gave up further search. The
probability, therefore, is, that the island upon which

we found the wreck is the Osnaburgh of Captain Carteret; and as it is equally probable, from what has been said, that the remains are those of the Matilda, it will be proper henceforward to affix to it the names of both Osnaburgh and Matilda.

A doubt might have arisen with respect to the island discovered to the southward being Osnaburgh Island, had Captain Carteret not expressly said in his journal, that the island he saw was to the *south* of him; but this bearing put such a supposition out of the question, as in that case he must have seen the island to the northward also. I have, in consequence, considered it a new discovery, and honoured it with the name of Cockburn Island, in compliment to the Right Honourable Sir George Cockburn, G.C.B., one of the lords of the Admiralty.

After we gave up the search to the eastward for the island of Captain Carteret, we pursued the same parallel of 22° S. some distance to the westward without being more successful, and then steered for the Lagoon Island of Captain Bligh, which was seen the following day. On our approach several large fires were kindled in different parts. The natives were darker than those of Lagoon Island of Cook, were nearly naked, and had their hair tied in a knot on the top of the head; they were all provided with stones, clubs, and spears. As the sea ran very high, we did not land, and consequently had no further communication with them. The island is larger than is exhibited upon Arrowsmith's Charts, but agrees in situation very closely with the position assigned to it by Captain Bligh.

Two days afterwards we discovered a small island in lat. 19° 40′ S. and long. 140° 29′ W., which, as it was not before known, I named Byam Martin Island, in compliment to Sir Thomas Byam Martin, K.C.B., the Comptroller of the Navy.

As we neared the shore the natives made several fires. Shortly afterwards three of them launched a canoe, and paddled fearlessly to the barge, which brought them to the ship. Instead of the deep-coloured uncivilized Indians inhabiting the coral islands in general, a tall well-made person, comparatively fair, and handsomely tattooed, ascended the side, and, to our surprise, familiarly accosted us in the Otaheitan manner. The second had a hog and a cock tattooed upon his breast—animals almost unknown among the islands of Eastern Polynesia; and the third wore a turban of blue nankeen. Either of these were distinctions sufficient to excite considerable interest, as they convinced us they were not natives of the island before us, but had either been left there, or drifted away from some other island : the latter supposition was the most probable, as they described themselves to have undergone great privation and suffering, by which many of their companions had lost their lives, and their canoe to have been wrecked upon the island ; and that they and their friends on shore were anxious to embark in the ship, and return to Otaheite. A little suspicion was at first attached to this account, as it seemed impossible for a canoe to reach their present asylum without purposely paddling towards it; as Byam Martin Island, unlike Wateo, upon which Omai found his countrymen, is situated six hundred miles from Otaheite, in the direction of the trade-wind. We could not doubt, however, that they were natives of that place, as they mentioned the names of the missionaries residing there, and proved that they could both read and write.

To their solicitation to return in the ship to Otaheite, as their numbers on shore amounted to forty persons, I could not yield, and I pointed out to them the impossibility of doing so; but that we might learn the real history of their adventures, I offered a passage to the man who first ascended the side, as he appeared the most intelligent of the party. The poor fellow was at first quite delighted, but suddenly became grave, and inquired if his wife and children might accompany him, as he could on no account consent to a separation. Our compliance with this request appeared to render him completely happy; but still fearful of disappointment, before quitting the ship he sent to ask if I was in earnest.

The next morning, on landing, we found him, his wife, and family, with their goods and chattels, upon the beach, ready to embark, and all the islanders assembled to take leave of them; but as we wished to examine the island first, we postponed this ceremony until the evening. The little colony gave us a very friendly reception, and conducted us to their village, which consisted of a few low huts, similar to those at Barrow Island; but they had no fruit to offer us, excepting pandanus-nuts, which they disliked almost as much as ourselves, and told us they had been accustomed to better fare.

In their huts we found calabashes of water suspended to the roof, mats, baskets, and every thing calculated for a sea-voyage; and not far from them a plentiful store of fish, raised about four feet above the ground, out of the reach of the rats, which were very numerous. They had clothing sufficient for the climate, and were in every respect stout and healthy ; there was therefore no immediate necessity for removing them, though I offered to take them as far as the next island, which was larger and inhabited, and where — concluding, from what we saw, that these people were auxiliary missionaries—they would have an opportunity of prosecuting their pious intentions in the conversion of the natives. This proposal, however, after a little consultation, was declined, from an apprehension of being killed and eaten, as they supposed the greater part of the inhabitants of the eastern islands of Polynesia to be cannibals.

We very soon discovered that our little colony were Christians : they took an early opportunity of convincing us of this, and that they had both Testaments, hymn-books, &c. printed in the Otaheitan language : they also shewed us a black-lead pencil, and other materials for writing. Some of the girls repeated hymns, and the greater part evinced a reverence and respect for the sacred books, which reflects much credit upon the missionaries, under whose care we could no longer doubt they had at one time been.

Tuwarri, to whom I offered a passage, we found was not the principal person on the island, but that their chief was a man who accompanied him in the boat, with his legs dreadfully enlarged with the elephantiasis : it was he who directed their course, rebuilt their canoe after it had been stranded, and who appeared also to be their protector, being the only one who possessed fire-arms. His importance in this respect was, however, a little diminished by the want of powder and shot, and by an accident which had deprived him of the hammer of his gun —a misfortune he particularly regretted, as it had been given him by King Pomarree. His anxiety on this head was relieved by finding our armourer could supply the defect, and that we could furnish him with the necessary materials for the defence of his party.

The canoe in which this extraordinary voyage had been made was found hauled up at a different part of the island from that on which we landed, and placed under a shed very neatly built, with the repairs executed in a workmanlike manner, and in every respect ready for sea. She was a double canoe, upwards of thirty feet long by nine broad, and three feet nine inches deep; each vessel having three feet three inches beam: one was partly decked, and the other provided with a thatched shed: they were sharp at both ends, each of which was fitted for a rudder, and the timbers were sewed together with strong plaited cord, after the manner of the canoes of Chain Island, where they are brought to great perfection.

We remained the whole day upon the island, contributing to the comfort of the inhabitants by the distribution of useful presents; and at the same time making our own observations, and endeavouring to learn something of their history, and at sunset we assembled upon the beach to embark. Poor Tuwarri was quite overwhelmed at separating from his companions and fellow-sufferers. The whole village accompanied him to the boat, to the last testifying their regard by some little act of civility. When the moment of departure arrived, the men gathered about him, embraced him, shed abundance of tears, and took their leave in a solemn manner with very few words. The women, on the other hand, clung about his wife and children, and indulged a weakness that better became their sex.

The island upon which we found them is nearly an oval of three miles and three-quarters diameter. It is of coral formation, and has a lagoon and productions very similar to the other islands recently described. One species of coral not noticed before was seen in the lagoon, growing above water: it was a millepore extending itself in vertical plates parallel to the shore. Among the vegetable productions, the *polypodium vulgare*, seen at Whitsunday Island, was found here ; and also a small shrub, which we afterwards ascertained to be an achyranthus. From the pemphis we procured a large supply of firewood, to which use it is well adapted, as it burns a long time, gives great heat, and occupies comparatively little room. The wood of this tree is as hard as lignum vitæ, and equally good for tools; its specific gravity much greater than seawater: its colour is deep red, but the inner bark more strongly tinged ; and if properly prepared, would perhaps afford a good dye.

From Byam Martin Island we steered for Gloucester Island of Captain Wallis, and early the next morning were close to it. The appearance of the island has been accurately described by its discoverer, but its present form and extent differ materially. At the S. E. angle of the island we noticed a morai built of stones, but there were no inhabitants upon the shore. In passing to windward of the island, the current unexpectedly set so strong upon it, that the ship was for a considerable time in imminent danger of being thrown upon the rocks, and her escape is entirely attributable to the rapid descent of the coral reef, which at times was almost under her bottom. She, however, fortunately cleared the reef, and was immediately in safety. After collecting the necessary information, we steered for Bow Island, which was seen from the mast-head at three o'clock the same afternoon.

CHAPTER VIII.

CHAP.
VIII.

Feb.
1826.

Bow Island was discovered by M. Bougainville
in 1768, and the following year was visited by
Captain Cook, who gave it its present name from
the resemblance its shape bore to a bow. Its figure
protracted upon paper, however, is very irregular,
and bears but small resemblance to the instrument
after which it was named ; but to a person viewing
it as Captain Cook did, the mistake is very likely to
occur. It is of coral formation, thirty-four miles
long, and ten broad ; well wooded on the weather
side, but very scantily so on the other ; and so low
in this half, that the sea in places washes into the
lagoon. We sailed close along what may be considered the string of the bow, while the barge navi-

gated the arch; and thus, between us, in a few hours made the circuit of the island.

Previous to quitting England, Captain Charlton, the consul at the Sandwich Islands, among other useful matter which he obligingly communicated, informed me of an opening through the coral reef of this island into the lagoon; and as I was desirous, at this period of the survey, of having a point astronomically fixed to correct the chronometrical measurements, I determined, if possible, to enter the lagoon with the ship. When we reached the supposed opening, a boat was lowered to examine it; and Tuwarri was sent in her to conciliate the natives, should any be seen in the course of the service. As she drew near the shore, several men were observed among the trees; and the officer in charge of the boat, acting under my general orders of being always prepared for an attack, desired the muskets to be loaded. Tuwarri, who had probably never possessed much courage, at the sight of these preparations wished himself anywhere else than in his present situation, and, to judge from his countenance, calculated at least upon being killed and eaten by cannibals: he was in the greatest agitation as the boat advanced, until she came within speaking distance of the strangers, when, instead of the supposed monsters ready to devour him, he recognised, to his surprise, his own brother and several friends whom he had left at Chain Island three years before, all of whom had long given him up as lost, and whom he never expected to see again.

The two brothers met in a manner which did credit to their feelings, and after the first salutation sat down together upon the beach with their hands firmly locked, and entered into serious conversation, consisting no doubt of mutual inquiries after friends and relations, and Tuwarri's account of his perilous adventure. They continued with their hands grasped until it was time for the boat to return to the ship, when they both came on board. This affecting interview increased our impatience to have the mystery which overhung the fate of our passenger cleared up, and an opportunity fortunately happened for doing so.

The gig, on entering the lagoon, had been met by a boat from an English brig (the Dart, employed by the Australian Pearl Company) at anchor there, with a number of divers, natives of Chain Island, hired into her service : among these men there was one who acted as interpreter, and who was immediately engaged to communicate to us the particulars of Tuwarri's adventures, which possess so much interest that the reader will not, I am sure, regret the relation of them.

Tuwarri was a native of one of the low coral formations discovered by Captain Cook in his first voyage, called Anaa by the natives, but by him named Chain Island, situated about three hundred miles to the eastward of Otaheite, to which it is tributary. About the period of the commencement of his misfortunes old Pomarree the king of Otaheite died, and was succeeded by his son, then a child. On the accession of this boy several chiefs and commoners of Chain Island, among whom was Tuwarri, planned a voyage to Otaheite, to pay a visit of ceremony and of homage to their new sovereign. The only conveyance these people could command was double canoes, three of which of the largest class were prepared for the occasion. To us, accustomed to navigate the seas in ships of many tons burthen, provided with a compass and the necessary instruments to determine our position, a canoe with only the stars for her guidance, and destined to a place whose situation could be at the best but approximately known, appears so frail and uncertain a conveyance, that we may wonder how any persons could be found sufficiently resolute to hazard the undertaking. They knew, however, that similar voyages had been successfully performed, not only to mountainous islands to leeward, but to some that were scarcely six feet above the water, and were situated in the opposite direction; and as no ill omens attended the present undertaking, no unusual fears were entertained. The canoes being accordingly prepared, and duly furnished with all that was considered necessary, the persons intending to proceed on this expedition were embarked, amounting in all to a hundred and fifty souls. What was the arrangement of the other two canoes is unknown to us, but in Tuwarri's there were twenty-three men, fifteen women, and ten children, and a supply of water and provision calculated to last three weeks.

On the day of departure all the natives assembled upon the beach to take leave of our adventurers; the canoes were placed with scrupulous exactness in the supposed direction which was indicated by certain marks upon the land, and then launched into the sea amidst the good wishes and adieus of their countrymen. With a fair wind and full sail they glided rapidly over the space without a thought of the possibility of the miseries to which they were afterwards exposed.

It happened, unfortunately, that the monsoon that year * began earlier than was expected, and blew with great violence; two days were, notwithstanding, passed under favourable circumstances, and the adventurers began to look for the high land of Maitea, an island between Chain Island and Otaheite, and to anticipate the pleasures which the successful termination of their voyage would afford them, when their progress was delayed by a calm, the precursor of a storm, which rose suddenly from an unfavourable quarter, dispersed the canoes, and drove them away before it. In this manner they drifted for several days; but on the return of fine weather, having a fortnight's provision remaining, they again resolutely sought their destination, until a second gale drove them still farther back than the first, and lasted so long that they became exhausted. Thus many days were past; their distance from home hourly increasing; the sea continually washing over the canoe, to the great discomfiture of the women and children; and their store of provision dwindled to the last extremity. A long calm, and, what was to them even worse, hot dry weather, succeeded the tempest, and reduced them to a state of the utmost distress. They described to us their canoe, alone and becalmed on the ocean; the crew, perishing with thirst beneath the fierce glare of a tropical sun, hanging exhausted over their paddles; children looking to their parents for support, and mothers deploring their inability to afford them assistance. Every means of quenching their thirst were resorted to; some drank the sea water, and

* In the South Pacific the monsoons are occasionally felt throughout all the islands of Eastern Polynesia.

others bathed in it, or poured it over their heads;
but the absence of fresh water in the torrid zone
cannot be compensated by such substitutes. Day
after day, those who were able extended their gourds
to heaven in supplication for rain, and repeated their
prayers, but in vain; the fleecy cloud floating high
in the air indicated only an extension of their suffering: distress in its most aggravated form had at
length reached its height, and seventeen persons fell
victims to its horrors.

The situation of those who remained may readily
be imagined, though their fate would never have
been known to us, had not Providence at this critical moment wrought a change in their favour. The
sky, which for some time had been perfectly serene,
assumed an aspect which at any other period would
have filled our sufferers with apprehension; but, on
the present occasion, the tropical storm, as it approached, was hailed with thankfulness, and welcomed as their deliverer. All who were able came
upon the deck with blankets, gourds, and cocoa-nut
shells, and held them toward the black cloud, as
it approached, pouring down torrents of rain, of
which every drop was of incalculable value to the
sufferers; they drank copiously and thankfully, and
filled every vessel with the precious element. Thus
recruited, hope revived; but the absence of food
again plunged them into the deepest despair. We
need not relate the dreadful alternative to which
they had recourse until several large sharks rose to
the surface and followed the canoe; Tuwarri, by
breaking off the head of an iron scraper, formed it
into a hook, and succeeded in catching one of them,
which was instantly substituted for the revolting
banquet which had hitherto sustained life.

Thus refreshed, they again worked at their paddles or spread their sail, and were not long before
their exertions were repaid with the joyful sight of
land, on which clusters of cocoa-nuts crowned the
heads of several tufts of palm-trees : they hurried
through the surf and soon reached the much wished
for spot, but being too feeble to ascend the lofty
trees, they were obliged to fell one of them with
an axe.

On traversing the island to which Providence had
thus conducted them, they discovered by several
canoes in the lagoon, and pathways intersecting the
woods, that it had been previously inhabited; and
knowing the greater part of the natives of the low
islands to be cannibals, they determined to remain no
longer upon it than was absolutely necessary to recruit their strength, imagining that the islanders,
when they did return, would not rest satisfied with
merely dispossessing them of their asylum.

It was necessary, while they were allowed to remain, to seek shelter from the weather and to exert
themselves in procuring a supply of provision for
their further voyage; huts were consequently built,
pools dug for water, and three canoes added to those
which were found in the lake. Their situation by
these means was rendered tolerably comfortable, and
they not only provided themselves with necessaries
sufficient for their daily consumption, but were able
to dry and lay by a considerable quantity of fish for
sea stock.

After a time, finding themselves undisturbed, they
gained confidence, and deferred their departure till
thirteen months had elapsed from the time of their
landing. At the expiration of which period, being
in good bodily health and supplied with every re-

quisite for their voyage, they again launched upon the ocean in quest of home.

They steered two days and nights to the north-west, and then fell in with a small island, upon which, as it appeared to be uninhabited, they landed, and remained three days, and then resumed their voyage. After a run of a day and a night they came in sight of another uninhabited island. In their attempt to land upon it their canoe was unfortunately stove, but all the party got safe on shore. The damage which the vessel had sustained requiring several weeks to repair, they established themselves upon this island, and again commenced storing up provision for their voyage. Eight months had already passed in these occupations, when we unexpectedly found them thus encamped upon Byam Martin Island; with their canoe repaired, and all the necessary stores provided for their next expedition. The other two canoes were never heard of.

Several parts of this curious history strongly favoured the presumption that the island upon which the party first landed and established themselves was Barrow Island: and, in order to have it confirmed, the piece of iron that had been brought from thence, and had fortunately been preserved, was produced. Tuwarri, when he saw it, immediately exclaimed that it was the piece of iron he had broken in two to form the shark-hook, which was the means of preserving the lives of his party, and said that the tree we found cut down with some sharp edged tool was that which his party felled before their strength enabled them to climb for the fruit; and hence the huts, the pools of water, the canoes, &c. were the remains of their industry.

This curious discovery enabled us to form a tolerably accurate idea of the distance the canoe had been drifted by the gale, as Barrow Island is 420 miles in a direct line from Chain Island, their native place; and if to this be added 100 miles for the progress they made during the first two days toward Maitea, and the distance they went on their return before they reached Barrow Island, the whole cannot amount to much less than 600 miles.

Before Tuwarri could be restored to his home, we visited in succession several low islands to which he was a stranger. While we were cruising among them he entertained the greatest apprehension lest we had lost our way, and perhaps pictured to himself a repetition of his disastrous voyage. He could not imagine our motive for pursuing so indirect a course, and frequently inquired if we were going to his native island, and if we knew where it was, occasionally pointing in the direction of it. He always boasted of a knowledge of the islands lying between Bow Island (He-ow) and Chain Island (Anaa), but never informed us right when we came to any of them. He had, it is true, reason to be anxious; for his wife, almost the whole of the passage, was very sea-sick, which gave him great concern; and when the sea was much agitated he appeared inconsolable. When he at length arrived within sight of Chain Island, his joy at the certainty of again setting foot on his native soil, and meeting friends who had long supposed him lost, may readily be imagined. His gratitude to us for having given him a passage, and for our attention to his comfort, was expressed in tears of thankfulness; and he testified his regret at parting in a manner which showed him to be sincere: and as he was going away, he expressed his sorrow that the ship would not remain long enough off the island for him to send some little token of his gratitude. These feelings, so highly creditable to Tuwarri, were not participated by his wife, who, on the contrary, showed no concern at her departure, expressed neither thanks nor regrets, nor turned to any person to bid him farewell; and while Tuwarri was suppressing his tears, she was laughing at the exposure which she thought she should make going into the boat without an accommodation-ladder. Tuwarri while on board showed no curiosity, knew nothing of our language, or evinced any desire to learn it; took very little interest in any thing that was going forward, and was very dull of comprehension. He appeared to be a man whose energies had been worn down by hardship and privation, and whom misfortune had taught to look on the worst side of every thing. But with all these weak points, he had many good qualities. He lent a willing hand to pull at a rope, was cleanly and quiet, punctually attended church on Sundays, and had a strong sense of right and wrong, which, as far as his abilities enabled him, governed his actions. He had a warm heart, and his attachment to his wife and children amounted even to weakness. He had a tolerable knowledge of the relative situation of the islands of the archipelago, and readily drew a chart of them, assigning to each its name, though, as I have said before, he never could recognise them. Some of these we were able to identify, and perhaps should have done so with others, had there not been so much sameness in all the coral islands.

Mr. Belcher, who was in command of the barge

which put him on shore, says, he was not received by his countrymen with the surprise and pleasure which might have been expected; but this may, perhaps, be explained by there being no one on the beach to whom he was particularly attached. Before the barge quitted the island, he put on board some shells as a present, in gratitude for the assistance which had been rendered him.

Reverting to the occurrences of the ship off Bow Island: Mr. Elson, the officer who was sent to examine the channel into the lagoon, returned with the supercargo of the Dart, Mr. Hussey, and made a favourable report of the depth of water in the passage, but said its width was so very contracted that it could not be passed without hazard. The exact distance from reef to reef is 115 feet, and there is a coral knoll in the centre; the trade-wind does not always allow a ship to lie well through it, and there is, at times, a tide running out at the rate of four knots an hour. It was, however, necessary to incur this risk; and, on the information of Mr. Hussey that the morning was the most favourable time for the attempt, shortly after daylight on the next day (15th), under Mr. Elson's skilful pilotage, we shot through the passage, at the rate of seven knots, and were instantly in a broad sheet of smooth water. We found the lagoon studded with coral knolls, which it was necessary to avoid by a vigilant look out from aloft, as the lead gave no warning of their vicinity; we beat among them at some risk, and at ten o'clock anchored at the N. E. angle of the lake, in ten fathoms water, on a broad patch of sand, about a quarter of a mile from the shore, and in as secure a harbour as could be required.

Nearly opposite to our anchorage, the natives,

about fifty in number, had erected temporary huts
during the stay of the Dart, their permanent residences being at the opposite end of the island.
They were in appearance the most indolent ill-look-
ing race we had yet seen; broad flat noses, dull
sunken eyes, thick lips, mouths turned down at the
corners, strongly wrinkled countenances, and long
bushy hair matted with dirt and vermin. Their
stature was above the middle size, but generally
crooked; their limbs bony, their muscles flaccid, and
their only covering a maro. But hideous as the
men were, their revolting appearance was surpassed
by the opposite sex of the same age. The males
were all lolling against the cocoa-nut trees, with
their arms round each other's necks, enjoying the
refreshing shade of a thick foliage of palm-trees;
while the women, old and young, were labouring
hard in the sun, in the service of their masters, for
they did not merit the name of husbands. The
children, quite naked, were placed upon mats, crying and rolling to and fro, to displace some of the
myriads of house-flies, which so speckled their bodies that their real colour was scarcely discernible.

Amidst this scene I was introduced to the chief,
who was distinguished from his subjects by his superior height and strength, and probably maintained
his authority solely by those qualities. He gave me
a friendly reception, and suffered us to cut down
what wood we wanted, confining us only to those
trees which produced no edible fruits. In return
for some presents made him, he drew from his canoe
several pearl fishing hooks and bundles of turtle-shell,
and begged my acceptance of them; but his extreme
poverty was such, that I could not bring myself to

do so, though I do not know to what material use the last mentioned article could be applied by him.

We availed ourselves of the areghe's permission, and sent a party to cut as many trees as we required, consisting principally of the pemphis acidula, as at Byam Martin Island. Mr. Marsh endeavoured to engage some of the natives in this employment, by offering shirts, tobacco, &c.; but, notwithstanding the munificence of the reward, the areghe alone could be roused from his lethargy; and even he quitted the axe before the first tree was felled.

A party of seamen was at the same time sent, under the direction of Lieutenant Wainwright, to dig wells; in which their success was so satisfactory, that in less than three days we procured thirty tons of fresh water. The wells were about four feet deep, dug through the sand into the coral rock. Into two of these the water flowed as fast as we could fill the casks; and when allowed to stand, rose eighteen inches. This water was drunk by all the ship's company for several weeks, and proved tolerably good, though it did not keep as well as spring water.* It is important to navigators to know, that even as good water as this may be procured on the

* Mr. Collie observes, in his Journal, that a " solution of nitrate of soda detected in it a moderate proportion of muriatic acid, most likely embodied in the soda. It had no brackish taste. With an alcoholic solution of soap it formed a copious white precipitate: with oxalate of ammonia it formed slowly, but after some time, a dense white cloud: with nitrate of silver an abundant purplish-white precipitate: it remained unchanged with nitrate of barytes. Thus showing that it contained no sulphuric acid, but that it was impregnated with muriatic acid and magnesia, most likely muriate of soda and magnesia, the component parts of sea water."

coral islands by means of wells. In digging them, the choice of situation should be given to the most elevated part of the island, and to a spot distant from the sea; perhaps in the vicinity of cocoa-nut trees. It is a curious fact that, in Bow Island, the water that flowed into holes dug within a yard of the sea was fresh enough to be drunk by the sailors, and served the purpose of the natives while they remained in our vicinity; though I do not think Europeans could have used it long with impunity.

Not far from the temporary residence of the natives, there was a level- spot of ground, overgrown with grass, upon which the observatory was erected; and I had in consequence frequent intercourse with them, and, through the medium of the interpreter of the Dart, learned many interesting particulars concerning them. By this account they have not long desisted from cannibalism. On questioning the chief, he acknowledged himself to have been present at several feasts of human bodies, and on expatiating on the excellence of the food, particularly when it was that of a female, his brutal countenance became flushed with a horrible expression of animation. Their enemies, those slain in battle, or those who die violent deaths, and murderers, were, he said, the only subjects selected for these feasts; the latter, whether justified or not, were put to death, and eaten alike with their victims. They have still a great partiality for raw food, which is but one remove from cannibalism; and when a canoe full of fish was brought one day to the village, the men, before it could be drawn to the shore, fell upon its contents, and devoured every part of the fish except the bones and fins. The women, whose business it was to unload the boat, did the best they could with one of them between their teeth, while their hands were employed portioning the contents of the canoe into small heaps. But even in this repast we were glad to observe some indication of feeling in putting the animal speedily out of torture by biting its head in two, the only proof of humanity which they manifested. In like manner, cleanliness was not overlooked by them, for they carefully rinsed their mouths after the disgusting meal.

It appeared that the chief had three wives, and that polygamy was permitted to an unlimited extent; any man of the community, we were told, might put away his wife whenever it was his pleasure to do so, and take another, provided she were disengaged. No ceremony takes place at the wedding; it being sufficient for a man to say to a woman, " You shall be my wife ;" and she becomes so.

The offspring of these unions seemed to be the objects of the only feelings of affection the male sex possessed, as there were certainly none bestowed on the women. Indeed the situation of the females is much to be pitied ; in no part of the world, probably, are they treated more brutally. While their husbands are indulging their lethargic disposition under the shade of the cocoa-nut trees, making no effort toward their own support, beyond that of eating when their food is placed before them, the women are sent to the reefs to wade over the sharp-pointed coral in search of shell-fish, or to the woods to collect pandanus-nuts. We have seen them going out at daylight on these pursuits, and returning quite fatigued with their morning toil. In this state, instead of enjoying a little repose on reaching their homes, they

are engaged in the laborious occupation of preparing
what they have gathered for their hungry masters,
who, immediately the nuts are placed before them,
stay their appetites by extracting the pulpy substance contained in the outside woody fibres of the
fruit, and throw the remainder to their wives, who
further extract what is left of the pulp for their own
share, and proceed to extricate the contents of the
interior, consisting of four or five small kernels about
the size of an almond. To perform this operation,
the nut is placed upon a flat stone endwise, and
with a block of coral, as large as the strength of the
women will enable them to lift, is split in pieces,
and the contents again put aside for their husbands.
As it requires a considerable number of these small
nuts to satisfy the appetites of their rapacious rulers,
the time of the women is wholly passed upon their
knees pounding nuts, or upon the sharp coral collecting shells and sea eggs. On some occasions the
nuts are baked in the ground, which gives them a
more agreeable flavour, and facilitates the extraction
of the pulp; it does not, however, diminish the
labour of the females, who have in either case to
bruise the fibres to procure the smaller nuts.

The superiority of sex was never more rigidly
enforced than among these barbarians, nor were the
male part of the human species ever more despicable.
On one occasion an unfortunate woman who was
pounding some of these nuts, which she had walked
a great distance to gather, thinking herself unobserved, ate two or three of the kernels as she extracted them; but this did not escape the vigilance
of her brutal husband, who instantly rose and felled
her to the ground in the most inhuman manner

with three violent blows of his fist. Thus tyran-
nised over, debased, neglected by the male sex, and
strangers to social affection, it is no wonder all those
qualities which in civilised countries constitute the
fascination of woman are in these people wholly
wanting.

The supercargo of the Dart, to forward the service
he was engaged in, had hired a party of the natives
of Chain Island to dive for shells. Among these
was a native missionary,* a very well-behaved man,
who used every effort to convert his new acquaintances to Christianity. He persevered amidst much
silent ridicule, and at length succeeded in persuad-
ing the greater part of the islanders to conform to
the ceremonies of Christian worship. It was interesting to contemplate a body of savages, abandoning
their superstitions, silently and reverently kneeling
upon the sandy shore, and joining in the morning
and evening prayers to the Almighty. Though
their sincerity may be questioned, yet it is hoped
that an impression may be made upon these neo-
phytes, which may tend to improve their moral
condition.

Previous to the arrival of the missionary, every
one had his peculiar deity, of which the most common was a piece of wood with a tuft of human hair
inserted into it ; but that which was deemed most
efficacious, when it could be procured, was the thigh
bone of an enemy, or of a relation recently dead.
Into the hollow of this they inserted a lock of the
same person's hair, and then suspended the idol to a

* We were told that at Chain Island there were thirteen houses
of prayer under the direction of native missionaries.

tree. To these symbols they address their prayers as long as they remained in favour; but, like the girl in China, who, when disappointed by her lover, pulled down the brazen image and whipped it, these people when dissatisfied with their deity, no longer acknowledged his power, and substituted some other idol. There were times, however, when they feared its anger, and endeavoured to appease it with cocoanuts; but I did not hear of any human sacrifices being offered. They appeared to entertain the Pythagorean doctrine of the transmigration of the soul, and supposed the first vessel which they saw to be the spirit of one of their relations lately deceased. The compartments allotted to the dead are here tabooed; and the bodies, first wrapped in mats, are placed under ground. As the soul is supposed for a time to frequent these places, provision and water are placed near the spot for its use; and it would be thought unkind, or that some evil would befal the person whose business it is to provide them, if these supplies were neglected.

The manufactures of these people are the same with those of all the other islanders, and are only such as nature renders necessary, consisting of mats, maros, baskets, fishing-tackle, &c. They have no occupation beyond the manufacture of these few articles, and providing for their daily support. On interrogating the chief how he passed the day, he said he rose early and ate his breakfast; he then invoked his deity; sometimes he went to fish or catch turtle; but more generally passed his time under the shade of the cocoa-nut trees: in the evening he ate again, and went to sleep.

The natives of this island, according to information obtained by the interpreter on board the Dart, amount altogether to about a hundred souls.

As my stay at the island was limited to four days, my time was much occupied at the observatory, and I am indebted to the journals of the officers for many interesting particulars relating to other parts of it, and to its natural productions.

By our trigonometrical survey, Bow Island is thirty miles long by an average of five miles broad. It is similar to the other coral islands already described, confining within a narrow band of coral a spacious lagoon, and having its windward side higher and more wooded than the other; which indeed, with the exception of a few clusters of trees and heaps of sand, is little better than a reef. The sea in several places washes into the lagoon, but there is no passage even for a boat, except that by which the ship entered, which is sometimes dangerous to boats, in consequence of the overfalls from the lagoon, especially a little after the time of high water. It is to be hoped that the rapid current which sets through the channel will prevent the growth of the coral, and leave the lagoon always accessible to shipping. It lies at the north side of the island, and may be known by two straggling cocoa-nut trees near it, on the western side, and a clump of trees on the other.

The bottom of the lagoon is in parts covered with a fine white sand, and it is thickly strewed with coral knolls; the upper parts of which overhang the lower, though they do not at once rise in this form from the bottom, but from small hillocks. We found comparatively few beneath the surface, though there are some; at the edge of such as are exposed, there is usually six or seven fathoms water; receding from it, the lead gradually descends to the general level, of about twenty fathoms. The lagoon contains an abundance of shell-fish, particularly those of the pearl-oyster kind. The party in the employ of the Dart sometimes collected seventeen hundred of these shells in one day.

The height of water in the lagoon is subject to the variations of the tides of the ocean; but it suffers so many disturbances from the waves which occasionally inundate the low parts of the surrounding land, that neither the rise of the tide nor the time of high water can be estimated with any degree of certainty. Were the communication between the lake and the sea larger, so as to admit of the water finding its level, the period of low water might be determined, as there is a change of tide in the entrance.

The strip of low land enclosing the lagoon is nearly seventy miles in extent, and the part that is dry is about a quarter of a mile in width. On the inner side, a few yards from the margin of the lake, there is a low bank formed of finely broken coral; and, at the outer edge, a much higher bank of large blocks of the same material, long since removed from the reach of the waves, and gradually preparing for the reception of vegetation. Beyond this high bank there is a third ridge, similar to that skirting the lagoon; and outside it again, as well as in the lagoon, there is a wide shelf three or four feet under water, the outer one bearing upon its surface huge masses of broken coral; the materials for an outer bank, similar to the large one just described. These appearances naturally suggest the idea of the island having risen by slow degrees. Thus the sand dispersed over the lagoon indicates a period when the sea rolled entirely over the reef, tore up blocks of coral from its margin, and by constant trituration ground them to powder, and finally deposited the particles where they now rest. The bank near the lake must have originated at a subsequent period, when the outer edge becoming nearer to the surface, moderated the strength of the waves, and the wash of the sea reached only far enough to deposit the broken coral in the place described. At a still less distant period, when the island became dry, and the violence of the sea was wholly spent upon its margin, the coral, which had before escaped by being beneath the surface, gave way to the impetuous wave, and was deposited in broken masses, which formed the high ridge. Here the sea appears to have broken a considerable time, until a second ledge gradually extending seaward, and approaching the surface, so lessened the effect of the waves upon this ledge also, that they were again only capable of throwing up an inferior heap similar to the one first mentioned. In process of time this outer ledge will become dry, and the many large blocks of coral now resting near its edge will, probably, form another heap similar to the large one; and thus the island will continue to increase by a succession of ledges being brought to the surface, while, by the same process, the lagoon will gradually become more shallow and contracted.

The ridges are particularly favourable to the formation of a soil, by retaining within them whatever may be there deposited until it decays, and by protecting the tender shrubs during their early growth. Near our observatory the soil had attained a depth of about eight inches before we came to broken coral.

" In the central and sheltered parts of the plain between the ridges the pandanus spreads its divergent roots and rears its fruitful branches; the pemphis also takes root in the same situation.   The loose dry stones of the first ridge are penetrated by the hard roots of the tefano, which expands its branches into a tall spreading tree, and is attended by the fragrant suriana, and the sweet-scented tournefortia, in the shelter of whose foliage the tender achyranthus and lepidium seem to thrive the best.   Beyond the first high and stony ridge the hardy scævola extends its creeping roots and procumbent verdure towards the sea, throwing its succulent leaves round the sharp coral stones."

" On the windward side, wherever the pandanus was devoid of the protection of the more hardy trees, the brown and decayed leaves showed it had advanced beyond its proper boundary."*

We quitted Bow Island on the 20th of February, and continued the survey of the archipelago, until the period had arrived when it was necessary to proceed direct to Otaheite, to prepare the ship for her voyage to the northward. We were greatly retarded toward the close of our operations by the rainy season, which was attended with calms, and hot, sultry, wet weather, and perhaps, had we continued at sea, would have prevented any thing more being done.   The dysentery about this time began to make its appearance among the ship's company, owing no doubt to the rains and closeness of the atmosphere, combined with the harassing duty arising from the navigation of a sea so thickly strewed with

* Mr. Collie's Journal.

islands, and to the men having been a long time on a reduced allowance of salt provisions.

The islands which were visited between Bow Island and Otaheite were all of the same character and formation as those already described, and furnished us with no additional information beyond the correct determination of their size and position; which, with some remarks that may be useful to navigation, are given in the Appendix to the 4to. ed. Among the number there were two which were previously unknown; the largest of these, which was also the most extensive of our discoveries in the archipelago, I named Melville Island, in honour of the first lord of the Admiralty; and the other, Croker Island, in compliment to the right honourable secretary.

The discoveries of Cook and Wallis in this track are relatively correctly placed; but those of the latter are as much as forty miles in error in longitude, and several miles in latitude, which has occasioned two of them to be mistaken for each other by Bellinghausen, and one to be considered as a new discovery by Captain Duperrey. It would not have been easy to detect these errors, had we not visited the discoveries of Wallis in succession, beginning with Whitsunday and Queen Charlotte's Islands, which are so situated that no mistake in them could possibly occur. Moreover, we always searched the vicinity narrowly for the existence of other islands.

The mistakes have arisen from placing too much confidence in the longitude of the early navigator. The true place of Cumberland Island lying much nearer the alleged position of Wallis's Prince William-Henry Island than any other, has occasioned Bellinghausen's mistake; and the true position of

Prince William-Henry being so remote from any of Wallis's discoveries, as placed by himself, has made Captain Duperrey think the one which he saw could not possibly be one of them, and he in consequence bestowed upon it the new name of L'Ostange.

There can be no doubt that the island which I consider Prince William-Henry Island is the L'Ostange of Captain Duperrey, as we had an opportunity of comparing longitudes with him at Moller Island; and it is equally certain that this island is the same with that discovered by Wallis, as its distance from Queen Charlotte's Island and his other discoveries to the eastward, each of which we visited, exactly coincides. Wallis has certainly erred ten miles in latitude, but it should be recollected that the position of the island was fixed by reckoning from noon, the island having been seen at daybreak "far to windward;" and it should not be overlooked that his latitude at Cumberland Island the day before was eight miles in error the same way, which makes it very probable that either his observations were indifferent, or that he had incorrect tables of declination.

In forming this conclusion, I am aware that I am depriving Captain Duperrey of the merit of a discovery, but he will, it is hoped, admit the justice of my opinion.

All the islands seen by Cook, Wallis, and Carteret, lying within the limit of our survey, have been found to be accurately described, excepting that their size has always been overrated; a mistake very likely to arise with low strips of land deficient in familiar objects to direct the judgment where actual measurement is not resorted to.

The discoveries of Mr. Turnbull are so loosely related in his entertaining Voyage, that their situation cannot be entertained; and unless some better clue to them is given, they will always be liable to be claimed by subsequent navigators.

Of the thirty-two islands which have thus been visited in succession, only twelve are inhabited, including Pitcairn Island, and the amount of the population altogether cannot possibly exceed three thousand one hundred souls; of which one thousand belong to the Gambier groupe, and twelve hundred and sixty to Easter Island, leaving eight hundred and forty persons only to occupy the other thirty islands.

All the natives apparently profess the same religion; all speak the same language, and are in all essential points the same people. There is a great diversity of features and complexion between those inhabiting the volcanic islands and the natives of the coral formations, the former being a taller and fairer race. This change may be attributed to a difference of food, habits, and comfort; the one having to seek a daily subsistence upon the reefs, exposed to a burning sun and to the painful glare of a white coral beach, while the other enjoys plentifully the spontaneous produce of the earth, reposes beneath the genial shade of palm or bread-fruit groves, and passes a life of comparative ease and luxury.

It has hitherto been a matter of conjecture how these islands, so remote from both great continents, have received their aborigines. The intimate connexion between' the language, worship, manners, customs, and traditions of the people who dwell upon them, and those of the Malays and other in-

habitants of the great islands to the westward, leaves
no doubt of frequent emigrations from thence; and
we naturally look to those countries as the source
from which they have sprung. The difficulty, however, instantly presents itself of proceeding so vast
a distance in opposition to the prevailing wind and
current, without vessels better equipped than those
which are in possession of the above-mentioned people.
This objection has so powerfully influenced the minds
of some authors that they have had recourse to the circuitous route through Tartary, across Beering's Strait,
and over the American continent, to bring the emi-
grants to a situation whence they might be drifted by
the ordinary course of the winds to the lands in question. But had this been the case, a more intimate resemblance would surely be found to exist between the
American Indians and the natives of Polynesia.

All have agreed as to the manner in which these
migrations between the islands have been effected,
and some few instances have actually been met
with; but they have been in one direction only,
and have rather favoured the opinion of migration
from the eastward. The accident which threw in
our way Tuwarri and his companions, who, it may
be recollected, were driven six hundred miles in a
direction contrary to the trade-wind in spite of their
utmost exertions, has fortunately enabled us to remove the objections which have been urged against
the general opinion. The fact being so well at-
tested, and the only one of the kind upon record, is,
consequently, of the highest interest, both as regards its singularity, and as it establishes the *possibility* of the case. Though this is the only instance that has come to our knowledge, there is no

reason why many other canoes may not have shared a similar fate; and some few of many thousands, perhaps, may have drifted to the remotest islands of the archipelago, and thus peopled them.

The navigation of canoes between islands in sight of each other was, and is still, very general; and it was not unusual, in early times, for warriors, after a defeat, to embark, careless of the consequences, in order to escape the persecution of their conquerors. To remain, was certain death and ignominy; to fly, was to leave their fate to chance.

The temporary obstruction of the trade wind in these seas, by the westerly monsoons, has not been duly considered by those who represent the difficulties as insurmountable. At the period of the year corresponding with our spring these gales commence, and blow with great violence during the rainy season. As they arise very suddenly, any canoes at sea must have difficulty in escaping them, and would, in all probability, be driven so far, as never to be able to regain their native country, or be drifted to islands upon which their crews might be contented to dwell, in preference to encountering farther risks.

The traces of inhabitants upon almost all the islands of the low archipelago, many of which are at present uninhabited, show both the frequency with which these migrations have occurred, and the extent to which they have been made: some of these isolated spots where remains have been found, Pitcairn Island for instance, are 400 miles from any land whence inhabitants were likely to be derived; and the circumstance of their having abandoned that island is a fair presumption that the people who landed there

knew of other lands which there was a probability of their reaching, and which certainly could
not be the coast of America, at least 2000 miles
against the trade-wind.

I shall now bring together a few facts connected
with the formation of these islands, which it is hoped
may be useful to those persons who are interested in
the subject, observing, in extenuation of the absence
of more detailed information, that our time did not
admit of more than was actually essential to the purposes of a correct delineation of their outline, and
that in general the islands were so surrounded by
breakers that it was dangerous to approach the shore,
in the ship in particular, which alone was calculated
to obtain very deep soundings.   To windward this
could not be done of course, and to leeward there
was not unfrequently a heavier swell setting upon
the island than in other parts of it.

In speaking of the coral islands hereafter, my
observations will be confined to the thirty-two islands
already stated to have fallen under our examination.
The largest of them was thirty miles in diameter,
and the smallest less than a mile : they were of
various shapes; were all formed of living coral,
except Henderson's Island, which was partly surrounded by it ; and they all appeared to be increasing their dimensions by the active operations
of the lithophytes, which appeared to be gradually
extending and bringing the immersed parts of their
structure to the surface.

Twenty-nine of the number had lagoons in their
centres, which is a proportion sufficiently large, when
coupled with information supplied from other parts
of the globe where such formations abound, to ren-

der it almost certain that the remainder also had
them in the early period of their formation, and that
such is the peculiar structure of the coral islands.
And, indeed, these exceptions can scarcely be considered objections, as two of them—Thrum Cap,
which is only seventeen hundred yards long by
twelve hundred broad; and Queen Charlotte's Island,
which is not more than three quarters of a mile wide
in its broadest part, and less than half a mile in other
places—are so circumstanced, that, had their lagoons
existed, they would have been filled in the course
of time with the masses of coral and other substances
which the sea heaps upon such formations as they
rise above the surface; they have, besides, long been
wooded and inhabited, though deserted at the present moment, both of which would tend to efface
the remains of a lagoon of such small dimensions.
The sea, however, prevented our boats from landing
upon either of these islands, to ascertain the fact of
the early existence of lagoons. The other exception,
Henderson's Island, though of coral formation, appears to have been raised to its present height above
the sea by a subterraneous convulsion, and has its
centre so incumbered and overgrown with bushes
that we could not determine whether it ever had a
lagoon.

In the above-mentioned twenty-nine islands the
strips of dry coral enclosing the lagoons, divested of
any loose sandy materials heaped upon them, are
rarely elevated more than two feet above the level
of the sea; and were it not for the abrupt descent
of the external margin, which causes the sea to break
upon it, these strips would be wholly inundated:
this height of two feet is continued over a small

portion only of the width of the island, which slopes
on both sides, by an almost imperceptible inclination
to the first ledge, where, as I said before, its descent
is very steep ; but this is greatly altered by circumstances, and the growth or age of the island. Those
parts of the strip which are beyond the reach of the
waves are no longer inhabited by the animals that
reared them, but have their cells filled with a hard
calcareous substance, and present a brown rugged
appearance. The parts still immersed, or which
are dry at low water only, are intersected by small
channels, and are so full of hollows, that the tide as
it recedes leaves small lakes of water upon them.
The width of the plain or strip of dead coral, in the
islands which fell under our observation, in no instance exceeded half a mile from the usual wash of
the sea to the edge of the lagoon, and in general was
only about three or four hundred yards. Beyond
these limits, on the lagoon side in particular, where
the coral was less mutilated by the waves, there was
frequently a ledge, two or three feet under water at
high tide,* thirty to fifty yards in width ; after
which the sides of the island descended rapidly,
apparently by a succession of inclined ledges formed
by numerous columns united at their capitals, with
spaces between them in which the sounding-lead
descended several fathoms. This formation, though
not clearly established as applying to all the islands,
was so conspicuous in some as to justify the conclusion with regard to others. At Bow and Matilda
Islands, I have been tolerably minute in my descriptions of them, and it will be unnecessary here to
repeat what has been said there ; but these two, as

---

* At Bow Island, on the sea side, it was more.

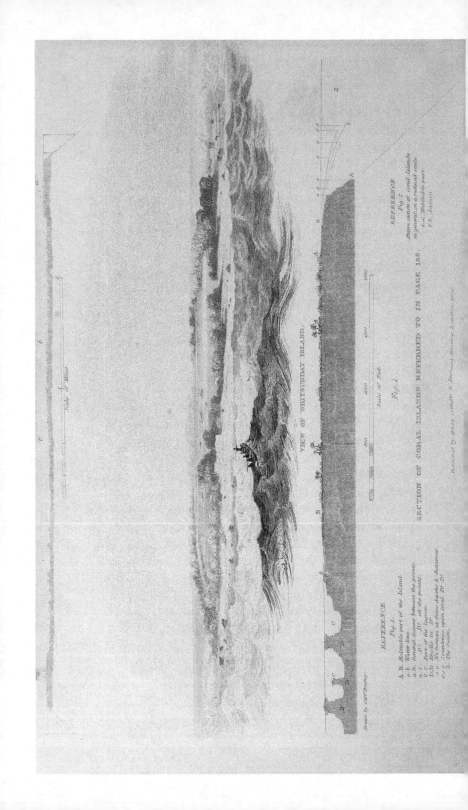

Drawn by Capt. Beechey.

REFERENCE

Fig. 1.

A B Habitable part of the Island.
a b Water line.
a a. General descent between the points
a f. D.º D.º off the points
C C Bars or the Lagoon.
D D Beaches.
v o No bottom at these depths & fathoms
v v g Soundings upon Coral D.º D.º
Z The Ocean.

VIEW OF WHITSUNDAY ISLAND.

Fig. 1.

Scale of Feet.

SECTION OF CORAL ISLANDS REFERRED TO IN PAGE 189.

REFERENCE

Fig. 2.

Entire section of Coral Islands
in general at a reduced scale.
a a. Habitable part.
b b. Lagoon.

Fig. 2.

Scale of Feet.

Scale of Miles.

Published by Henry Colburn & Richard Bentley, London 1831.

also Henderson's Island, afford good examples of what I have been describing. To enable the reader more readily to comprehend the nature of these singular formations, I subjoin a sketch and a section of a coral Island, with the slope of the sides of several of them, laid down according to the soundings and the depths at which attempts were made to reach the bottom.

All these islands are situated within the trade-wind, with the exception of Oeno, which is only on the verge of it, and follow one general rule in having their windward sides higher and more perfect than the others, and not unfrequently well wooded, while the opposite ones are only half-drowned reefs, or are wholly under water. At Gambier and Matilda Islands this inequality was very conspicuous, the weather side of both being wooded, and of the former, inhabited, while the other sides were from twenty to thirty feet under water, where they might be perceived equally narrow, and well defined. It is on the leeward side also that the entrances into the lagoons generally occur, though they are sometimes situated in a side that runs in the direction of the wind, as at Bow Island; but I do not know of any one being to windward. The fact, if it be found to be general with regard to other coral islands, is curious, and is not fully accounted for by the continued operation of the trade-wind upon its side, as the coincidence would suggest. After the reef has arrived at the surface of the sea, it is easy to conceive what would be the effect of the trade-wind; but it does not seem possible that its influence could be felt so far under water as some of the reefs are situated.

All the points or angles of these islands descend into the sea with less abruptness than the sides, and, I think, with more regularity. The wedge-shaped space that the meeting of the two sides would form in the lagoon is filled up by the ledges there being broader; in such places, as well as in the narrow parts of the lake, the coralline are in greater numbers, though, generally speaking, all the lagoons are more or less incumbered with them. They appear to arise to the surface in the form of a truncated cone, and then, their progress being arrested, they work laterally, so that if several of them were near each other they would unite and form a shelf similar to that which has been described round the margins of some of the lagoons.

The depth of these lagoons is various: in those which we entered it was from twenty to thirty-eight fathoms, but in others, to which we had no access, by the light-blue colour of the water it appeared to be very small. It is, however, tolerably certain that the coral forms the bases of them, and consequently, unless depositions of sand or other substances, obnoxious to the coral insects, take place, their depth must depend upon their age.

Very little offered itself to our notice, by which we could judge of the rapidity of the growth of the coral, as the islands which we examined had never been described with the accuracy necessary for this purpose; and there were, consequently, no means of comparing the state in which they were found by us, with that which was presented to our predecessors; but from the report of the natives, the coral bordering the volcanic islands does not increase very fast, as we never heard of any channels being filled up; but, on the contrary, that the passages through the reefs were apparently always in the same condition. The only direct evidence, however, which I could obtain of this fact was that of the Dolphin reef off Point Venus in Otaheite. This reef, when first examined by Captain Wallis in 1769, had "two fathoms water upon it." Cook sounded upon it a few years afterwards, and gave its depth fifteen feet. In our visit to this place, we found, upon the shallowest part of it, thirteen feet and a half. These measurements, though at variance, from the irregularity of the surface of the reef, are sufficiently exact to warrant the conclusion that it has undergone no very material alteration during an interval, it should be recollected, of fifty-six years. But the Dolphin, as well as the above-mentioned reefs and channels, are within the influence of rivers, which, in my opinion, materially retard their increase, and their growth must not be taken as a criterion of that of the islands of which I have been speaking. With regard to them, there is one fact worthy of consideration, and upon which every person must form his own judgment. I allude to the remains of the Matilda, a ship which a few pages back is stated to have been cast away upon one of these coral islands. In my description of Matilda Island, it is stated, that one of the anchors of this ship, a ton in weight, a four-pounder gun, her boilers and iron-work, are lying upon the top of the reef, two hundred yards from the present break of the sea, and are dry at low water.* The nature of these articles and the quantity of iron bolts and other materials lying with them renders it probable

* The rise of the tide is about two feet.

that the vessel went to pieces in that spot, for had the sea been heavy enough to wash the anchor from deeper water, the boiler must have been carried much beyond it; and the question is, whether the hull of a vessel of the Matilda's tonnage could be washed upon a reef dry at low water, and be deposited two hundred yards within the usual break of the sea. The circumstance of the hatches, staves of casks, and part of the vessel, being deposited in parts of the dry land not far distant, and scarcely more than four feet from the present level of the sea, offers a presumption that the sea did not rise more than that height above its ordinary level, or it would have washed the articles further and left them in the lagoon, whence they would have been carried to sea by the current.

The materials were not in the least overgrown with coral, nor had they any basin left round them by which the progress of the coral could be traced; and yet, in other parts of this reef, we noticed the chama gigas of seven or eight inches in diameter so overgrown by it, that there was only a small aperture of two inches left for the extremity of the shell to open and shut.

When the attention of men of science was called to these singular formations by the voyages of Captain Cook, one opinion, among others respecting their formation was, that they sprung from a small base, and extended themselves laterally as they grew perpendicularly towards the surface of the sea; and that they represented upon a large scale the form which is assumed by some of the corallines. In particular this theory was entertained by Mr. John R. Forster, who accompanied Captain Cook on his second voyage and visited several of the coral islands, and was founded, no doubt, upon the experience which he had derived upon that voyage. But considering the extent of some of these islands, it is evident that if this be their form, the lythophites, the animals which construct them, must commence their operations at very great depths, a fact which is doubted by naturalists. The general opinion now is, that they have their foundations upon submarine mountains, or upon extinguished volcanoes, which are not more than four or five hundred feet immersed in the ocean; and that their shape depends upon the figure of the base whence they spring. It would be immaterial which of these theories were correct, were it not that in the latter instance the lagoon that is formed in all the islands of this description might be occasioned by the shape of the crater alone, whereas, in the former, it must result from the propensity of the coral animals, and this, if true, forms a remarkable and interesting feature in their natural history. Mr. Forster* thought this peculiarity might arise from the instinct of the animalcules forming the reefs, which from a desire to shelter their habitation from the impetuosity of the winds, and the power and rage of the ocean, endeavoured to construct a ledge, within which was a lagoon entirely screened against the power of the elements, and where a calm and sheltered place was by these means afforded to the animals in the centre of the island.

Another reason why the consideration of the nature of their foundation is not immaterial is, that if the form of the islands arose from the peculiar shape

* Forster's Observations, 4to, page 150.

of the craters, and it be admitted that the lithophytes are unable to exist at greater depths than those above-mentioned, we shall have examples of craters of considerably larger dimensions, and more complete in their outline, than any that are known upon the land, which, if true, is a curious fact. Until the voyage of the Blossom, it was not generally known that the lagoons in these islands were of such depths, or that the wall of coral which encircles them was so narrow and perfect, as in almost every instance it has been found; nor that the islands were of such dimensions, as they were designated groups, or chains of islands, in consequence of the wall being broken by channels into the lagoon; but on examination, the chain is found continuous under water; and as in all probability it will in time reach the surface and become dry, the whole group may be considered as one island.

In the plans which I have delivered into the Admiralty, the figure and extent of thirty coral islands, out of the many which exist in the Pacific, are carefully delineated, and a reference to them will more fully explain their nature than any description I can give here. One of these plans * being of particular interest, I have inserted it in the present work, as it exhibits, not only the coral chain enclosing the lagoon, which is the common character of the coral islands; but, also, an example of several volcanic islands rising within it; and likewise the peculiarity of the inequality in the sides of the chain mentioned in page 189.

The subject of the formation of these islands is

* See the plan of Gambier groupe.

one of great interest, and will require a numerous and careful collection of facts before any entirely satisfactory conclusion can be arrived at. I regret that my time did not permit me to inquire more particularly into this curious matter; but having to survey about fifty islands, some of which were of great extent, in the space of about four months, I could not accomplish more than was absolutely necessary to the purposes of a safe navigation of the Archipelago. We were, however, not inattentive to the subject, and when opportunity offered, soundings were tried for at great depths, and the descent of the islands was repeatedly ascertained as far as the common lines would extend. Some of these experiments are given in the annexed plate, representing a section of a coral island from actual measurement.

In considering the subject of these coral formations, my attention was drawn to the singularity of the occurrence of openings in them, either opposite to, or in the direction of some stream of fresh water from the mountains; and on searching several charts, I find so many corroborations of the fact, that I have no doubt of the truth of it: as far as my own observations extended, it was always so. The aversion of the lithophytes to fresh water is not singular, as, independent of its not being the natural element of those animals, it probably supplies no materials with which they can work.

It has been suggested, that these openings being opposite to valleys, the continuation of them under water is the cause of the break in the reef. But when we consider the narrowness of these openings, compared with the width of the valleys, and that the latter are already filled up to the surface and fur-

nished with a smooth sandy beach, many obstacles
will be found to the confirmation of such an opinion;
and it appears to me more reasonable to attribute it
to the nature of the element. The depth of these
channels rarely exceeds twenty-five feet, the greatest limit probably to which the influence of fresh
water would be felt.

Henderson Island, one of the exceptions mentioned in the early part of this discussion, is among
the rare instances of its kind in these seas. It is an
island composed of dead coral, about eighty feet
above the sea, with perpendicular cliffs nearly all the
way round it, as if after being formed in the ocean
it had been pushed up by a subterraneous convulsion. These cliffs are undermined at the base, as
though the sea had beaten against them a considerable time in their present position. There are no
marks upon them indicative of the island having
risen by degrees; but, on the contrary, a plain surface indicating its ascent by one great effort of nature.
On examining the volcanic islands near Henderson
Island, no traces appeared of the sea having retired;
and we may, therefore, presume it to have risen as
described. Its length is five miles, and breadth one
mile; it is nearly encompassed by a reef of living
coral, so wide that the cliffs, which were at first subjected to the whole force of the waves, are now
beyond the reach even of their spray.

The navigation of this archipelago was made at a
period of the year when the westerly monsoon was
about to commence, and toward the end of which it
had actually begun, and materially retarded our operations; but previous to that time, or about the beginning of March, the trade was fresh and steady,

blowing between S.E. by E. and E.N.E., which is
more northerly than the direction of the same trade
between corresponding parallels in the Atlantic. In
consequence of this opposition to the trade wind the
currents were very variable, sometimes setting to the
eastward, and at others in the opposite direction;
and on the whole, the body of water at that period
is not drifted to the westward with the same rapidity
that it is in other parts of the ocean within the influence of the tropical winds. The mean temperature for the above-mentioned period, the weight and
humidity of the atmosphere, with other meteorolo-
gical observations, are given in the Appendix to the
4to ed. under their respective heads.

For the information of persons who may traverse
this archipelago, it is evident from the account of
Tuwarri, that there is a small island situated about
half way between Byam Martin and Barrow Islands,
which was not seen by us; and hence it is possible
that there are other low islands lying between the
tracks of the Blossom which were not seen; and
ships ought in consequence to keep a vigilant look-
out during the night, or adopt the precaution of
lying to when the weather is dark or thick. The
lead is no guide whatever in these seas, and the
islands are so low that in the night the white line
of the surf or the roar of the breakers would give
the first warning. Fallacious as the appearance of
birds is generally considered, and in some parts of
the globe justly so, in this archipelago, when seen
in flocks, it is an almost certain indication of land.
They range about forty miles from the islands, and
consist principally of black and white tern. This,
however, applies particularly to uninhabited islands;

for when they become peopled, the birds generally
quit them, and resort to those where they are less
molested.

At day-light on the 15th the Island of Maitea was
seen in the north-west, and soon afterwards the
mountains of Otaheite appeared five minutes above
the horizon at the distance of ninety miles, from
which its height may be roughly estimated at 7000
feet. As we passed Maitea we had an opportunity
of verifying its position and ascertaining its height
to be 1432 feet. Baffling winds prevented us from
reaching our port until the evening of the 18th,
when, at the suggestion of Captain Charlton, his
Majesty's consul for the Society and Sandwich Islands, from whom we had the pleasure of receiving a
visit, we anchored in the outer harbour of Toanoa,
about four miles to the westward of Matavai Bay.

CHAPTER IX.

Proceedings at Otaheite—The Ship visited by the Queen Regent, the Royal Family, and several Chiefs—Short Account of the former since Captain Cook's Visit—Successful Issue of a Dispute with the Government respecting the Detention of a trading Vessel—Visit to the Queen Regent's House—Present Condition of the Chiefs and of the Inhabitants—Superstitions—Trial of Natives for Theft of the Ship's Stores—The King visits the Ship —Lake and Morai of Mirapaye—Dance exhibited by a Party of New Zealanders—Considerations on the Effect of the Introduction of Christianity.

THE diversity of feature of the romantic Island of Otaheite formed a strong contrast with the monotonous appearance of the coral formations; the variety of hill and valley, and of woods and rivers in the one, after the sameness of flat, sterile, parched-up surface in the other; and the glassy smoothness of the harbours around us, opposed to the turbulent shores we had recently quitted, were gratifying in the extreme, and impressed us most forcibly with the truth of the observations of our predecessors, who have spoken of the scenery of this island in the highest terms of commendation.

As I proposed to remain here a few weeks to recruit the health of the crew, who were somewhat debilitated, and to prepare the ship for her voyage to the northward, she was moved to an inner anchorage opposite a small village called Toanoa, and there secured by a cable fastened to some trees on one

CHAP.
IX.

March,
1826.

side, and by a bower anchor dropped at the edge of a coral reef on the other. This reef forms one side of the harbour; which, though small, possesses several advantages over the more spacious one of Papiete generally resorted to, and of which the superior freshness and salubrity of its atmosphere are not the most inconsiderable.

Previous to entering upon a relation of our proceedings with the natives, it must be understood that the short time we remained, and our various occupations necessarily rendered our intercourse with them very limited compared with that of many of our predecessors. Still, it is hoped, the remarks which I shall offer will be sufficient to present a candid and faithful picture of the existing state of society in the island; a feature by no means unimportant in the history of the country, which is otherwise complete. To exceed this, by dwelling upon the beauties of the scenery, the engaging manners of the inhabitants, their mythology, superstitions, and legends, &c. would be only to recapitulate what has been detailed in the interesting voyages of Wallis, Cook, Vancouver, Wilson, Turnbull, and others, and very recently by Mr. Ellis, in his valuable work entitled " Polynesian Researches," compiled after ten years' residence in the Pacific, and from the journals of other missionary gentlemen in those parts. In this useful work Mr. Ellis has traced the history of some of the islands through all their various stages; he has explained the origin of many of their barbarous customs, has elucidated many hitherto obscure points, and has shown the difficulties which opposed themselves to the introduction of Christianity; the hardships, dangers, and privations, which were endured by himself and his brethren, who, actuated by religious motives, were induced to sacrifice their own health, comfort, and worldly advantages in the attempt to ameliorate the condition of their fellow-creatures. But complete as that work is in many respects, it is nevertheless deficient in some essential points. The author, with a commendable feeling of charity, consonant with his profession, has by his own admission in the account of the biography of Pomarree, glossed over the failings and dwelt upon the better qualities of the subject of his memoir; and pursuing the same course throughout, he has impressed the reader with a more elevated idea of their moral condition and with a higher opinion of the degree of civilization to which they have attained, than they deserve; or, at least, than the facts which came under our observation authorise. There seems to be no doubt that he has drawn the picture, generally, as it was presented to him; but he has unconsciously fallen into an error almost inseparable from a person of his profession, who, when mixing with society, finds it under that restraint which respect for his sacred office and veneration for his character create. As in our intercourse with these people they acted more from the impulse of their natural feelings, and expressed their opinions with greater freedom, we were more likely to obtain a correct knowledge of their real disposition and habits.

To convey to the reader, who has not perused the above-mentioned work, an idea of the political state of the island, in which there has been a material alteration since the period alluded to in the early voyages, it will be necessary to state briefly that since 1815 a code of laws has been drawn up oy Pomarree II., with the assistance of the missionaries, which has subsequently been extended from time to time; and that since 1825 a house of parliament has been established, to which representatives of the several districts in the island are returned by popular election. The penalties proposed by Pomarree were very severe, but that of death has as yet been enforced upon four culprits only.

The limit thus imposed on the arbitrary power of the monarch, and the security thus afforded to the liberties and properties of the people, reflect credit upon the missionaries, who were very instrumental in introducing these laws: at the same time, had they been better informed in the history of mankind, they would have been less rigid upon particular points, and would have more readily produced those benefits which they no doubt hoped would ensue. Magistrates are appointed to try cases, and conduct their judicial proceedings in open court, and the police are continually on the alert both day and night to prevent irregularities, and to suppress the amusements of the people, whom, from mistaken views of religion, they wish to compel to lead a life of austere privation.

We found the consul in possession of a small but comfortable house opposite the anchorage, which had been hastily run up by the natives for his use; and we took the earliest and most favourable opportunity of impressing the importance of his situation upon the inhabitants, by the salute due to his rank. Besides the missionary gentlemen, we found that several other Europeans were residing in our vicinity; and as some of these, as well as the consul, had

their wives and female relatives with them, we
looked forward to the pleasure of varying our intercourse with the uncouth natives by more agreeable
society—an anticipation which was fully realised by
their unremitting attention, especially on the part
of the consul, whose house was the general resort
of all the officers.

Our arrival was immediately communicated,
through the proper channel, to the queen regent,
who lived about a mile from the anchorage, and we
received an intimation of her intention of paying an
early visit to the ship.

The arrival of a man-of-war at Otaheite is still an
event of much interest, and brings a number of the
inhabitants from the districts adjoining the port,
some in canoes, others on foot. The little hamlet
opposite the ship was almost daily crowded with
strangers, and a vast number of canoes skimmed the
smooth surface of the harbour, or rather the narrow
channel of water which is tied to the shores of this
luxuriant island by reefs of living coral. A remarkable exception to this scene of bustle occurred
on the day of our arrival, which, although Saturday,
according to our mode of reckoning, was here observed as the Sabbath, in consequence of the missionaries having proceeded round by the Cape of
Good Hope, and having thereby gained a day upon
us. Next morning, however, a busy scene ensued.
Canoes laden with fruit, vegetables, and articles of
curiosity, thronged as closely round the ship as their
slender outriggers would allow, while such of the inhabitants as wanted these means of approaching us
awaited their harvest on the shore.

We soon found that the frequent intercourse of

Europeans with the islanders had effected an alteration in the nature of the currency, and that those
tinselled ornaments with which we had provided
ourselves were now objects of desire only as presents;
the more substantial articles of clothing and hard
dollars being required for the purposes of the market,
except, perhaps, where a ring or a jew's harp happened for the moment to attract the attention of
some capricious individual. However gratified we
might be to observe this advance towards civilization, we experienced considerable inconvenience
from its effects; for on leaving the coast of Chili,
very few of us had provided dollars, under an impression that they would not be necessary; and those
which we had were principally of the republican
coinage, and as useless in the Otaheitan market as
they would have been in New Zealand. No dollars
bear their full value here, unless the pillars on the
reverse are clearly distinguishable, and a greater degree of value is attached to such as are bright than
to others. So ignorant, indeed, were these simple
people of the real worth of the coin, that it was not
unusual for them to offer two that were blemished
in exchange for one that was new, and in the market
a yard of printed calico, a white shirt, new or old,
provided it had not a hole in it (even a threadbare
shirt that is whole being whimsically preferred to one
which might have been eaten through by a mouse),
or a Spanish dollar that had two pillars upon it, were
in the ordinary way equivalent to a club, a spear, a
conch shell, a paddle, or a pig. Deviations, of course,
occurred from this scale, founded on the superior
quality or size of the article, and occasionally on the
circumstances of the vendor, who, when he antici-

pated a better bargain, would accommodate his price to his preconceived opinion of the disposition of the purchaser. We were not more conveniently circumstanced in regard to the clothing which we could offer in exchange, as we had a long voyage before us, and little to spare without subjecting ourselves to future inconvenience. We, consequently, found ourselves at first surrounded with plenty, without the means of purchase, or obliged to part in payment with what we could very ill spare: and we incurred the additional risk of being charged with parsimony, which the good people of Otaheite are very apt to attach to those who may not meet their ideas of generosity. "Taata paree," or stingy people, is an epithet which they always affix to such persons, with a feeling of contempt, although they are themselves equally open to the charge, never offering a present without expecting a much larger one in return. It is very desirable to secure a favourable impression by liberality on your first arrival at this island; it being a constant custom with the natives to mark those who have any peculiarity of person or manner by a nickname, by which alone the person will be known as long as any recollection of his visit may remain. Among the many instances which occurred of this, was one of a brother officer, who, when we quitted England, begged to be remembered to his old acquaintances in Otaheite; but we found they had lost all memory of his name, and we at last only brought him to their recollection by describing his person, and mentioning that he had lost an eye by a wound received in service; on which they at once exclaimed "Tapane Matapo!" or "Captain Blind-eye." We were the more anxious to avoid acquiring a distinc-

tion of this kind for ourselves, as a Russian ship had
just preceded us, the crew of which, according to the
natives, purchased every thing that was offered without regard to price, at whom they laughed heartily,
because one of the officers had given a blue jacket in
exchange for a pearl which had been ingeniously
made out of an oyster-shell.

Some of us, therefore, had recourse to the European residents, and fortunately obtained what cloth
and specie we wanted; while others preferred bartering such portions of their wardrobes as they
considered unnecessary for their approaching change
of climate.

On the Monday succeeding our arrival, all the
stores of the ship that required removal were landed
and placed under a shed; the observatory was
erected close to the consulate; a rope-walk was constructed, and the forge was put up under the shade of
some trees. Thus, as the shore was so near, all the
duties of the ship were carried on under our own
immediate superintendence far more expeditiously
than the confined space on board would have allowed. The sick were also landed, and provided
with a place better adapted to their situation.

The state of our provisions rendered it necessary
to observe the strictest economy, for we had been
confined to our own resources during several months,
and Otaheite afforded nothing except beef and pork,
nor had we any certainty of an opportunity of replenishing them. The bread fruit was, fortunately,
at this time excellent, and was substituted for the
daily allowance of flour, at first in moderate proportions, that no bad effects might arise from such
a change of diet; but, latterly, the crew were al-

lowed as much as they could consume, by which necessary piece of economy we saved during our stay about 2,000 pounds of flour, the most valuable article of sea store ; a measure which subsequently proved of the utmost importance to us. I do not think that this fruit, though very delicious and more farinaceous than potatoes, is a satisfactory substitute for bread, but it is by no means a bad one.

Foreseeing the possibility of being obliged to cure our own meat, we fortunately provided a quantity of salt for that purpose at Chili, an article which we found very scarce at Otaheite ; and the consul made arrangements for salting both beef and pork for our future use, which succeeded uncommonly well ; and he materially forwarded the object of our voyage by exerting himself to satisfy all our demands, so far as the resources of the island would admit. Before our arrival articles of food were sufficiently cheap ; but the great demand which we occasioned materially enhanced their prices, and there appeared to be a great dislike to competition. The resources of the island, fruit excepted, are considerably diminished from what they formerly were, notwithstanding the population at one time exceeded its present amount twenty-fold.

On the day appointed for the visit of the royal party, the duty of the ship was suspended, and we were kept in expectation of their arrival until four o'clock in the afternoon, when I had the honour of receiving a note, couched in affectionate terms, from the queen regent, to whom, as well as to her subjects, the loss of time appears to be immaterial, stating her inability to fulfil her engagement, but

that she would come on board the following day.
Scarcely twenty minutes had elapsed, however, from
the receipt of this note, when we were surprised by
the appearance of the party, consisting of the queen
regent, the queen dowager and her youthful husband, and Utamme and his wife. Their dress was
an incongruous mixture of European and native
costumes; the two queens had wrappers of native
cloth wound loosely round their bodies, and on their
heads straw poked bonnets, manufactured on the
island, in imitation of some which had been carried
thither by European females, and trimmed with
black ribands. Their feet were left bare, in opposition to the showy covering of their heads, as if
purposely to mark the contrast between the two
countries whose costumes they united; and neatly
executed blue lines formed an indelible net-work
over that portion of the frame which in England
would have been covered with silk or cotton.
Utamme, who, without meaning any insinuations
to the disadvantage of the queen, appeared to be on
a very familiar footing with her majesty, (notwithstanding he was accompanied by his own wife), was
a remarkably tall and comely man; he wore a straw
hat, and a white shirt, under which he had taken
the necessary precaution of tying on his native
maro, and was provided with an umbrella to screen
his complexion from the sun. This is the common
costume of all the chiefs, to whom an umbrella is
now become almost as indispensable as a shirt; but
by far the greater part of the rest of the population
are contented with a mat and a maro.

It may be desirable, in this early period of our
communications with the court of Otaheite, to state

the relationship which exists between the reigning
family and Otoo, who was king of the larger peninsula at the period of Captain Cook's last visit.

Otoo, after Cook's departure, was surnamed Pomarree, from a hoarseness that succeeded a sore
throat which he caught in the mountains, and this
afterwards became the royal patronymic. His son,
Pomarree II., who was a child at that period, succeeded him in 1803, and reigned until December,
1821, when, having effected many most important
changes in the customs of the island, and having,
under the zealous exertions of the missionaries, converted the chief part of the population to Christianity, he expired in a fit of apoplexy, accelerated, no
doubt, by frequent excesses. Of this man it may be
lamented that his exertions in the cause of Christianity were not seconded in the fullest extent by a
rigid adherence to its precepts in his own person.
He had two wives, or rather a wife and a mistress,
who were sisters, named Terre-moe-moe, and Pomarree Waheine. This woman, daughter of the
King of Ulietea, had been sent for from Huaheine
to be married to the king, but being accompanied
by her sister, Terre-moe-moe, who was very superior
in personal attractions, the latter captivated his
majesty at first sight, and received the honour of
his hand, while Pomarree Waheine was retained in
the more humble capacity of mistress. Each sister
bore a child, Terre-moe-moe giving birth to Pomarree III., and the mistress presenting him with a
daughter named Aimatta, the present queen. Pomarree III. was only six years old at the time we
arrived, and the regency was administered by his
aunt Pomarree Waheine, who I suppose was con-

sidered a more fit person to manage the affairs of the
state than her sister, who had doubtless the greater
claim to the office.   We found that the queen
mother, widow of Pomarree II., had married a chief
of Bora Bora, a fine-looking lad of ten or eleven
years of age, and that Aimatta was united to a chief
of Huaheine, a short corpulent person, who, in consequence of his marriage, was allowed to bear the
royal name of Pomarree, to which, however, in allusion to his figure, and in conformity with their
usual custom, they had added the appropriate but
not very elegant surname of "Aboo-rai," or big-
belly.

We treated the royal party with the few good
things which remained, and they landed at night,
highly delighted with a display of fire-works purposely prepared for them.   Next morning the party
repeated their visit, somewhat better dressed, and
accompanied by Aimatta and Aboo-rai.   They were
followed by a large double canoe and many small
single ones, bearing upon their gunwales heaps of
fruit and roots, and four enormous hogs, at the imminent risk of upsetting the whole.   The double
canoe was the "last of her race," and had been used
for the nobler purposes of war, but, like the inhabitants, was now devoted to humbler but more useful
occupations.

As soon as the queen reached the deck she ten-
dered the present to me in the name of the young
king, then at the missionary school at Eimeo, and I
returned the compliment that was due to her for this
mark of her attention, as well as for the munificence
of the gift.   As soon as the remainder of the party
were assembled, it was proposed that we should

adjourn to the breakfast prepared in the cabin; but the regent desired that every part of the present should previously be set out on a particular part of the deck, pigs and all, in order to impress us more fully with an idea of her liberality; and when the whole was collected, she led me to the pile, and expatiated on the superior quality of the fruit.

Having at length assembled at breakfast, which by this time was cold, a difficulty arose, I was informed, in consequence of Aimatta, the king's sister, being unwilling to relinquish the distinction she had enjoyed under the former custom of the island, which rendered it indecorous for some of her countrywomen, who were of the party, to presume to eat in the presence of so exalted a personage. As these distinctions, however, had been removed upon the introduction of Christianity, there was an evident apprehension of giving offence to the assembled chiefs by such a display of ambition on the present occasion. The inconvenience which it was suggested would attend the observance of the custom in this instance, and the opposition afforded by the precepts of the missionaries to any such mode of displaying the royal prerogative, relieved us from our dilemma. A cloud of discontent hung for a short time on the countenance of our royal guest, but it was dispelled by the first breeze of mirth, and the party appeared to enjoy greatly the remainder of their visit.

It is by no means surprising that the chiefs should wish to adhere to such of their old customs as constituted the principal if not the only distinction between them and their vassals. Should they be deprived of these, and should the superstitions, by means of which they awed the lower classes of the community, be brought into contempt, they would be left with no other superiority than that conferred by bodily strength; for in education, and not unfrequently even in wealth, their advantages were very limited. Pomarree, in framing his laws to meet the new circumstances of his subjects, seems to have been too zealous in pressing his reforms in this as well as in many other points. It would be ridiculous to advocate the perpetuation of customs fit only for the darkest ages of barbarism; but it might probably not be unwise to retain in the earlier progress towards improvement such as are least objectionable; particularly in a country like Otaheite, where their observance had been enforced with the greatest rigour. The effect produced by the abolition of that most detestable of all their pagan rites, human sacrifice, is noticed by Mr. Ellis in his Polynesian Researches, to have endangered the royal authority.*

In the course of the day several chiefs came on board, dressed in white shirts and straw hats; and were all remarkable for their extraordinary height and noble appearance. Whether this superiority of stature is the result of the better quality of their food, or whether, by the commission of infanticide, their parents have preserved only the largest or most healthy children, and bestowed upon them a more careful nursing than may have fallen to the lot of their vassals, I cannot say, but it is beyond a doubt that the advantage which their chiefs enjoyed in this respect had a strong influence on the minds of

* He says (vol. ii. p. 378.) that " many, free from the restraint it (human sacrifice) had imposed, seemed to refuse almost all lawful obedience and rightful support to the king."

the simple Otaheitans, who were with difficulty convinced that the size of the purser (who was the largest man in the ship) did not confer on him the best
claim to be the Ratira-rai, or captain of the Blossom.

The arrival of the chiefs was an event very favourable to the wishes of the consul, who availed
himself of the opportunity it afforded of urging,
with some prospect of success, the repeal of an order issued by the regent, which had occasioned serious mischief to one of our merchant ships; and
which, if not speedily rescinded, must have endangered not only the property, but even the lives of
individuals trading to these islands. The consul
had already appealed against the obnoxious decree,
but it was at a time when he was not supported by
the presence of a king's ship; and the short-sighted
policy of the regent did not anticipate the probability of the consul soon receiving such a powerful
support to his negotiation. She had ventured,
therefore, to dismiss his remonstrance, intimating
that she was fully aware of his defenceless situation.
The case under discussion was as follows.

The queen, seeing the estimation in which the
pearl oyster-shells were held by Europeans, imagined that by levying a duty on them she would greatly increase her revenue. Orders were accordingly
issued to all the tributary islands to seize every vessel trading in shells, which had not previously obtained the royal licence to procure them. The
Chain Islanders, who, from their enterprising and
marauding habits, may be considered the buccaneers of the eastern South Sea archipelago, were
too happy to find themselves fortified with a plea
for a proceeding of this nature, and instantly sent

one of their double canoes to Tiokea, where they found the Dragon, an English brig, taking in pearl shells. These people behaved in a very friendly manner to her crew, and allowed her quietly to take her cargo on board; but the Dragon was no sooner ready to put to sea, than several of the islanders went on board with the ostensible purpose of taking leave, but suddenly possessed themselves of the vessel, overpowering the master and crew, binding their hands, and sending them on shore as prisoners. A general plunder of the vessel ensued, in which every thing moveable was carried away. The natives, after this atrocious act, went to church to return thanks for their victory, and to render their prayers more acceptable, transferred the bell of the ship to their place of worship. During several days they detained the master bound hand and foot, and debated whether he should not be put to death and eaten; a fate which we were informed he would in all probability have encountered but for the interference of one of their chiefs, for the Tiokeans are still reputed to be cannibals, notwithstanding they have embraced the christian religion. The crew, more fortunate than their commander, very soon obtained their release, upon condition of fitting the brig for sea, the natives imagining they could navigate her themselves. The vessel being ready, the master, under some pretext, obtained permission to go on board, and having speedily established an understanding with his crew, he cut the cables and carried her out to sea.

The stolen property was of course never recovered, and the vessel was so plundered of her stores that the object of her voyage was lost. When she reached

Otaheite the master stated the case to the consul, whose representation of the outrage to the queen was, as has already been said, treated with derision. The consul availed himself of the present occasion to obtain restitution of the stolen property, or remuneration for the owners, and a repeal of the objectionable order, the execution of which it is evident could not be safely confided to a barbarous people, at all times too prone to appropriate to themselves whatever might fall within their reach. Her majesty was exceedingly unwilling to abandon this source of revenue, and strenuously urged her indubitable right to levy taxes within her own dominions, maintaining her arguments with considerable shrewdness, and appealing finally to the chiefs. Finding them, however, disposed to accede to the demands of the consul, she burst into tears; but at length consented, by their advice, to send a circular to the Pamoutas, or Low Islands, directing that no molestation should be offered to any vessels trading in shells, or touching at those islands for refreshment; but on the contrary that all necessary aid and assistance should be afforded to them; and that in the event of any dispute, the matter should be referred to the authorities at Otaheite.

This concession destroyed the complacency of the queen for some time, but she recovered her spirits in the course of the afternoon, and amused herself much by listening to the drum, which she begged might be permitted to play on the upper deck. As this species of music, however, was not very agreeable in the confined space of a ship, it was proposed that the instrument should be removed to the shade of some tall trees on the shore, whither the whole party repaired; the drummer continuing his performance, and marching to and fro, until he became heartily tired, to the infinite delight of the assembled populace, who crowded round, and even scaled the loftiest trees, to obtain a glimpse of him.

A few days after this visit the queens came again to Toanoa, and I invited them into the tent we had pitched on shore, with the view of making a present to each of them, and of confiding to their care the presents intended for Pomarree Aboo-rai, Aimatta, and Utamme, who were absent. The present for the king, which consisted of a handsome double-barrelled gun inlaid with silver, with some broadcloth and other valuables, I reserved until I should have an opportunity of seeing him. The other parcels were apportioned according to what I considered to be the rank of the parties, and the name of each person was placed on his destined share. The regent, however, opened them all, and very unceremoniously transferred a portion of each to her own, and huddling the whole together, she sent them off to her canoe. Then finishing half a bottle of brandy between them, the regent and her sister despatched the remainder of the spirits after the presents, and took their leave.

In the course of the day we received an invitation to pass the evening at the regent's house at Papiete, a very romantic spot about a mile from the place where the ship was anchored. After a delightful walk along the shore in the refreshing coolness which succeeds a tropical day, we arrived at the royal residence, which was in one of those spacious sheds frequently mentioned by my predecessors. It was about a hundred feet in length, by thirty-five in width, of an oval form, with a thatched roof, supported upon small poles placed close together. By the light of the moon we discovered a small door about mid-way between the extremities, which we entered, and immediately found ourselves in darkness. On groping our way, our shins came in contact with several bamboo partitions dividing the area into various compartments. In one of these we distinguished by the rays of moonlight which fell through the interstices of the dwelling, that it was occupied by toutous, or common people, of both sexes. We, therefore, turned to the opposite direction, which soon led us to the royal saloon, which we found illuminated by a yellow and melancholy light proceeding from a rag hung over the edge of a broken cocoa-nut shell half filled with oil. The apartment, to our surprise, was quite still; but we were soon greeted with the salutation of " Euranna-poy" (How do you do?) from a number of athletic men, her majesty's favourites, as they awoke in succession from their nap.

We at length discovered the queen regent extended upon a mat spread upon dried grass, with which the whole apartment was strewed; around her, upon mats also, were several interesting young females; and occupying a wooden bedstead, placed against a slight partition, which contained numerous cases filled with cocoa-nut oil, we found Pomarree Aboo-rai, and Aimatta. Our entry threw this numerous party into a state of activity and bustle, some to procure a second light, and some to accommodate us with mats; while Pomarree, drawing his tappa round him, led forward his princess, Aimatta, and extended his politeness much beyond what we

could possibly have anticipated from so young a
husband.

Fearful that we might have misunderstood the
morning invitation, or that we were later than we
had been expected, we began to offer apologies, and
to excuse ourselves for breaking in upon the repose
of the party; but the indisposition of the queen
appeared to be the cause, as she was suffering from
repletion, and, forgetting all about the invitation,
had retired earlier than usual. She had scarcely had
sufficient rest when we arrived to engage in any
amusement herself, but gave us a friendly reception,
and desired that a dance might be performed for our
entertainment. This was an indulgence we hardly
expected, such performances being prohibited by
law, under severe penalties, both against the performers, and upon those who should attend such
exhibitions; and for the same reason it was necessary
that it should be executed quietly, and that the *vivo*,
or reed pipe, should be played in an under tone, that
it might not reach the ears of an aava, or policeman,
who was parading the beach, in a soldier's jacket,
with a rusty sword; for even the use of this melo-
dious little instrument, the delight of the natives,
from whose nature the dance and the pipe are inseparable, is now strictly prohibited. None of us had
witnessed the dances of these people before they
were restrained by law; but in that which was
exhibited on the present occasion, there was nothing
at which any unprejudiced person could take offence;
and it confirmed the opinion I had often heard expressed, that Pomarree, or whoever framed the laws,
would have more effectually attained his object had
these amusements been restricted within proper

limits, rather than entirely suppressed. To some of us, who had formed our opinion of the native dance of this island from the fascinating representation of it by Mr. Webber, who accompanied Captain Cook, that which we saw greatly disappointed our expectation, and we turned from it to listen to the simple airs of the females about the queen, who sang very well, and were ready *improvisatrices*, adapting the words of the song to the particular case of each individual.

While these amusements engaged the attention of our party, scenes of a very different nature were passing in the same apartment, which must have convinced the greatest sceptic of the thoroughly immoral condition of the people; and if he reflected that he was in the royal residence, and in the presence of the individual at the head of both church and state, he would have either concluded, as Turnbull did many years before, that their intercourse with Europeans had tended to debase rather than to exalt their condition, or that they were wilfully violating and deriding laws which they considered ridiculously severe.

In our intercourse with the chiefs and middle classes of society, the impression left by this night's entertainment was in some measure removed; and especially as regards the former, who are, on the whole, a well-behaved class of men, though they are much addicted to intemperance. A party of them, among which were Utammee and Pa-why, came on board one day, and having received a present of a bottle of rum from the cabin, went to pay a visit to the gun-room officers, who politely offered them a glass of wine, but evincing some reluctance to this beverage, rum was placed upon the table, upon which the chiefs manifested their approbation, and Utammee seizing the bottle requested it as a present, and then emptying their glasses, which had been filled with wine, to the toast of *Euranna poy*, they bowed politely and withdrew. This partiality for spirits seems to be an incorrigible vice, and it is a fortunate circumstance that their means of indulging in it are so very limited. Some of them have materially benefited by the residence of the missionaries, and, in particular, two who resided at Matavai, about four miles to the eastward of our anchorage. They piqued themselves on their imitation of European customs, and had neat little cottages, built after the European style, with whitewashed fronts, which, peeping through some evergreen foliage, had a most agreeable effect, and being the only cottages of this description upon the island in the possession of the natives, were the pride of their owners. The apartments contained chests, chairs, a table, and a knife and fork for a guest; and nothing gave these chiefs greater pleasure than the company of some of the officers of the ship. Each of them could read and write their own language, and the elder, Pa-why, had, I believe, been useful to the missionaries in translating some part of the Scriptures. He was the more learned of the two brothers; but Hetotte was the more esteemed, and was an exception to almost all his countrymen in not asking for what was shown to him. His inquiries concerning the use of every thing which offered itself to his notice, on coming on board the ship, surprised and interested us; while his amiable disposition and engaging manners won him the

esteem of almost all on board. An anecdote illus-
trative of his character will be read with interest.
The missionaries had for several years endeavoured
to produce a change of religion in the island, by
explaining to the natives the fallacy of their belief,
and assuring them that the threats of their deities
were absurd. Hetotte at length determined to put
their assertions to the test, by a breach of one of the
strictest laws of his religion, and resolved either to
die under the experiment or embrace the new faith.

A custom prevailed of offering pigs to the deity,
which were brought to the morai and placed upon
whattas, or fautas, for the purpose. From that moment they were considered sacred, and if afterwards
any human being, the priests excepted, dared to
commit so great a sacrilege as to partake of the offering, it was supposed that the offended god would
punish the crime with instant death. Hetotte
thought a breach of this law would be a fair criterion of the power of the deity, and accordingly
stole some of the consecrated meat, and retired with
it to a solitary part of the wood to eat it, and perhaps to die. As he was partaking of the food, he expected at each mouthful to experience the ven-
geance he was provoking; but having waited a considerable time in the wood in awful suspense, and
finding himself rather refreshed, than otherwise, by
his meal, he quitted the retreat and went quietly
home. For several days he kept his secret, but
finding no bad effects from his transgression, he disclosed it to every one, renounced his religion, and
embraced Christianity. Such instances of resolution
and good sense, though they have been practised
before, are extremely rare in Otaheite, and in this

sketch of the two brothers a highly favourable picture is presented of the class to which they belong; though there are others, particularly Taate, the first and most powerful chief upon the island, who are equally deserving of favourable notice.

Of the rest of the population, though their external deportment is certainly more guarded than formerly, in consequence of the severe penalties which their new laws attach to a breach of decorum, yet their morals have in reality undergone as little change as their costume. Notwithstanding all the restrictions imposed, I do not believe that I should exceed the bounds of truth in saying, that, if opportunity offered, there is no favour which might not be obtained from the females of Otaheite for the trifling consideration of a Jew's harp, a ring, or some other bauble.

Their dwellings, with the exception of doors to some, and occasionally latches and locks, are precisely what they were when the island was first discovered. The floor is always strewed with grass, which they are not at all careful to preserve clean or dry, and it consequently becomes extremely filthy and disagreeable; and when it can be no longer endured, it is replaced by fresh material. Their household furniture has been increased by the introduction of various European articles; and a chest, or occasionally a bedstead, may be seen occupying the corner of an apartment; but these are not yet in great demand, the natives having little to put into the former, and esteeming such of the latter as have found their way to Otaheite scarcely more desirable places of repose than their mats spread upon straw. The extreme mildness of the climate, however, sufficiently accounts for the contented state of the population in this respect.

Their occupations are few, and in general only such as are necessary to existence or to the gratification of vanity. In our repeated visits to their huts we found them engaged either in preparing their meals, plaiting straw-bonnets, stringing the smallest kinds of beads to make rings for the fingers or the ears, playing the Jew's harp, or lolling about upon their mats; the princess excepted, whose greatest amusement consisted in turning a hand-organ. The indolence of these people has ever been notorious, and has been a greater bar to the success of the missionaries than their previous faith. The fate of the experiment on the cotton in Eimeo is an exemplification of this. It is well known that the land was cleared, and the cotton planted and grown, but the perseverance to clean the crop, to make it marketable, was wanting; and finding no sale for the article in its rude state, they forbore to cultivate it the next year. A small portion, however, was picked by way of experiment: the missionaries taught the girls to spin, and even furnished them with a loom, and instructed them in the use of it, upon condition that they should weave fifty yards of cloth for the king, and fifty for themselves. The novelty of the employment at first brought many pupils, but they would not persevere, and not one was found who fulfilled the engagement. The proportion due to the king was wove, but not as much more as would make a single gown, and the pupils, after a dispute regarding their wages, abandoned the employment about the period of our arrival. " Why should we work?" they would say to us;

" have we not as much bread-fruit, cocoa-nuts, bananas, vee-apples, &c. as we can eat? It is very good for you to work who require fine clothes and fine ships; but," looking around their apartment with evident satisfaction, " we are contented with what we possess." And in disposition they certainly appeared to be so; for a more lively, good-natured, inoffensive people it is impossible to conceive. The only interruption to their general serenity appears to be occasioned by the check which the laws have placed upon their amusements; a feeling which became very apparent the moment the missionaries were mentioned. They have in general, however, a great respect for those gentlemen, and are fearful of the consequences of offending them.

Some of the natives had an indistinct notion of this philanthropic society, and were not a little surprised at being told that we were not missionaries; and in answer to their inquiry " King George missionary?" their astonishment was greatly increased at being informed that he was not; for as they had an idea that King George was at the head of the missionary society, they naturally imagined that his officers must of course also belong to it. This misconception had been so generally entertained before our arrival, that we were told they had threatened to complain to the society of the master of a merchant ship who had by some means incurred their displeasure.

The Otaheitans were always a very superstitious people, and notwithstanding their change of religion still entertain most absurd notions on several points. Though they have ceased to give credit to any

recent prophecies, many firmly believe they have
seen the fulfilment of some of the predictions that
were made before their conversion to Christianity,
of which the invasion of the island by the natives of
Bora Bora was one. This event was foretold by a
little bird called Oomamoo, which had the gift of
speech, and used to warn persons of any danger
with which they were threatened. On many occasions, when persons have taken refuge in the mountains to avoid a mandate for a victim for the morai,
or to escape from some civil commotion, this little
bird has been their guardian spirit, has warned them
when danger was near, and directed them how to
escape pursuit. I used to laugh at Jim, our interpreter, a good-natured intelligent fellow, for his
belief in these tales; but he was always very earnest
in his relation of them, and never allowed himself
to join in our ridicule. Though he confessed that
this little monitor had been dumb since the introduction of Christianity, yet it would evidently have
been as difficult to make him believe it never had
spoken, as that the danger of which it warned him
had never existed; and this feeling is, I believe,
common to all his countrymen. Nothing is more
difficult than the removal of early impressions, particularly when connected with superstitions. I was
one evening returning with him round the shore
of the bay from Papiete, a favourite route, and
was conversing on the superstitions of his countrymen, when we came to a retired spot crowned
with tall cocoa-nut trees, with a small glen behind
it. Night was fast approaching, and the long
branches of the palm, agitated by the wind, produced a mournful sound, in unison with the subject

of our conversation. As we passed I observed Jim endeavouring to get on the outside, and latterly walking in the wash of the sea ; and found that he never liked to pass this spot after dark for fear of the spirits of his unfortunate countrymen who were hanged there between the cocoa-nut trees. The popular belief, before the introduction of our faith, was, that the spirit of the deceased visited the body for a certain time, and for this reason many of them would on no account approach this place in the night time.

A few days after our arrival some offenders were brought to trial, and as we were desirous of witnessing the proceedings of the court, it was removed from its usual site, to the shade of some trees in our immediate vicinity. The court was ranged upon benches placed in successive rows under the trees, with the prisoners in front, under the charge of an officer with a drawn sabre, and habited in a volunteer's jacket and a maro. The aava-rai of the district in which the crimes had been committed took his place between the court and the prisoners, dressed in a long straw mat, finely plaited, and edged with fringe, with a slit cut in it for the head to pass through ; a white oakum wig, which, in imitation of the gentlemen of our courts of law, flowed in long curls over his shoulders, and a tall cap surmounting it, curiously ornamented with red feathers, and with variously coloured tresses of human hair. His appearance without shoes, stockings, or trousers, the strange attire of the head, with the variegated tresses of hair mingling with the oakum curls upon his shoulders, produced, as may be imagined, a ludicrous effect ; and I regret that the

limits of this work prevent my subjoining an ad-
mirable representation of it by Mr. Smyth.

The prisoner being brought up, the aava read
certain passages from the penal code, and then accused the prisoner of having stolen a gown from a
European resident. He instantly pleaded guilty to
the charge, and thereby saved a great deal of trouble.
He was then admonished against the repetition of
evil practices, and fined four hogs, two to the king,
and two to the person from whom the property had
been stolen. Bail is not necessary in Otaheite; and
the prisoner, consequently, was allowed to go where
he pleased, which of course was to such of his friends
as were most likely to supply him with a hog.
Three other persons were then put to the bar, and
fined for a breach of our seventh commandment.
The young lady, who had sinned with several persons, but two of whom only were detected, smil-
ingly heard herself sentenced to make twenty yards
of cloth, and the two men to furnish six posts each,
for a building that was about to be erected at Papiete. In default of payment transgressors are con-
demned to labour.

Before we sailed, a more serious theft was committed on the stores of the ship, which had been
placed under a shed, and likewise on the wearing
apparel of one of the officers who was ill on shore.
Immediately the aavas (policemen) heard of it, they
were on the alert, and arrested two men, on whom
suspicion fell, from their having slept in the place
the night of the robbery, and absconded early in the
morning. The news of the offence spread with its
accustomed rapidity among uncivilized tribes; and
various were the reports in circulation, as to the

manner in which I intended to visit the misde-
meanour. The prisoners at first acknowledged their
guilt, but afterwards denied it; and declared they
had been induced to make the confession from the
threats of the aavas who apprehended them. Nothing was found upon them, and no person could be
brought forward as a direct witness of the fact; so
that their guilt rested on circumstantial evidence
alone. I was, however, anxious to bring the offenders to trial, as all the sails and the stores of the
ship were on shore, and at the mercy of the inhabitants; and unless severe measures were pursued in
this instance, successive depredations would in all
probability have occurred. The chiefs were in consequence summoned, and at an early date the prisoners were brought to trial opposite the anchorage.
As it was an extraordinary case, I was invited to the
tribunal, and paid the compliment of being allowed
to interrogate the prisoners; but nothing conclusive
was elicited, though the circumstantial proof was so
much against them that five out of six of the chiefs
pronounced them guilty. The penalty in the event
of conviction in a case of this nature is, that the cul-
prit shall pay fourfold the value of the property
stolen: in this instance, however, as the articles could
not be replaced, and the value was far beyond what
the individuals could pay, I proposed, as the chiefs
referred the matter to me, that, by way of an
example, and to deter others from similar acts, the
prisoners should suffer corporal punishment. Their
laws, however, did not admit of this mode of punishment, and the matter concluded by the chiefs making
themselves responsible for the stores, and directing
Pa-why to acquaint the people that they had done
so, promising to make further inquiry into the mat-

ter; which was never done, and the prisoners escaped: but the investigation answered our purpose
equally well, as the stores afterwards remained un-
touched. The various reports which preceded the
trial, the assembling of the chiefs, and other circumstances, had brought together a great concourse of
people. Pa-why, raising himself above the multitude, harangued them in a very energetic and apparently elegant manner, much to the satisfaction of
the inhabitants, who all dispersed and went quietly
to their homes. The consideration which the chiefs
gave to the merits of this question, and the pains
they took to elicit the truth, reflect much credit
upon them. The case was a difficult one, and Hetotte, not being able to make up his mind to the
guilt of the prisoners, very honestly differed from
his colleagues; and his conduct, while it afforded a
gratifying instance of the integrity of the man,
showed a proper consideration for the prisoners,
which in the darker ages would have been sacrificed
to the interested motive of coinciding in opinion
with the majority. If we compare the fate which
would have befallen the prisoners, supposing them
innocent, had they been arraigned under the early
form of government, with the transactions of this
day, we cannot but congratulate the people on the
introduction of the present penal code, and acknow-
ledge that it is one of the greatest temporal blessings
they have derived from the introduction of Christianity. At the same time it is just to observe, that
had a similar depredation been committed under
those circumstances, there is every reason to believe
from former experience, that the real offender would
have been detected, and the property restored.

On the 3d April the young king landed at Otaheite from Eimeo, and was received with the most enthusiastic shouts of his subjects, who were assembled in great numbers on the beach to welcome his arrival. The following day he paid a visit to the ship, attended by the queen, a numerous retinue, and Mr. Pritchard, the principal missionary upon the island. I saluted the king on the occasion with nine guns, much to the delight of his subjects; and presented him with the fowling-piece which was sent out by the government for that purpose. The stock was inlaid with silver, and the case handsomely lined, and fitted up in a manner which made a deep impression on the minds of the Otaheitans, who are extremely fond of display, and who expressed their approbation by repeated exclamations of " My-tie! mia my-tie Pretannee!" as each article was exhibited. The king was a well-behaved boy, of slender make, uniting with the rudiments of an European education much native shrewdness; and the chiefs were considerably interested in him, as they considered his education would give him advantages over his predecessors; and his succession to the throne would remove the reins of government from the hands of the present possessor, whose measures were not always the most disinterested or beneficial to her country; and who, in consequence of her influence with the Boo-ratiras, the most powerful body of men upon the island, often carried her plans into execution in spite of the wishes of the chiefs to the contrary. But the object of their hopes unfortunately died the following year, and the sceptre passed to the hands of Aimatta, his sister, of whom the missionaries speak well.

Before we sat down to dinner, I was amused at

Jim, the interpreter, bringing me the queen dowager's compliments, and "she would be much obliged
by a little rum," to qualify a repast she had been
making on raw fish, by way, I suppose, of provoking
an appetite for dinner. We had missed her majesty
a few minutes before from the cabin, and on looking
over the stern of the ship, saw her seated in a native
boat finishing her crude repast.

A few days previous to this visit Lieutenant
Belcher was despatched in the barge to Mirapaye,
in the district of Papara, to bring round a quantity
of beef which had been prepared there for the ship's
use by Mr. Henry, the son of one of the early
missionaries. In this district there is a lake and a
morai, of which it will be proper to give a short
notice, as the former is considered curious, and
foreigners are often led, by the exaggerated account
of the natives, to visit the place, which really does
not repay the trouble it involves. To convey some
idea of the difficulty of reaching this lake, Lieutenant Belcher and Mr. Collie, who accompanied
him, crossed a stream which ran through the valley
leading to it twenty-nine times in their ascent,
sometimes at a depth considerably above their knees;
and after it was passed it was necessary to climb the
mountain upon hands and knees, and to maintain
their position by grasping the shrubs in their way,
which indeed were, for the most part, weak and
treacherous, consisting principally of the *musa sapientum, spondias dulcis*, and some ferns.

"In this manner," says Mr. Collie, "after tracing a zigzag and irregular course, and losing our
way once or twice, we reached the highest part of
the acclivity; and then descending a short distance,

the puny lake burst upon our disappointed view."
Its dimensions were estimated at three quarters of
a mile in circumference; and it was stated by the
natives to be fourteen fathoms deep. The water of
the lake was muddy, and appeared to receive its
supplies from several small streams from the mountains, and the condensation of the vapour around,
which fell in a succession of drops, and, bounding
off the projecting parts of the cliff, formed here and
there thin and airy cascades. Though there is a
constant accession of water, there has not yet been
found any outlet to the lake; and what renders it
still more curious is, that when heavy rains descend,
the water, instead of rising and overflowing its margin, is carried off by some subterraneous channel.
The natives say, when these rains occur there is a
great rush of water from a large cavern beneath the
bed of the lake. The temperature of the lake at
seven A. M. was 72°, and that of the atmosphere 71°.
During a shower of rain it rose to 74°: a thermometer at the level of the sea at the same time stood
at 77°. One side of the lake was bounded by lofty
perpendicular precipices, the other by a gentle slope
covered with the varied verdure of trees, shrubs,
and ferns, with a few herbaceous plants. The general appearance of the country suggested the idea of
an enormous avalanche, which stopped up the valley,
and intercepted the streams that heretofore found
their way along its bed to the sea.

The lake was estimated at 1500 feet above the
level of the sea, and the cliffs from which this avalanche appeared to have been precipitated were considered to be eight hundred feet more. Though at
so great a height, and so far from any large tract of

land, this extraordinary basin is said to abound in fresh-water eels of an enormous size. On the margin of the basin, blocks of columnar basalt, with porous and vesicular lava, were heaped in great confusion.

On the eastern side, Mr. Belcher found great quantities of vesicular shaggy lava, which led him to suppose a volcano had existed in the vicinity; and he remarks that many persons who have visited the lake were of opinion that it was a crater filled with water. In other parts he collected some very perfect crystals of basaltic hornblend, and found one or two of olivine on the surface of the vesicular lava. The lake appeared to be falling rapidly when they saw it; at a place where Mr. Belcher was obliged to cross it there were eighteen inches of water; some time after, at sunset, there were only six inches; and the next morning the rock was dry. On examining this place he noticed a large chasm beneath a rock, through which it appeared the water had found an outlet; and favoured the opinion of the basin being caused by an avalanche.

The morai is the same as that exhibited in the voyage of Mr. Wilson and mentioned by Captain Cook. Its measurements have been given in those voyages, and perhaps more correctly than the present dilapidated state of the edifice admits. But its history is interesting, as it was told by a descendant of the chief who erected it, and whose family, as well as himself, were priests of the god to whom it was consecrated. It differs in several respects from the account given by Mr. Ellis; but I insert it as related to Mr. Belcher by the chief.

The great-grandfather of Taati, the present chief,

whose name holds a conspicuous place in the wars of
Pomarree, was defeated in a pitched battle by the
king. The chief, incensed at the god under whose protection he fought, went to Ulietea, and by devotion,
presents, and promises, induced the god of that place,
Oroo, to accompany him to Otaheite. On his return, the new and, as it was supposed, powerful god,
so inspired the refugee party with courage, that they
again rallied around their chief, and so forcibly did
the superstition of those dark ages operate, that the
king, before victorious, was now repeatedly beaten
and driven to the opposite side of the island. The
chief, having secured tranquillity to his district, began to construct the morai above alluded to, which
was of such magnitude as to require two years for
its completion. It was then dedicated to the god
whose presence had achieved for him such repeated
victories.

The change effected in the circumstances of the
chief of Papara by the introduction of this new god,
acquired for the deity a reputation beyond any thing
that had been known in Otaheite ; and the king determined to obtain possession of it. By bribing the
priests, he was allowed to pay his devotions to the deity, and afterwards to fight under its auspices, which
he did so successfully that he ultimately obtained
possession of the idol. A morai was then built for it
in the valley of Atehuru, situated between Mirapaye
and Papiete ; memorable as the place where the last
battle was fought which decided the cause between
Christianity and paganism, and crowned with success
the labours of the missionaries, who for eighteen
years had been unremitting in their endeavours to
accomplish this great end ; this valley is also cele-

brated in consequence of a strong-hold on an eminence near it, where the old men and women used
to retire in all cases of attack upon the district. In
this last and important battle Taati's brother lost his
life, supporting to the last the cause of idolatry.
Taati himself had been converted to the new faith,
and was joined with Pomarree in opposition to his
relation.

While we were at anchor, a whale-ship arrived
from New Zealand, with a party of natives of that
country on board, whom the master permitted to
exhibit their war-dance for our diversion. After
the duty of the day was over, the party assembled
in front of the consul's house, and the Otaheitans,
anxious for an opportunity of comparing the dances
of other countries with their own, crowded round in
great numbers to witness the performance.

The exhibition took place by torch-light, and
began by the party being drawn up in a line with
their chief in advance, who regulated their motions ;
which, though very numerous, were all simultaneous,
and showed that they were well practised in them.
They began by stamping their feet upon the ground,
and then striking the palms of the hands upon the
thighs for about a minute, after which, they threw
their bodies into a variety of contortions, twisted their
heads about, grinned hideously, and made use of all
kinds of imprecations and abuse on their supposed
enemy, as if to defy him to battle : having at length
worked themselves into a complete frenzy, they ut-
tered a yell, and rushed to the conflict ; which, from
what we saw represented, must in reality be horrible ;
the effect upon the peaceable Otaheitans was such
that long before they came to the charge some of them

ran away through fear, and all, no doubt, congratu-
lated themselves that there was so wide an expanse
of water between their country and New Zealand.
A dirge over the fallen enemy concluded the performance, which it is impossible adequately to describe. We learned from the whaler, that Shonghi,
the New Zealand chief who was educated in England, was availing himself of the superiority he had
acquired, and was making terrible ravages among
his countrymen, whose heads, when dried, furnished
him with a lucrative trade.

On the 24th we prepared for our departure : during our stay we visited the natives almost daily in
their habitations, and became well acquainted with
their habits and manner of living ; but in this intercourse there was so little novelty, that, considering
how many volumes have been written upon the
country, by persons whose stay far exceeded ours, it
would be both tiresome and useless to detain the
reader with their description. The conclusion generally arrived at was, that the people retain much of
their original character and many of their habits, and
appear to have been particularly described by Turnbull ; but if early historians err not, they have lost
much of their cheerfulness, and the women a great
deal of their beauty.

Considering the advances the country had made
toward the formation of a government by the election of a parliament, and by the promulgation of
laws, we certainly expected to find something in
progress to meet approaching events, yet in none of
our excursions did we see any manufactures beyond
those which were in use when the island was first
discovered, but on the contrary, it was evident that

they had neglected many which then existed. We were sorry to find that none of those in operation could be materially useful to the state; that there were no dawnings of art, nor did there appear to be any desire on the part of the people to improve their condition; but so far from it, we noticed a feeling of composure and indifference which will be the bane of their future prosperity.

The island is nevertheless imperceptibly entering into notice: it is advantageously situated for various purposes of commerce, and, consequently, in the event of a war between England and other powers, it might be subjected to many annoyances from the most insignificant force—from any armed vessel indeed which might think it her duty to annoy the island on the ground of its reputed alliance with England. There are no works of defence to obviate such a possibility: the natives have not yet thought of the precaution, much less have they commenced any preparation, and the island throughout is in a perfectly defenceless condition. The weapons with which their battles were formerly fought are now in disuse, and the inhabitants have lost the skill necessary to employ them to advantage. A number of muskets distributed amongst the population creates an imaginary security, but the bad condition of the arms, and the want of powder, would render them unavailable. At all events they are deficient in an organised body of men; a species of defence which seems necessary for the security of every country that does not wish her shores to be invaded, or to have her internal tranquillity disturbed by feuds; which in Otaheite have frequently occurred, and are very likely to do so again, either from the differences of opinion in the affairs of the government, or from the jealousies between the chiefs and the great landholders, the *Boo Ratiras*. Their tranquillity besides may have hitherto depended upon their obscurity, or on the equally defenceless condition of their neighbours with themselves; but the extension of navigation has removed the one, and an advancement of civilization and of power has destroyed the balance of the other.

Religious books are distributed among the huts of such of the natives as are converted, or who are, as they term themselves, *missi-narees;* but many of the inhabitants are still tooti-ouris or bad characters, an old expression signifying literally rusty iron, and now indiscriminately used for a dissenter from the Christian religion and a low character. These persons are now of no religion, as they have renounced their former one, and have not embraced that which has been recently introduced.

Ignorance of the language prevented my obtaining any correct information as to the progress that had been made generally towards a knowledge of the Scriptures by those who were converted; but my impression was, and I find by the journals of the officers it was theirs also, that it was very limited, and but few understood the simplest parts of them. Many circumstances induced me to believe that they considered their religious books very much in the same light as they did their household gods; and in particular their conduct on the occasion of a disturbance which arose from some false reports at the time of the robbery on the stores of the ship, when they deposited these books in the mission, and declared themselves to be indifferent about their lives and property, so long as the sacred volume, which could be replaced at any time for a bamboo of oil, was in safety. In general those who were *missi-narees* had a proper respect for the book, but associating with it the suppression of their amusements, their dances, singing, and music, they read it with much less good will than if a system had been introduced which would have tempered religion with cheerfulness, and have instilled happiness into society.

The Otaheitans, passionately fond of recreation, require more relaxation than other people; and though it might not have been possible at once to clear the dances from the immoralities attending them, still it would have been good policy to sanction these diversions under certain restrictions, until laws which were more important began to sit easy on the shoulders of the people. Without amusements, and excessively indolent, they now seek enjoyment in idleness and sensuality, and too much pains cannot be bestowed to arouse them from their apathy, and to induce them to emerge from their general state of indifference to those occupations which are most essential to their welfare. Looking only to the past, they at present seem to consider that they can proceed in the same easy manner they have hitherto done; forgetting that their wants, formerly gratified by the natural produce of the earth, have lately been supplied by foreign commodities, which, by indulgence, have become essential to their comfort; and that as their wants increase, as in all probability they will, they will find themselves at a loss to meet the expenses of the purchase. They forget also that being dependent upon the casual arrival of merchant vessels for these supplies, they are liable to be deprived of them suddenly by the occurrence of a war or of some other contingency, and this at a period perhaps when by disuse they will not have the power of falling back upon those which have been discontinued.

The country is not deficient in productions adapted to commerce. The sugar-cane grows so luxuriantly that from two small enclosures five tons of white sugar are annually manufactured under the superintendence of an Englishman; cotton has been found to succeed very well; arrow-root of good quality is plentiful: they have some sandal-wood, and other ornamental woods suitable for furniture, and several dyes. Besides these, coffee and other grain might no doubt be grown, and they might salt down meat, which, with other articles I have not mentioned, would constitute a trade quite sufficient to procure for the inhabitants the luxuries which are in a gradual course of introduction, and to make it desirable for merchant vessels to touch at the island. It is not from the poverty of the island, therefore, from which they are likely to feel inconvenience, but from their neglect to avail themselves of its capabilities, and employ its productions to advantage.

It seemed as if the people never had these things revealed to them, or had sunk into an apathy, and were discouraged at finding each year burthened with new restrictions upon their liberties and enjoyments, and nothing in return to sweeten the cup of life. I cannot avoid repeating my conviction that had the advisers of Pomarree limited the penal code at first, and extended it as it became familiar to the people; had they restricted instead of suppressed

the amusements of the people, and taught them
such parts of the Christian religion as were intelligible to their simple understandings, and were
most conducive to their moral improvement and
domestic comfort, these zealous and really praise-
worthy men would have made greater advances
towards the attainment of their object.

If in offering these remarks it should be thought
I have been severe upon the failings of the people,
or upon the conduct of the missionary gentlemen, I
have only to say, that I have felt myself called upon
to declare the truth, which I trust has been done
without any invidious feeling to either; indeed, I
experienced nothing during my stay that could
create such a feeling, but very much to the contrary,
as both my officers and myself received every possible kindness from them. And if I have pour-
trayed their errors more minutely than their virtues,
it has been done with a view to show, that although
the condition of the people is much improved, they
are not yet blessed with that state of innocence and
domestic comfort of which we have read. It would
have been far more agreeable to have dwelt on the
fair side of their character only, but that has already
been done, and by following the same course I
should only have increased the general misconception.

At the time of our arrival, the rainy season, which
had been somewhat protracted, was scarcely over.
Its proper period is December, January, and February. So much wet weather in the height of
summer is always the occasion of fevers, and together with the abundance of vee-apples (*spondias
dulcis*), which ripen about that period, produce dy-

sentery and sickness among the poorer class of inhabitants, several of whom were labouring under these
and other complaints during our stay. Miserable
indeed was the condition of many of them. They
retired from their usual abode and the society of
their friends, and erected huts for themselves in the
woods, in which they dwelt, until death terminated
their sufferings. The missionaries and resident Europeans strove as much as was in their power to alle-
viate these distresses; but the natives were so im-
provident and careless that the medicine often did
them harm rather than good, and many preferred their
own simple pharmacopœia, and thus fell victims to
their ignorance. Our own ship's company improved
upon the abundance of fish and vegetable diet; but
from what afterwards occurred, I am disposed to
think the change from their former food, to so much
vegetable substance was very injurious. Regard to
this subject ought not to be overlooked in vessels
circumstanced as the Blossom was.

The winds during our stay were principally from
the eastward, freshening in the forenoon and mode-
rating toward sun-set to a calm, or giving place to a
light breeze off the land, which sometimes prevailed
through the night. This effect upon the trade-
wind, by comparatively so small a tract of coast,
shows the powerful influence of the land upon the
atmosphere.

In the height of summer, or during the rainy season, the winds fly round to the W. and N. W. and
blow in gales or hard squalls, which it is necessary
to guard against in anchoring upon the north-western coast, particularly at Matavai Bay, which is
quite open to those quarters. The mean tempera-

ture of the atmosphere during our stay was 79°, 98, the minimum 75°, and maximum 87°.

The many excellent ports in Otaheite have been enumerated by Captain Cook, though he only made use of one, Matavai Bay, and that which was most exposed; in consequence, probably, of the facility of putting to sea. Those on the north-western coast are the most frequented, as some difficulty of getting out and in attends most of the others, particularly those in the south-western side of the island, which are subject to a constant heavy swell from the higher latitudes, and in the long calms that prevail under the lee of the island, are apt to endanger vessels approaching the reefs. Of the four on the north-western coast, viz. Matavai, Papawa, Toanoa, and Papiété, the last is the most common anchorage, and were it not that it is subject to long calms and very hot weather, in consequence of being more to leeward than the others, it would certainly be the best. Toanoa is very small, but conveniently adapted to the refit of one or two ships. The best port however lies between this anchorage and Matavai, and is called Papawa; several ships may anchor there in perfect safety quite close to the shore, and if a wharf were constructed, might land their cargoes upon it without the assistance of boats. It may be entered either from the east or west, and it has the additional advantage of having Matavai Bay for a stopping place, should circumstances render it inconvenient to enter at the moment; but this channel which communicates with Matavai Bay must be approached with attention to two coral knolls that have escaped the notice of both Cook and Bligh. I have given directions for avoiding them in my nautical remarks.

The tides in all harbours formed by coral reefs are very irregular and uncertain, and are almost wholly dependant upon the sea-breeze. At Toanoa it is usually low water about six every morning, and high water half an hour after noon. To make this deviation from the ordinary course of nature intelligible, it will be better to consider the harbour as a basin, over the margin of which, after the breeze springs up, the sea beats with considerable violence, and throws a larger supply into it than the narrow channels can carry off in the same time, and consequently during that period the tide rises. As the wind abates the water subsides, and the nights being generally calm, the water finds its lowest level by the morning.

CHAPTER X.

Departure from Otaheite—Arrival at Woahoo, Sandwich Islands
—Contrast between the two Countries—Visit the King and
Pitt—Departure—Oneehow—Passage to Kamschatka—Petro-
paulski—Beering's Island—St. Lawrence Island—Esquimaux
—King's Island—Diomede Islands—Pass Beering's Strait—
Arrive in Kotzebue Sound—Anchor off Chamisso Island—Ice
Formation in Escholtz Bay.

ON the 26th of April, we left this delightful
island, in which we had passed many very pleasant
days, in the enjoyment of the society of the residents, and of the scenery of the country   We put
to sea in the morning, and about noon reached the
low Island of Tethoroa, the watering place of the
Otaheitans.   It is a small coral island, distant about
seven leagues from Otaheite; from the hills of
which it may be distinctly seen, and is abundantly
provided with cocoa-nut trees.   The salubrity of
this little island, which was formerly the resort of
the chiefs, arreoys, and others, for the purpose of
recruiting their health after their debaucheries, is
still proverbial at Otaheite.   Spare diet and fresh
air were the necessary consequences of a visit to this
place, and for a good constitution were the only

CHAP.
X.

April,
1826.

restoratives required ; and, as these seldom failed in their effects, it obtained a reputation in Otaheite, no less famous than that of the celebrated spring of eternal youth, which Ponce de Leon so long sought in vain. From the proximity of the islands of Tethoroa, Otaheite, and Eimeo, we were enabled to connect them trigonometrically. Upon the latter there is a peak with a hole through it, to which a curious history is attached, connected with the superstition of early times. It is asserted that the great god, Oroo, being one day angry with the Tii, or the little God of Eimeo, he threw his spear across the water at him, but the activity of the Tii evaded the blow, and the spear passed through the mountain, and left the hole which we saw. The height of this peak is 4,041 feet.

On the 27th, we were within six miles of the situation in which Arrowsmith has placed Rogge-wein's high Island of Recreation ; but nothing was in sight from the mast-head. In all probability this island, which answers so well in its description, ex-cepting as to its size, is the Maitea of Mr. Turnbull, situated nearly in the same latitude. From this time we endeavoured to get to the eastward, and to cross the equator in about 150° W. longitude, so that when we met the N. E. tradewind, we might be well to windward. There is, otherwise, some difficulty in rounding Owyhee, which should be done about forty miles to the eastward to ensure the breeze.

The passage between the Society and Sandwich groupes differs from a navigation between the same parallels in the Atlantic, in the former being ex-empt from long calms which sometimes prevail about the equator, and in the S. E. trade being more easterly. The westerly current is much the same in both; and if not attended to in the Pacific, will carry a ship so far to leeward, that by the time she reaches the parallel of the Sandwich Islands, she will be a long way to the westward, and have much difficulty in beating up to them.

Soon after leaving Otaheite, the officers and ship's company generally were afflicted with dysentery, which, at one time, assumed an alarming appearance. On the 3d of May, we had the misfortune to lose Mr. Crawley, one of the midshipmen, a young gentleman of very good abilities, and much regretted by all who knew him; and on the 6th, William Must, my steward, sunk under the same complaint: on the 7th, great apprehensions were entertained for Mr. Lay, the naturalist; but fortunately his complaint took a favourable turn, and he ultimately recovered. The disease, however, continued among us some time, threatening occasionally different portions of the ship's company.

As we approached the Sandwich Islands, our view was anxiously directed to the quarter in which Owyhee* was situated, in the hope of obtaining a sight of the celebrated Mouna Roa; but the weather was so unfavourable for this purpose, that the land at the foot of the hills was the only part of the coast which presented itself to our view On the 18th, we passed about thirty miles to windward of the eastern points of the island; and in the afternoon of the following day, as it was too late to fetch the

* More recently written Hawaii.

anchorage off Woahoo, we rounded to under the lee of Morotoi, the next island. The following morning we came to an anchor in nineteen fathoms outside the reefs of Honoruro, the principal port of the Sandwich Islands, and the residence of the king. This anchorage is very much exposed, and during the N. W. monsoon, unsafe; but as there is great difficulty attending a large ship going in and out of the harbour, it is the general stopping place of such vessels as make but a short stay at the island.

Our passage from Otaheite to this place had been so rapid, that the contrast between the two countries was particularly striking. At Woahoo, the eye searches in vain for the green and shady forests skirting the shore, which enliven the scene at Otaheite. The whole country has a parched and comparatively barren aspect; and it is not until the heights are gained, and the extensive ranges of taro plantations are seen filling every valley, that strangers learn why this island was distinguished by the name of the garden of the Sandwich Islands.

The difference between the appearance of the natives of Woahoo and Otaheite is not less conspicuous than that of the scenery. Constant exposure to the sun has given them a dark complexion and a coarseness of feature which do not exist in the Society Islands, and their countenances moreover have a wildness of expression which at first misleads the eye; but this very soon wears off, and I am not sure whether this manliness of character does not create a respect which the effeminacy of the Otaheitans never inspires.

As we rowed up the harbour, the forts, the cannon, and the ensign of the Tamahamaha, displayed upon the ramparts of a fort mounting forty guns, and at the gaff of a man-of-war brig, and of some other vessels, rendered the distinction between the two countries still more evident; and on landing, the marked attention to etiquette and the respect shown by the subjects to their chiefs offered a similar contrast. In every way this country seemed far to surpass the other in civilization—but there were strong indications of a close connexion between the natives of both.

It was not long since Lord Byron in the Blonde had quitted these islands; the appearance of a man-of-war was, therefore, no novelty; but the beach was thronged to excess with people of all distinctions, who behaved in a very orderly manner, helped us out of the boats, and made a passage as we advanced. In our way nothing more strikingly marked the superiority of this country over that we had recently quitted than the number of wooden houses, the regularity of the town laid out in squares, intersected by streets properly fenced in, and the many notices which appeared right and left, on pieces of board on which we read " an Ordinary at one o'clock, Billiards, the Britannia, the Jolly Tar, the Good Woman," &c. After a short walk we came to a neatly built wooden house with glass windows, the residence of Krimakoo, or, as he was commonly called, Pitt, whom I found extended upon the floor of his apartment, suffering under a dropsical complaint, under which he ultimately sunk. This disease had so increased upon him of late that he had undergone five operations for it since the departure of the Blonde. Though unable to rise from his bed, his mind was active and unimpaired; and when the conversation turned upon the affairs of the island he was quite energetic, regretting that his confinement prevented his looking more into them, and his greatest annoyance seemed to be his inability to see every thing executed himself. He expressed his attachment strongly to the British government, and his gratitude for the respect that had been shown the descendant of his illustrious patron, and his queen, by sending their bodies to the Sandwich Islands in so handsome a manner, and also for the footing upon which the affairs of the state had been placed by Lord Byron in command of the Blonde. He was anxious to requite these favours, and pressed his desire to be allowed to supply all the demands of the ship himself, in requital for the liberality with which his countrymen were treated in England. I could not accede to this effusion of the chief's gratitude, as the expense attending it would have been considerably felt by him, and more particularly as Lord Byron had previously declined the same offer.

The young king, who had been taking an airing, arrived at this moment, and repeated the sentiments of his protector; making at the same time many inquiries for his friends in the Blonde. Boki was absent at Owyhee attending his sister, who was dangerously ill. Madam Boki, Kuanoa, Manuia, and the other chiefs who were of the party in England, were all anxious to show us civility; and spoke of England in such high terms, that they will apparently never forget the kind treatment they experienced there: but they had a great dread of the diseases of our country, and many of them considered it very unhealthy. My impression was, that those who had already been there had had their curiosity satisfied to feel in no way disposed to risk another visit. The want of their favourite dish Poe was, besides, so serious an inconvenience that when allusion was made to England, this privation was always mentioned.

Our reception was friendly in the extreme; all our wants, as far as possible, were supplied, but unfortunately there was this year a scarcity of almost every kind of production; the protracted rainy season and other circumstances having conspired to destroy or lessen the crops, and the whole population was in consequence suffering from its effects. There was also a scarcity of dried provision, our visit having preceded the usual time of the arrival of the whalers, who discharge all they can spare at this place previous to their return home. Our expectation of replenishing the ship's provisions was consequently disappointed, and it therefore became necessary to reduce the daily allowance of the ship's company, and to pursue the same economical system here, with regard to taro and yams, as was done at Otaheite with the bread-fruit.

The few days I had to remain here were devoted to astronomical and other observations, and I had but little opportunity of judging of the state of the island; but from a letter which I received from Boki, it was evident that he did not approve of the system of religious restraint that had been forced into operation, which was alike obnoxious to the foreigners residing upon the island and to the natives.

At the time of our departure the health of Mr. Lay was by no means restored, and as it appeared to me that his time during the absence of the ship could

be more profitably employed among the islands of
the Sandwich groupe than on the frozen shores of
the north, he was left behind, under the protection
of Pitt, whose kindness on the occasion nothing
could exceed. Mr. Collie took upon himself the
charge of naturalist, and acquitted himself in a
highly creditable manner.

On the 31st of May we took our leave of Woahoo,
and proceeded to Oneehow, the westernmost island
of the Sandwich groupe, famous for its yams, fruit,
and mats. This island is the property of the king,
and it is necessary, previous to proceeding thither, to
make a bargain with the authorities at Woahoo for
what may be required, who in that case send an
agent to see the agreement strictly fulfilled. On the

1st of June we hauled into a small sandy bay on the
western side of the island, the same in which Vancouver anchored when he was there on a visit of a
similar nature to our own ; and I am sorry to say
that like him we were disappointed in the expected
supplies ; not from their scarcity, but in consequence
of the indolence of the natives.

Oneehow is comparatively low, and, with the exception of the fruit trees, which are carefully cultivated, it is destitute of wood. The soil is too dry to
produce taro, but on that account it is well adapted
to the growth of yams, &c. which are very excellent
and of an enormous size. There is but one place in
this bay where the boat of a man of war can effect a
landing with safety when the sea sets into the bay,
which is a very common occurrence ; this is on its
northern shore, behind a small reef of rocks that lies
a little way off the beach, and even here it is necessary to guard against sunken rocks ; off the western

point these breakers extend a mile and a half.   The
soundings in the bay are regular, upon a sandy
bottom, and there is good anchorage, if required,
with the wind from the eastward ; but it would not
be advisable to bring up under any other circumstances. The natives are a darker race of people
than those of Woahoo, and reminded us strongly of
the inhabitants of Bow Island.   With the exception
of the house of the Earee, all the huts were small,
low, and hot ; the one which we occupied was so
close that we were obliged to make a hole in its side
to admit the sea breeze.

We took on board as many yams as the natives
could collect before sun-set, and then shaped our
course for Kamschatka. In doing this I deviated
from the tracks of both Cook and Clerke, which I
think was the occasion of our passage being shorter
than either of theirs. Instead of running to the
westward in a low latitude, we passed to the eastward of Bird Island, and gained the latitude of 27° N.
In this parallel we found the trade much fresher,
though more variable, and more subject to interruption, than within the tropic ; we had also the advantage of a more temperate climate, of which we stood
in need, as the sickness among the ship's company
was so far from being removed, that on the 13th we
had the misfortune to lose one of the marines. On
this day we spoke the Tuscan, an English whale-
ship, and found that on quitting the Sandwich
Islands her crew had suffered in the same way as our
own, but had since quite recovered. In all probability the sudden change of diet from the usual
seafare to so much vegetable food, added to the heat
and humidity of the atmosphere at the season in

which our visits were made to those islands, was the cause of the sickness of both vessels. The master of the Tuscan informed me that the preceding year his ship's company had been so severely afflicted with disease that he found it necessary to put into Loo Choo, where he was well received, and his people were treated with the greatest kindness. He was supplied with fresh meat and vegetables daily, without being allowed to make any other payment than that of a chart of the world, which was the only thing the natives would accept. It was, however, not without the usual observance of narrow-minded Chinese policy, that himself and his invalid crew were allowed to set their feet on shore, and even then they were always attended by a party of the natives, and had a piece of ground bordering on the beach fenced off for them. The salute which the Alceste and Lyra had fired on the 25th of October was well remembered by these people, and they had an idea that it was an annual ceremony performed in commemoration of something connected with the king of England. On the return of this day, during the Tuscan's visit, they concluded that the ship would observe the same ceremony, and looked forward with such anxiety and delight to the event, that the master of the whaler was obliged to rub up his four patereros, and go through the salute without any intermission, as the Loochooans counted the guns as they were fired.

A few hours after we parted with the Tuscan, we fell in with two other whale ships, neither of which could spare us any provisions. These ships were no doubt fishing down a parallel of latitude, which is a common custom, unless they find a continued scarcity of whales. The 30th degree, I believe, is rather a favourite one with them.

Ten days after our departure from Oneehow we lost the trade wind in latitude 30° N. and longitude 195° W.; it had been variable before this, but had not fairly deserted us: its failure was of little consequence, as in three days afterwards we were far enough to the westward to ensure the remainder of the passage; and indeed from the winds which ensued, a course might as well have been shaped for Kamschatka on the day we lost the wind.

On the 3d of June, the day after leaving Oneehow, in latitude 25° N. and longitude 163° 15′ W., we saw large flocks of tern and noddies, and a few gannets and tropic birds, also boneta, and shoals of flying-fish; and on the 5th, in latitude 28° 10′ N. and longitude 172° 20 W., we had similar indications of the proximity of land. Though such appearances are by no means infallible, yet as so many coral islands have recently been discovered to the W. N. W. of the Sandwich Islands, ships in passing these places should not be regardless of them. On this day we observed an albatross (*diomedia exulans*), the first we had seen since quitting the coast of Chili. It is remarkable that Captain King in his passage to Kamschatka first met these birds within thirty miles of the same spot. We noticed about this time a change in the colour of the wings of the flying-fish, and on one of the species being caught it was found to differ from the common *exocætus volitans*. We continued to see these fish occasionally as far as 30° N., about which time the tern also quitted us. In 33° N. we first met the birds of the northern regions, the *procellaria puffinus*, but it was not until we were within a hundred miles of the coast of Kamschatka that we saw the lumme, dovekie, rotge, and other alca, and the shag. The tropic birds accompanied us as far as 36° N.

On the 18th and 19th, in latitude 35° N., longitude 194° 30′ W., we made some experiments on the temperature of the sea at intermediate depths, as low as 760 fathoms, where it was found to be twenty-eight degrees colder than at the surface; two days afterwards another series was obtained, by which it appeared that the temperature at 180 fathoms was as cold as that at 500 fathoms on the former occasion, and it was twenty degrees colder at 380 fathoms on this, than it was at 760 fathoms on the other. Between these experiments we entered a thick fog, which continued until we were close off the Kamschatka coast; and we also experienced a change of current, both of which no doubt contributed towards the change of temperature of the sea, which was much greater than could have been produced by the alteration in the situation of the ship: the fog by obstructing the radiation of heat, and the current by bringing a colder medium from higher latitudes. About this period we began to see drift wood, some of which passed us almost daily The sea was occasionally strewed with moluscous animals, principally beroes and nereis, among which on the 19th were a great many small crabs of a curious species. Whether it was that these animals preferred the foggy weather, or that we more narrowly scrutinized the small space of water around us to which our view was limited, I cannot say, but it appeared to us that they were much more numerous while the fog lasted than before or afterwards.

In the afternoon of the 23d, in latitude 44° N., the wind, which had been at S. W., drew round to the west, and brought a cold atmosphere in which the thermometer fell fourteen degrees; it is remarkable that sixteen hours before this change occurred, the temperature of the sea fell six degrees, while that of the atmosphere was affected only four hours previous. In my remarks on our passage round Cape Horn, I have mentioned the frequency with which the temperature of the surface of the sea was affected before that of the atmosphere when material changes of wind were about to occur.

On the 26th, in latitude 49° N., after having traversed nearly seven hundred miles in so thick a fog that we could scarcely see fifty yards from us, a north-east wind cleared the horizon for a few hours: this change again produced a sensible diminution of the temperature, which was thirty-one degrees lower than it had been thirteen days previous. The next day we had the satisfaction of seeing the high mountains of Kamschatka, which at a distance are the best guides to the port of Awatska. The eastern mountain, situated twenty-five miles from Petropaulski, is 7·375 feet high by my trigonometrical measurement; another, which is the highest, situated N. 5° E. from the same place, and a little to the northward of a short range upon which there is a volcano in constant action, is 11·500 feet high. At eight o'clock we distinguished Cape Gavarea, the southern point of a deep bay in which the harbour of Petropaulski is situated, and the same evening we were becalmed within six miles of our port. Nothing could surpass the serenity of the evening or the magnificence of the mountains capped with perennial

snows, rising in majestic array above each other.
The volcano emitted smoke at intervals, and from a
sprinkling of black dots on the snow to leeward of
the crater we concluded there had been a recent
eruption.

At two o'clock the following afternoon we anchored off the town of Petropaulski, and found lying in
the inner harbour his imperial majesty's ship Mo-
deste, commanded by Baron Wrangel, an enterprising officer, well known to the world as the commander of a hazardous expedition on sledges over the
ice to the northward of Schelatskoi Noss.

I found despatches awaiting my arrival, communicating the return of the expedition under Captain
Parry, and desiring me to cancel that part of my instructions which related to him. The officers on
landing, at the little town of Petropaulski, met with
a very polite reception from the governor, Stanitski,
a captain in the Russian navy, who, during our short
stay in port, laid us under many obligations for
articles of the most acceptable kind to seamen after
a long voyage. I regretted extremely that confinement to my cabin at this time prevented my having
the pleasure of making either his acquaintance or
that of the pastor of Paratounka, of whose ancestor
such honourable mention has been made in the voyages of Captain Cook, a pleasure which was reserved
for the following year. The worthy pastor, in strict
compliance with the injunctions of his grandfather,
that he should send a calf to the captain of every
English man of war that might arrive in the port,
presented me with one of his own rearing, and sent
daily supplies of milk, butter, and curds. Had our
stay in this excellent harbour permitted, we should

have received a supply of oxen, which would have
been most acceptable to the ship's company ; but
the animals had to be driven from Bolcheresk, and,
pressed as we were for time, too great a delay would
have been incurred in waiting for them. The colony at this time was as much distressed for provisions as ourselves, and was even worse off, in consequence of the inferior quality of the articles.

On the 1st of July we weighed and attempted to
put to sea, but after experiencing the difficulties
of which several navigators have complained, were
obliged to anchor again, and that at too great a distance from the town to have any communication.

On the 2nd, as well as on the 3rd, we also weighed,
but were obliged to anchor as before ; and it was
not until the 5th, after weighing and anchoring
twice that morning, to prevent going ashore, that
the ship reached the outside of the harbour ; this
difficulty arises from counter currents which prevent
the steerage of the ship. After clearing the harbour
there was a strong wind against us, but it soon died
entirely away, and left us exposed to a heavy swell,
which rolled with great violence upon the shore ; so
much so, that for some time the boats were insuffi-
cient to prevent the ship nearing the land, and there
was no anchorage, in consequence of the great depth
of water : fortunately towards night a light air favoured our departure, and we succeeded in getting
an offing.

My object was now to make the best of my way
to Kotzebue Sound, as there were but fourteen days
left before the arrival of the appointed time of rendezvous there, and every effort was directed towards
that end. As we sailed across the wide bay in which

Petropaulski is situated, we connected the capes at its extremities with the port and intermediate objects, by which it appears that Cape Gavarea has hitherto been erroneously placed with regard to Chepoonski Noss; but I shall not here interrupt the narrative by the insertion of the particulars of the operations.

At day-light the following morning Chepoonski Noss was seen N. 19° W and in the afternoon of the next day high land was discerned from the masthead in the direction of Krotnoi Mountain. This was the last view we had of Kamschatka, as a thick fog came on, and attended us to Beering's Island.

At day-light on the 10th a high rock was seen about nine miles off, and shortly afterwards Beering's Island appeared through the fog. When we had reached close in with the land the mist partially dispersed, and exposed to our view a moderately high island armed with rocky points. The snow rested in ridges upon the hills, but the lower parts of the island were quite bare, and presented a green mossy appearance, without a single shrub to relieve its monotony. Its dreary aspect, associated with the recollection of the catastrophe that befel Beering and his shipmates, who were cast upon its shores on the approach of winter with their own resources exhausted, produced an involuntary shudder. The bay in which this catastrophe occurred is on the north side of the island, on a part of the coast which fortunately afforded fresh water, and abounded in stone foxes, sea otters, and moor-hens; and where there was a quantity of drift wood washed upon the shore, which served for the construction of huts; but notwithstanding these resources, the commander

Vitus Beering, and twenty-nine of the crew, found their graves on this desolate spot.   The island is now visited occasionally by the Russians for the skins of the sea otter and black fox.   The highest part of the island which we saw was towards its N. W. extremity, from whence the shore slopes gradually to the coast, and is terminated by cliffs.   At the foot of these there are low rocky flats, which can only be seen when quite close to them, and outside again are breakers.   Off the western point these reefs extend about two miles from the shore, and off the northern, about a mile and a quarter, so that on the whole it is a dangerous coast to approach in thick weather. The rock first seen was situated five miles and a half off shore, and was so crowded with seals basking upon it, that it was immediately named Seal Rock.*

To the northward of this there were several small bays in the coast, which promised tolerably good anchorage, particularly one towards the eastern part of the indentation in the coast line, off which there was a small low island or projecting point of land.   This, in all probability, is the harbour alluded to by Krenitzen, as there were near it " two small hillocks like boats with their keels upwards."

We did not see the south-eastern part of this island, as it was obscured by fog, but sailed along the southern and western shores as near as circumstances permitted until seven in the evening, when we got out of the region of clear weather, which

* Kotzebue observes in his narrative that " this rock has not been laid down in any chart:" I presume he alludes to those which are modern, as on a reference to the map of Captain Krenitzen's discoveries in 1768, it will be found occupying its proper place.

usually obtains in the vicinity or to leeward of land
in these seas, and entered a thick fog. With the
summer characteristics of this latitude—fine weather and a thick fog—we advanced to the northward, attended by a great many birds, nearly all of
the same kind as those which inhabit the Greenland
Sea, sheerwaters, lummes, puffins, parasitic gulls,
stormy petrel, dusky albatross, a larus resembling
the kittiwake, a small dove-coloured tern, and shags.
In latitude 60° 47′ N. we noticed a change in the
colour of the water, and on sounding found fifty-
four fathoms, soft blue clay. From that time until
we took our final departure from this sea the bottom
was always within reach of our common lines. The
water shoaled so gradually that at midnight on the
16th, after having run a hundred and fifty miles, we
had thirty-one fathoms. Here the ground changed
from mud to sand, and apprized us of our approach
to the Island of St. Lawrence, which on the following morning was so close to us that we could hear
the surf upon the rocks. The fog was at the same
time so thick that we could not see the shore; and
it was not until some time afterwards, when we had
neared the land by means of a long ground swell,
for it was quite calm, that we discerned the tops of
the hills.

It is a fortunate circumstance that the dangers in
these seas are not numerous, otherwise the prevalence of fogs in the summer time would render the
navigation extremely hazardous. About noon we
were enabled to see some little distance around us;
and, as we expected, the ship was close off the
western extremity of St. Lawrence Island. In this
situation the nearest hills, which were about five

hundred feet above the sea, were observed to be
surmounted by large fragments of rock having the
appearance of ruins. These hills terminate to the
southward and south-westward in bold rocky cliffs,
off which are situated three small islands; the hills
have a gradual slope to the coast line to the northward and westward; but at the north-western extremity of the island they end in a remarkable
wedge-shaped promontory — particulars which may
be found useful to navigators in foggy weather.
The upper parts of the island were buried in snow;
but the lower, as at Beering's Island, were bare and
overgrown with moss or grass. We stood close into
a small bay at the S. W angle of the island, where
we perceived several tents, and where, from the
many stakes driven into the ground, we concluded
there was a fishing-station. The natives soon afterwards launched four baidars,* of which each contained eight persons, males and females. They paddled towards the ship with great quickness, until
they were within speaking distance, when an old
man who steered the foremost boat stood erect and
held up in succession nets, walrus teeth, skin shirts,
harpoons, bows and arrows, and small birds; he then
extended his arms, rubbed and patted his breast,†
and came fearlessly alongside. We instantly detected in these people the features of the Esquimaux, whom in appearance and manners also, and
indeed in every particular, they so much resembled,
that there cannot, I think, be the least doubt of

* This boat, called by the natives oomiac, is the same in every
respect as the oomiac, or woman boat of the Esquimaux. It is
here used by the men instead of by the women.

† This is the usual Esquimaux indication of friendship.

their having the same origin. They were if any thing less dirty, and somewhat fairer, and their implements were better made. Their dress, though Esquimaux, differed a little from it in the skin shirts being ornamented with tassels, after the manner of the Oonalashka people, and in the boots fitting the leg, instead of being adapted to the reception of either oil or infants.

The old men had a few gray hairs on their chins, but the young ones, though grown up, were beardless. Many had their heads shaved round the crown, after the fashion of the Tschutschi, the Otaheitans, or the Roman Catholic priesthood in Europe, and all had their hair cut short. Their manner of salutation was by rubbing their noses against ours, and drawing the palms of their hands over our faces; but we were not favoured as Kotzebue was, by their being previously spit upon. In the stern of one of the baidars there was a very entertaining old lady, who amused us by the manner in which she tried to impose upon our credulity. She was seated upon a bag of peltry, from which she now and then cautiously drew out a skin, and exhibited the best part of it, with a look implying that it was of great value; she repeatedly hugged it, and endeavoured to coax her new acquaintances into a good bargain, but her furs were scarcely worth purchasing. She was tattooed in curved lines along the sides of the cheek, the outer one extending from the lower jaw, over the temple and eyebrow.

Our visiters on board were not less accomplished adepts at bartering than the old woman, and sold almost every thing they had. With the men, " tawac," as they called our tobacco, was their great object; and with the women, needles and scissors; but with both, blue beads were articles highly esteemed. We observed, that they put some of these to the test, by biting them to ascertain whether they were glass; having, perhaps, been served with wax ones by some of their former visiters.

Their implements were so similar to those of the Esquimaux as to need no description; except that their bows partook of the Tschutschi form. They had a great many small birds of the alca crestatella, strung upon thongs of hide, which were highly acceptable to us, as they were very palatable in a pie. These birds are, I believe, peculiar to St. Lawrence Island, and in proceeding up the strait their presence is a tolerably certain indication of the vicinity of the island. They are very numerous, and must be easily taken by the natives, as they sold seven dozen for a single necklace of blue glass beads.

About seven o'clock in the evening, the natives quitted us rather abruptly, and hastened toward the shore, in consequence of an approaching fog which their experience enabled them to foresee sooner than us, who, having a compass to rely upon, were less anxious about the matter. We soon lost sight of every distant object, and directed our course along the land, trying the depth of water occasionally. The bottom was tolerably even; but we decreased the soundings to nine fathoms, about four miles off the western point, and changed the ground from fine sand, to stones and shingle. When we had passed the wedge-shaped cliff at the north-western point of the island, the soundings again deepened, and changed to sand as at first.

At night the fog cleared away for a short time,

and we saw the Asiatic coast about Tschukutskoi Noss; but it soon returned, and with it a light air in the contrary direction to our course. The next day, as we could make no progress, the trawl was put overboard, in the hope of providing a fresh meal for the ship's company; but after remaining down a considerable time, it came up with only a sculpen (*cottus scorpius*), a few specimens of moluscæ, and crustaceæ, consisting principally of maias. In the evening, Lieutenant Peard was more successful in procuring specimens with the dredge, which supplied us with a great variety of invertebral animals, consisting of asterias, holothurias, echini, amphitrites, ascidias, actinias, euryales, murex, chiton crinitus, nereides, maias, gammarus, and pagurus, the latter inhabiting chiefly old shells of the murex genus. This was in seventeen fathoms over a muddy bottom, several leagues from the island.

About noon the fog dispersed, and we saw nearly the whole extent of the St. Lawrence Island, from the N. W. cape we had rounded the preceding night to the point near which Cook reached close in with, after his departure from Norton Sound. The middle of this island was so low, that to us it appeared to be divided, and I concluded, as both Cook and Clerke had done before, that it was so; circumstances did not, however, admit of my making this examination, and the connexion of the two islands was left for the discovery of Captain Schismareff of the Russian navy. The hills situated upon the eastern part of the island, to which Cook gave the name of his companion Captain Clerke, are the highest part of St. Lawrence Island, and were at this time deeply buried in snow.

The current off here, on one trial, ran **N. E.** five-eighths of a mile per hour, and on another, **N. 60° E.** seven-eighths of a mile per hour: as observations on this interesting subject were repeatedly made, they will be classed in a table in the Appendix.

Favoured with a fair wind, on the 19th we saw King's Island; which, though small, is high and rugged, and has low land at its base, with apparently breakers off its south extreme.

We had now advanced sufficiently far to the northward to carry on our operations at midnight; an advantage in the navigation of an unfrequented sea which often precludes the necessity of lying to.

We approached the strait which separates the two great continents of Asia and America, on one of those beautiful still nights, well known to all who have visited the arctic regions, when the sky is without a cloud, and when the midnight sun, scarcely his own diameter below the horizon, tinges with a bright hue all the northern circle. Our ship, propelled by an increasing breeze, glided rapidly along a smooth sea, startling from her path flocks of lummes and dovekies, and other aquatic birds, whose flight could, from the stillness of the scene, be traced by the ear to a considerable distance. Our rate of sailing, however, by no means kept pace with our. anxiety that the fog, which usually succeeds a fine day in high latitudes, should hold off until we had decided a geographical question of some importance, as connected with the memory of the immortal Cook. That excellent navigator, in his discoveries of these seas, placed three islands in the middle of the strait (the Diomede Islands). Kotzebue, however, in passing them, fancied he saw a fourth, and conjectured that it must have been either overlooked by Cook and Clerke, or that it had been since raised by an earthquake.*

As we proceeded, the land on the south side of St. Lawrence Bay made its appearance first, and next the lofty mountains at the back of Cape Prince of Wales, then hill after hill rose alternately on either bow, curiously refracted, and assuming all the varied forms which that phenomenon of the atmosphere is known to occasion. At last, at the distance of fifty miles, the Diomede Islands, and the eastern Cape of Asia, rose above the horizon of our mast-head. But, as if to teach us the necessity of patience in the sea we were about to navigate, before we had determined the question, a thick fog enveloped every thing in obscurity We continued to run on, assisted by a strong northerly current, until seven o'clock the next morning, when the western Diomede was seen through the fog close to us.

In our passage from St. Lawrence Island to this situation, the depth of the sea increased a little, until to the northward of King's Island, after which it began to decrease; but in the vicinity of the Diomede Islands, where the strait became narrowed, it again deepened, and continued between twenty-five and twenty-seven fathoms. The bottom, until close to the Diomedes, was composed of fine sand, but near them it changed to coarse stones and gravel, as at St. Lawrence Island; transitions which, by being attended to, may be of service to navigators in foggy weather.

* Some doubt, it appears, was created in the minds of the Russians themselves as to this supposed discovery, as we understood at Petropaulski, that a large wager was depending upon it.

During the day we saw a great number of whales, seals, and birds; but none, I believe, that are not mentioned in Pennant's Arctic Zoology.

We noticed upon the island abreast of us, which we conjectured to be the westernmost Diomede, several tents and yourts, and, also, two or three baidars, hauled upon the beach. On the declivity of the hill were several frames, apparently for drying fish and skins, and depositing canoes and sledges upon. It was nearly calm when we were off this place, but the current, which still ran to the northward, carried us fast along the land. I steered for the situation of the supposed additional island, until by our reckoning we ought to have been upon it, and then hauled over towards the American shore. In the evening the fog cleared away, and our curiosity was at last satisfied. The extremities of the two great continents were distinctly seen, and the islands in the strait clearly ascertained to be only three in number, and occupying nearly the same situations in which they were placed in the chart of Captain Cook.

The south-eastern of the three islands is a high square rock; the next, or middle one, is an island with perpendicular cliffs, and a flat surface; and the third, or north-western, which is the largest, is three miles long, high to the southward, and terminates, in the opposite direction, in low cliffs with small rocky points off them. East Cape in almost every direction is so like an island, that I have no doubt it was the occasion of the mistake which the Russian navigator has committed.

For the sake of convenience, I named each of these islands. The eastern one I called Fairway

Rock, as it is an excellent guide to the eastern channel, which is the widest and best; the centre one I
named after the Russian Admiral Krusenstern; and
to the north-western island I transferred the name
of Ratmanoff, which had been bestowed upon the
supposed discovery of Kotzebue. We remarked
that the Asiatic shore was more buried in snow
than the American. The mountains in the one
were entirely covered; in the other, they were
streaked and partly exposed. The low land of both
on the coast was nearly bare.

Near the Asiatic shore we had a sandy bottom,
but on crossing over the strait, it changed to mud,
until well over on the American side, where we
passed a tongue of sand and stones in twelve
fathoms, which in all probability was the extremity
of a shoal, on which the ship was nearly lost the
succeeding year. After crossing it the water deepened, and the bottom again changed to mud, and
we had ten and a half fathoms within two and a
half miles of the coast.

We closed with the American shore, a few miles
to the northward of Cape Prince of Wales, and
found the coast low, with a ridge of sand extending
along it, on which we noticed several Esquimaux
habitations. Steering along this shore to the northward, in ten and eight fathoms water, a little before
noon we were within four and a half miles of
Schismareff Inlet. Here we were becalmed, and
had leisure to observe the broad sheet of water that
extends inland in an E. S. E. direction beyond the
reach of the eye.* The width of the inlet between

---

* Mountains were seen at the back of it, but the coast was not
visible—probably it is low.

the two capes is ten miles; but Saritcheff Island lies immediately before the opening, and we are informed by Kotzebue, that the channel, which is on the northern side of it, is extremely intricate and narrow, and that the space is strewed with shoals. The island is low and sandy, and is apparently joined under water, to the southward, to the strip of sand before mentioned as extending along the coast: we noticed upon it a considerable village of yourts, the largest of any that had as yet been seen. The natives appear to prefer having their dwellings upon this sandy foundation to the main land, probably on account of the latter being swampy, which is the case every where in the vicinity of this inlet and Kotzebue Sound. Several of them taking advantage of the calm came off in baidars, similar to those used by the St. Lawrence Islanders, though of inferior workmanship. The people, however, differed from them in many respects; their complexion was darker, their features were more harsh and angular, they were deficient in the tattooing of the face; and what constituted a wider distinction between them was, a custom, which we afterwards found general on the American coast, of wearing ornaments in their under lips. Our visiters were noisy and energetic, but good-natured, laughed much, and humorously apprized us when we were making a good bargain.

They willingly sold every thing they had, except their bows and arrows, which they implied were required for the chase on shore; but they could not resist "tawac" (tobacco) and iron knives, and ultimately parted with them. These instruments differed from those of the islanders to the southward,

in being more slender, but they were made upon
the same principle, with drift pine assisted with
thongs of hide, and occasionally with pieces of
whalebone placed at the back of them neatly bound
round with small cord. Their arrows were tipped
with bone, flint, or iron, and they had spears or
lances headed with the same materials. Their dress
was the same as that worn by the whole tribe inhabiting the coast. It consisted of a shirt which
reached half way down the thigh, with long sleeves
and a hood to it, made generally of the skin of the
reindeer, and edged with the fur of the gray or
white fox, and sometimes with dog's skin. The
hood is usually edged with a longer fur than the
other parts, either of the wolf or dog. They have
besides this a jacket made of eider drakes' skins
sewed together, which put on underneath their
other dress is a tolerable protection against a distant
arrow, and is worn in times of hostility. In wet
weather they throw a shirt over their fur dress made
of the entrails of the whale, which, while in their
possession, is quite water tight, as it is then, in
common with the rest of their property, tolerably well supplied with oil and grease; but after
they had been purchased by us and became dry,
they broke into holes and let the water through.
They are on the whole as good as the best oil-skins
in England. Besides the shirt, they have breeches
and boots, the former made of deer's hide, the latter
of seal's skin, both of which have drawing strings at
the upper part made of sea-horse hide. To the end
of that which goes round the waist they attach a
tuft of hair, the wing of a bird, or sometimes a fox's
tail, which, dangling behind as they walk, gives

them a ridiculous appearance, and may probably have occasioned the report of the Tschutschi, recorded in Muller, that the people of this country have " tails like dogs."

It was at Schismareff Inlet that we first saw the lip ornaments which are common to all the inhabitants of the coast thence as far as Point Barrow. These ornaments consist of pieces of ivory, stone, or glass, formed with a double head, like a sleeve-button, one part of which is thrust through a hole bored in the under lip. Two of these holes are cut in a slanting direction about half an inch below the corners of the mouth. The incision is made when about the age of puberty, and is at first the size of a quill; as they grow older the natives enlarge the orifice, and increase the dimensions of the ornament accordingly, that it may hold its place : in adults, this orifice is about half an inch in diameter, and will, if required, distend to three quarters of an inch. Some of these ornaments were made of granite, others of jade-stone, and a few of large blue glass beads let into a piece of ivory which formed a white ring round them. These are about an inch in diameter, but I afterwards got one of finely polished jade that was three inches in length, by an inch and half in width.

About noon, a breeze springing up, the natives quitted us for the shore, and we pursued our course to the northward without waiting to explore further this deep inlet, which has since been a subject of regret, as the weather afterwards in both years prevented it being done. I could not, however, consistently with my instructions, wait to examine it at this moment, as the appointed time of rendezvous

at Chamisso Island was already past.* While becalmed off it, we were carried slowly to the north-
eastward by a current which had been running in
that direction from the time of our quitting St. Lawrence Island. With a fair wind we sailed along the
coast to the northward, which was low and swampy,
with small lakes inland. The ridge of sand continued along the coast to Cape Espenburgh, and
there terminated.

We entered Kotzebue Sound early in the morning
of the 22d of July, and plied against a contrary
wind, guided by the soundings; the appearance of
the land was so distorted by mirage, and in parts so
obscured by low fog, that it was impossible to distinguish where we were. The naturalist who accompanied Kotzebue in his voyage particularly remarks this state of the atmosphere in the vicinity of
the sound, and suggests that it may be occasioned by
the swampy nature of the country; in which opinion I fully concur. When it cleared off we were
much surprised to find ourselves opposite a deep inlet in the northern shore, which had escaped the observation of Captain Kotzebue. I named it Hotham
Inlet, in compliment to the Hon. Sir Henry Hotham, K.C.B., one of the lords of the Admiralty.
We stood in to explore it, but found the water too
shallow, and were obliged to anchor in four fathoms
to prevent being carried away by a strong tide which
was setting out of the sound, the wind being light
and contrary.

As it would be necessary to remain three or four
days at Chamisso Island to increase our stock of
water, previous to proceeding to the northward, the

* It has since been surveyed by the Russians.

Drawn by Wᵐ Smyth.

BAIDARS OF HOTHAM INLET.

barge was hoisted out and sent to examine the inlet, with directions to meet the ship at Chamisso Island. She was again placed under the command of Mr. Elson, and equipped in every way necessary for the service required.

We were visited by several baidars, containing from ten to thirteen men each, whose object was to obtain articles in exchange. They were in every respect similar to the natives of Schismareff Inlet, though rather better looking, and were all, without exception, provided with labrets, either made of ivory and blue beads, as before described, of ivory alone, or of different kinds of stone, as steatite, porphyry, or greenstone; they readily disengaged these from their lips, and sold them, without minding the inconvenience of the saliva that flowed through the badly cicatriced orifice over the chin; but on the contrary derided us when we betrayed disgust at the spectacle, by thrusting their tongues through the hole, and winking their eyes. One or two had small strings of beads suspended to their ears.

The articles they brought off were, as before, skins, fish, fishing implements, and nic-nacs. Their peltry consisted of the skins of the seal, of the common and arctic fox, the common and musk-rat, the marten, beaver, three varieties of ermine, one white, one with a light brown back and yellow belly, and the third with a gray back spotted white and yellow; the American otter, the white hare, the polar bear, the wolf, the deer, and the badger. Their fish were salmon and herrings: their implements, lances, either of stone or of a walrus tooth fixed to the end of a wooden staff; harpoons precisely similar to the Esquimaux; arrows; drills; and an instrument, the use of which was at first not very evident. It was part of a walrus tooth shaped something like a shoe-horn, with four holes at the small end communicating with a trough that extended along the middle of the instrument and widened as it neared the broad part. From the explanation given of it by the natives, it was evidently used to procure blood from dying animals, by inserting the end with the holes into the wound, and placing the mouth at the opposite end of the trough to receive the liquid as it flowed. From the satisfaction that was evinced by the describer during the explanation, it is evident that the blood of animals is as much esteemed by these people as by the eastern Esquimaux.* On the outside of this and other instruments there were etched a variety of figures of men, beasts, and birds, &c., with a truth and character which showed the art to be common among them. The reindeer were generally in herds: in one picture they were pursued by a man in a stooping posture in snow-shoes; in another he had approached nearer to his game, and was in the act of drawing his bow. A third represented the manner of taking seals with an inflated skin of the same animal as a decoy; it was placed upon the ice, and not far from it a man was lying upon his belly with a harpoon ready to strike the animal when it should make its appearance. Another was dragging a seal home upon a small sledge; and several baidars were employed harpooning whales which had been previously shot with arrows; and thus by comparing one device with another a little history was obtained which gave us a better insight

* See Captain Parry's Second Voyage, 4to., p. 510.

into their habits than could be elicited from any signs or intimations.

The natives also offered to us for sale various other articles of traffic, such as small wooden bowls and cases, and little ivory figures, some of which were not more than three inches in length, dressed in clothes which were made with seams and edgings precisely similar to those in use among the Esquimaux.

The staves of the harpoons and spears were made of pine or cypress, in all probability from drift wood, which is very abundant upon the shores; and yet the circumstance of their having lumps of the resin in small bags favoured the supposition that they had access to the living trees. They had also iron pyrites, plumbago, and red ochre, with which the frame of the baidar was coloured.

The people themselves, in their persons as well as in their manners and implements, possessed all the characteristic features of the Esquimaux; large fat round faces, high cheek bones, small hazel eyes, eyebrows slanting like the Chinese, and wide mouths. They had the same fashion with their hair as the natives of Schismareff Inlet, cutting it close round the crown of the head, and thereby leaving a bushy ring round the lower part of it. Ophthalmia was very general with them, and obliged some to wear either some kind of shade or spectacles, made of wood, with a wide slit for each eye to look through. At Schismareff Inlet diseases of this nature were, also, prevalent among those who visited us.

The salutation of our visiters was, as before, by a contact of noses, and by smoothing our faces with the palms of their hands, but without any disgusting practice.

When they had parted with all they had for sale, they quitted the ship, well pleased with their excursion, and having pushed off to a little distance, clapped their hands, extended their arms, and stroked their bodies repeatedly; which we afterwards found to be the usual demonstration of friendship among all their tribe. They then pointed to the shore, and with one consent struck the water with their paddles, and propelled their baidars with a velocity which we were not prepared to witness. These boats are similar in construction to the oomiaks of Hudson's Bay; but the model differs in being sharp at both ends. They consist of a frame made of drift wood, covered with the skins of walruses which are strained over it, and are capable of being tightened at any time by a lacing on the inside of the gunwale; the frame and benches for the rowers are fastened with thongs, by which the boat is rendered both light and pliable; the skin, when soaked with water, is translucent; and a stranger placing his foot upon the flat yielding surface at the bottom of the boat fancies it a frail security; but it is very safe and durable, especially when kept well greased.

In Hudson's Strait the oomiak is principally used by the women and children; here it is the common conveyance of the men, who, without them, would not be able to collect their store of provisions for the winter. They are always steered by the elderly men, who have also the privilege of sitting in the stern of the boat when unemployed. The starboard paddles of those which we saw were stained with black stripes, and the larboard with red, as were also the frames of some of the baidars.

We formed a favourable opinion of our visiters from the strict integrity which they evinced in all their dealings, even when opportunities offered of evading detection, which I notice the more readily, as we afterwards experienced very different behaviour from the same tribe.

Light winds kept us at anchor for twenty-four hours, during which time the current ran almost constantly to the south-westward, at the rate of from two fathoms to two miles per hour; and the water was nearly fresh (1.0089 to 1.0096 sp. gr.): this stratum, however, was confined to a short distance from the surface, as a patent log, which was sunk for three hours at the depth of three fathoms, showed only a fifth of a mile in that time. These facts left no doubt of our being near the estuary of a considerable river, flowing, in all probability, through the large opening abreast of us, which the boat had been sent to explore.

We weighed in the afternoon of the 23d, but in consequence of light winds and counter currents made very little progress; indeed, a great part of the time the ship would not steer, even with a moderate breeze and two boats a-head, and it was necessary to keep carrying out the kedge anchors on the bow to maintain the ship's head in the right direction. This was occasioned by some large rivers emptying themselves into the sound, the fresh water of which remained at the surface, and flowed in a contrary direction to the tide of the ocean. Had this occurred in an intricate channel it might have been dangerous; but in Kotzebue Sound the bottom is quite even, and there is plenty of room to drift about.

At four o'clock in the morning of the 25th we reached our appointed rendezvous at Chamisso Island, ten days later than had been agreed upon by Captain Franklin and myself, but which, it appeared, was quite early enough, as there were no traces of his having arrived. On approaching the island we discovered, through our telescopes, a small pile of stones upon its summit ; and as every object of this kind which was likely to be the work of human hands was interesting, from the possibility that it might be the labour of the party we were in search of, it was not long in undergoing an examination ; there was nothing however to lead to its history, but conjecture attributed it to Captain Kotzebue, who visited that spot in 1816.

The ship was anchored nearly as far up in Kotzebue Sound as a vessel of her class can go, between Chamisso Island on the south, and Choris peninsula on the north, with Escholtz Bay on the east, and an open space in the west, in which the coast was too distant to be seen. The land about this part of the Sound is generally characterised by rounded hills from about six hundred feet to a thousand above the sea, with small lakes and rivers ; its surface is rent into deep furrows, which, until a very late period in the summer, are filled with water, and being covered with a thick swampy moss, and in some places with long grass or bushes, it is extremely tedious to traverse it on foot. Early in the summer myriads of moskitos infest this swampy shore, and almost preclude the possibility of continuing any pursuit ; but in August they begin to die off, and soon afterwards entirely disappear.

Chamisso Island, the highest part of which is 231

feet above the sea, is steep, except to the eastward, where it ends in a low sandy point, upon which are the remains of some Esquimaux habitations; it has the same swampy covering as the land just described, from which, until late in the summer, several streams descend, and are very convenient for procuring water. Detached from Chamisso, there is a steep rock which by way of distinction we named Puffin Island, composed of mouldering granite, which has broken away in such a manner that the remaining part assumes the form of a tower. During the period of incubation of the aquatic birds, every hole and projecting crag on the sides of this rock is occupied by them. Its shores resound with the chorus of thousands of the feathery tribe; and its surface presents a curiously mottled carpet of brown, black, and white.

In a sandy bay upon the western side of the peninsula we found a few Esquimaux who had hauled up their baidars, and erected a temporary hut; they were inferior in every respect to those we had seen before, and furnished us with nothing new. In this bay we caught enough salmon, and other fish, to give a meal to the whole of the ship's company, which was highly acceptable; but we had to regret that similar success did not attend our subsequent trials.

By my instructions, I was desired to await the arrival of Captain Franklin at this anchorage; but in a memoir drawn up by that officer and myself, to which my attention was directed by the Admiralty, it was arranged that the ship should proceed to the northward, and survey the coast, keeping the barge in shore to look out for the land party, and to erect posts as signals of her having been there, and also to leave directions where to find the ship.

I was also desired to place a small party in occupation of Chamisso Island during the absence of the ship; but this spot proved to be so different from what we imagined, being accessible in almost every quarter, instead of having only one landing place, that a small party would have been of no use if the natives were inclined to be hostile, and the numerical strength of the crew did not admit of a large detachment being spared from her. But in order that Captain Franklin should not want provision in the event of his missing the ship along the coast, and arriving at the island in her absence, a tight barrel of flour was buried upon Puffin Rock, which appeared to be the most unfrequented spot in the vicinity, and directions for finding it were deposited in a bottle at Chamisso Island, together with such other information as he might require, and the place where it was deposited was pointed out by writing upon the cliffs with white paint. It was further arranged, that a party should proceed over land in a northerly direction, in the hope of falling in with Captain Franklin, as it was possible the shore of the Polar Sea might lie more to the southward than the general trending of that part of its coast which had been explored led us to expect. But as the ship was likely to be absent several weeks, and we were unacquainted with the disposition of the people or with the country, further than that from its swampy nature, it seemed to present almost insurmountable difficulties to the journey, I deferred the departure of the party, and afterwards wholly abandoned the project, as the coast was found to extend so far to the northward as to render it quite useless.

As I wished to avail myself of the latitude afforded by this memoir, to survey and examine as much of the coast as possible before Captain Franklin arrived, no time was lost in preparing the ship for sea, which it required only a little time to effect.

On the 28th Mr. Elson returned from the examination of the opening we discovered on the north side of Kotzebue Sound, and reported the water at the entrance to be so shallow that the barge could not enter. The inlet was of considerable width, and extended thirty or forty miles in a broad sheet of water, which at some distance up was fresh. This was ascertained by landing in the sound to the eastward of the opening, at which place it was found that the inlet approached the sea within a mile and a half. The time to which it was necessary to limit Mr. Elson prevented his doing more than ascertaining that this opening was navigable only by small boats; and by the water being quite fresh, that it could not lead to any sea beyond.

The Esquimaux in the inlet were more numerous than we supposed, but were very orderly and well behaved. When the barge anchored off a low sandy point, on which they had erected their summer habitations and fishing stakes, she was surrounded by fourteen baidars, containing 150 men; which, considering the crew of the barge only amounted to eight men and two officers, was a superiority of strength that might well have entitled them to take liberties, had they been so disposed, armed as they usually are with bows and arrows, spears, and a large knife strapped to their thigh: but so far from this being the case, they readily consented to an arrangement, that only one baidar at a time should come alongside to dispose of her goods, and then make way for another:

the proposal was made while the baidars were assembled round our boat, and was received with a shout of general applause.

Blue beads, cutlery, tobacco, and buttons, were the articles in request, and with which almost any thing they had might have been purchased: for these they sold their implements, ornaments, and some very fine salmon; also a small caiac very similar to those of Greenland and Hudson's Strait.

While the duties of the ship were being forwarded under my first lieutenant, Mr. Peard, I took the opportunity to visit the extraordinary ice-formation in Escholtz Bay, mentioned by Kotzebue as being " covered with a soil half a foot thick, producing the most luxuriant grass," and containing an abundance of mammoth bones. We sailed up the bay, which was extremely shallow, and landed at a deserted village on a low sandy point, where Kotzebue bivouacked when he visited the place, and to which I afterwards gave the name of Elephant Point, from the bones of that animal being found near it.

The cliffs in which this singular formation was discovered begin near this point, and extend westward in a nearly straight line to a rocky cliff of primitive formation at the entrance of the bay, whence the coast takes an abrupt turn to the southward. The cliffs are from twenty to eighty feet in height; and rise inland to a rounded range of hills between four and five hundred feet above the sea. In some places they present a perpendicular front to the northward, in others a slightly inclined surface; and are occasionally intersected by valleys and watercourses generally overgrown with low bushes. Opposite each of these valleys, there is a projecting flat

piece of ground, consisting of the materials that
have been washed down the ravine, where the only
good landing for boats is afforded. The soil of the
cliffs is a bluish-coloured mud, for the most part
covered with moss and long grass, full of deep furrows, generally filled with water or frozen snow.
Mud in a frozen state forms the surface of the cliff
in some parts; in others the rock appears, with the
mud above it, or sometimes with a bank half way
up it, as if the superstratum had gradually slid
down and accumulated against the cliff. By the
large rents near the edges of the mud cliffs, they
appear to be breaking away, and contributing daily
to diminish the depth of water in the bay.

Such is the general conformation of this line of
coast. That particular formation, which, when it
was first discovered by Captain Kotzebue, excited
so much curiosity, and bore so near a resemblance
to an iceberg, as to deceive himself and his officers,
when they approached the spot to examine it, remains to be described. As we rowed along the
shore, the shining surface of small portions of the
cliffs attracted our attention and directed us where
to search for this curious phenomenon, which we
should otherwise have had difficulty in finding, notwithstanding its locality had been particularly described; for so large a portion of the ice cliff has
thawed since it was visited by Captain Kotzebue
and his naturalist, that only a few insignificant
patches of the frozen surface now remain. The
largest of these, situated about a mile to the westward of Elephant Point, was particularly examined
by Mr. Collie, who, on cutting through the ice in a
horizontal direction, found that it formed only a

casing to the cliff, which was composed of mud and gravel in a frozen state. On removing the earth above, it was also evident, by a decided line of separation between the ice and the cliff, that the Russians had been deceived by appearances. By cutting into the upper surface of the cliff three feet from the edge, frozen earth, similar to that which formed the face of the cliff, was found at eleven inches' depth; and four yards further back the same substance occurred at twenty-two inches' depth.

The glacial facing we afterwards noticed in several parts of the sound; and it appears to me to be occasioned either by the snow being banked up against the cliff, or collected in its hollows in the winter, and converted into ice in the summer by partial thawings and freezings—or by the constant flow of water during the summer over the edges of the cliffs, on which the sun's rays operate less forcibly than on other parts, in consequence of their aspect. The streams thus become converted into ice, either while trickling down the still frozen surface of the cliffs, or after they reach the earth at their base, in which case the ice rises like a stalagmite, and in time reaches the surface. But before this is completed, the upper soil, loosened by the thaw, is itself projected over the cliff, and falls in a heap below, whence it is ultimately carried away by the tide. We visited this spot a month later in the season, and found a considerable alteration in its appearance, manifesting more clearly than before the deception under which Kotzebue laboured.

The deserted village upon the low point consisted of a row of huts, rudely formed with drift-wood and turf, about six feet square and four feet in height.

In front of them was a quantity of drift-wood raised upon rafters; and around them there were several heaps of bones, and skulls of seals and grampuses, which in all probability had been retained conformably with the superstitions of the Greenlanders, who carefully preserve these parts of the skeleton.* A rank grass grew luxuriantly about these deserted abodes, and also about the edges of several pools of fresh water, in which there were some wild fowl. We returned to the ship late at night, and found her ready for sea.

* Crantz Greenland, Vol. I.

CHAPTER XI.

Quit Kotzebue Sound, and proceed to survey the Coast to the
Northward—Interviews with the Natives—Cape Thomson —
Point Hope—Current—Capes Sabine and Beaufort—Barrier of
Ice—Icy Cape—advanced Position of the Ship—Discover Cape
Franklin, Wainwright Inlet, Shoals off Icy Cape, &c.—Boat sent
on an Expedition along the Coast—Return of the Ship to Kot-
zebue Sound—Interviews with the Esquimaux—Boat rejoins
the ship—Important Results of her Expedition.

CHAP.
XI.

July,
1826.

On the 30th of July we weighed from Chamisso
Island attended by the barge, and steered out of the
sound. The day was very fine ; and, as we sailed
along the northern shore, the sun was reflected from
several parts of the cliff, which our telescopes discovered to be cased with a frozen surface similar to
that just described in Escholtz Bay. We kept at six
or seven miles distance from the land, and had a
very even bottom, until near Hotham Inlet, when
the soundings quickly decreased, and the ship struck
upon a shoal before any alteration of the helm had
materially changed her position. The water was
fortunately quite smooth, and she grounded so easily
that, but for the lead-lines, we should not have
known any thing had occurred. We found upon
sounding, that the ship had entered a bight in the
shoal, and that there was a small bank between her

and the deep water, so that it became necessary to carry out the stream anchor in the direction of her wake, by means of which, and a little rise of the tide, she was soon got off.

This shoal, which extends eight miles off the land, is very dangerous, as the soundings give very short warning of its proximity, and there are no good landmarks for avoiding it. The distance from the shore, could it be judged of under ordinary circumstances, would on some occasions be a most treacherous guide, as the mirage in fine weather plays about it, and gives the land a very different appearance at one moment from that which it assumes at another.

As soon as we were clear of the shoal, we continued our course for Cape Krusenstern, near which place we the next day buried a letter for Captain Franklin, and erected a post to direct him to the spot. The cape is a low tongue of land, intersected by lakes, lying at the foot of a high cluster of hills not in any way remarkable. The land slopes down from them to several rocky cliffs, which, until the low point is seen at the foot of them, appear to be the entrance to the sound, but they are nearly a mile inland from it. The coast here takes an abrupt turn to the northward, and the current sets strong against the bend; which is probably the reason of there being deep water close to the beach, as also the occasion of a shoal in a north-westerly direction from the point, which appears to have been thrown up by the eddy water.

The boat landed about two miles to the northward of this point, upon a shingly beach sufficiently steep to afford very good landing when the water is smooth; behind it there was a plain about a mile

wide, extending from the hills to the sea, composed
of elastic bog earth, intersected by small streams, on
the edges of which the buttercup, poppy, blue-bell,
pedicularis, vaccinium, saxifrages, and some cruci-
form plants* throve very well; in other parts, however, the vegetation was stinted, and consisted only
of lichens and mosses. There were here some low
mud cliffs frozen so hard that it required considerable labour to dig fifteen inches to secure the end of
the post that was erected.

Mr. Elson, in command of the barge, was now
furnished with a copy of the signals drawn up by
Captain Franklin and myself, and directed to proceed close along the shore to the northward, vigi-
lantly looking out for boats, and erecting posts and
landmarks in the most conspicuous places for Captain Franklin's guidance, and to trace the outline of
the beach.   He was also desired to explore the coast
narrowly, and to fill in such parts of it as could not
be executed in the ship, and instructed where to
rendezvous in case of separation.

We then steered along the coast, which took a
north-westerly direction, and at midnight passed a
range of hills terminating about four miles from the
sea, which must be the Cape Mulgrave of Captain
Cook, who navigated this part of the coast at too
great a distance to see the land in front of the hills,
which is extremely low, and after passing the Mulgrave Range, forms an extensive plain intersected
by lakes near the beach ; these lakes are situated so
close together that by transporting a small boat from
one to the other, a very good inland navigation, if

* The botany of this part of the coast is published in the Flora
Americana of Dr. Hooker.

necessary, might be performed. They are supplied by the draining of the land and the melting snow, and discharge their water through small openings in the shingly beach, too shallow to be entered by any thing larger than a baidar, one of them excepted, through which the current ran too strong for soundings to be taken.

On the 1st of August we did little more than drift along the coast with the current—which was repeatedly tried, and always found setting to the north-west—from half a mile to a mile and a half per hour. The Esquimaux, taking advantage of the calm, came off to the ship in three baidars, and added to our stock of curiosities by exchanging their manufactures for beads, knives, and tobacco.

On the 2d, being favoured with a breeze, we closed with a high cape, which I named after Mr. Deas Thomson, one of the commissioners of the navy.* It is a bold promontory 450 feet in height, and marked with differently coloured strata, of which there is a representation in the geological memorandum. As this was a fit place to erect a signal-post for Captain Franklin, we landed, and were met upon the beach by some Esquimaux, who eagerly sought an exchange of goods. Very few of their tribe understood better how to drive a bargain than these people ; and it was not until they had sold almost all they could spare, that we had any peace. We found them very honest, extremely good natured, and friendly. Their features, dress, and weapons were the same as before described in Kotzebue Sound, with the exception of some broad-headed

* A cape close to this has been named Cape Ricord by the Russians.

spears, which they had probably obtained from the Tschutschi. They had more curiosity than our former visiters, and examined very minutely every part of our dress; from which circumstance, and their being frightened at the discharge of a gun, and no less astonished when a bird fell close to them, we judged they had had a very limited intercourse with Europeans. The oldest person we saw among the party was a cripple about fifty years of age. The others were robust people above the average height of Esquimaux: the tallest man was five feet nine inches, and the tallest woman five feet four inches. All the women were tattooed upon the chin with three small lines, which is a general distinguishing mark of the fair sex along this coast; this is effected by drawing a blackened piece of thread through the skin with a needle, as with the Greenlanders. Their hair was done up in large plaits on each side of the head, as described by Captain Parry at Melville Peninsula. We noticed a practice here amongst the women, similar to that which is common with the Arabs, which consisted of blacking the edges of the eyelids with plumbago rubbed up with a little saliva upon a piece of slate. All the men had labrets, and both sexes had their teeth much worn down, probably by the constant application of them to hard substances, of which their dresses, implements, and canoes are made.

They had several rude knives, probably obtained from the Tschutschi, some lumps of iron pyrites, and pieces of amber strung round their neck; but I could not learn where they had procured them.

As soon as we finished the necessary observations with the artificial horizon, to the no small diversion

NATIVES OF THE COAST NEAR CAPE THOMSON

Pub.d by H.Colburn & R.Bentley. 1831.

and surprise of our inquisitive companions, we paid
a visit to the next valley, where we found a small
village situated close upon a fine stream of fresh
water flowing from a large bed of thawing snow.
The banks of the brook were fertile, but vegetation
was more diminutive here than in Kotzebue Sound;
notwithstanding which, several plants were found
which did not exist there. The tents were constructed of skins loosely stretched over a few spars
of drift-wood, and were neither wind nor water
tight. They were, as usual, filthy, but suitable to
the taste of their inhabitants, who no doubt saw nothing in them that was revolting. The natives testified much pleasure at our visit, and placed before
us several dishes, among which were two of their
choicest—the entrails of a fine seal, and a bowl of
coagulated blood. But, desirous as we were to
oblige them, there was not one of our party that
could be induced to partake of their hospitality.
Seeing our reluctance, they tried us with another
dish, consisting of the raw flesh of the narwhal nicely cut into lumps, with an equal distribution of black
and white fat; but they were not more successful
here than at first.

An old man then braced a skin upon a tambourine frame, and striking it with a bone gave the signal for a dance, which was immediately performed
to a chorus of Angna aya! angna aya! the tambourine
marking time by being flourished and twirled about
against a short stick instead of being struck. The
musician, who was also the principal dancer, jumped
into the ring, and threw his body into different attitudes until quite exhausted, and then resigned his
office to another, from whom it passed to a lad who

occasioned more merriment by his grimaces and ludicrous behaviour than any of his predecessors. His song was joined by the young women, who until then had been mute and almost motionless, but who now acquitted themselves with equal spirit with their leader, twisting their bodies, twirling their arms about, and violently rubbing their sides with their garments, which, from some ridiculous associations no doubt, occasioned considerable merriment.

Against an obscure part of the cliff near the village we noticed a broad iron-headed halberd placed erect, with several bows and quivers of arrows; and near them a single arrow, with a tuft of feathers attached to it, suspended to the rock. The Esquimaux were reluctant to answer our inquiries concerning this arrangement, and were much displeased when we approached the place. From the conduct of the natives at Schismareff Inlet toward Captain Kotzebue, it is not impossible that the shooting of this arrow may be a signal of hostility, as those people after eying him attentively and suspiciously, paddled quickly away, and threw two arrows with bunches of feathers fastened to them toward their habitations, whence shortly afterwards issued two baidars, who approached Captain Kotzebue with very doubtful intentions.

Upon an eminence beyond this cliff we found several dogs tethered to stakes; and all the little children of the village, who had perhaps been sent out of the way, and who, on seeing us, set up a general lamentation.

After viewing this village we ascended Cape Thomson, and discovered low land jetting out from the coast to the W. N. W. as far as the eye could reach.   As this point had never been placed in our charts, I named it Point Hope, in compliment to Sir William Johnstone Hope.

Having buried a bottle for Captain Franklin upon the eminence, we took leave of our friends, and made sail towards the ship, which, in consequence of a current, was far to leeward, although she had been beating the whole day with every sail set.   We continued to press the ship during the night, in order to maintain our position, that the barge might join; but the current ran so strong, that the next morning, finding we lost rather than gained ground, I bore away to trace the extent of the low point discovered from Cape Thomson.   On nearing it, we perceived a forest of stakes driven into the ground for the purpose of keeping the property of the natives off the ground ; and beneath them several round hillocks, which we afterwards found to be the Esquimaux yourts, or underground winter habitations.   The wind fell very light off this point, and I went in the gig to pay a visit to the village, leaving directions to anchor the ship in case the wind continued light.   After rowing a considerable time, we found a current running so strong that we did not make any progress, and it was as much as we could do to get back to the ship, which had in the mean time been anchored with the bower, having previously parted from the kedge.

The current was now running W. by N. at the rate of three miles an hour.   About five o'clock the next morning, however, it slackened to a mile and a half, and the boats were sent to creep for the kedge anchor, but it could not be found.   A thick fog afterwards came on, which kept us at anchor until the next day. During this time signal guns were fired every two hours, as well on account of Captain Franklin as of our own boat.

On the 5th we weighed, and set the studding-sails, but the ship would not steer, and came broadside to the tide, in spite of the helm and three boats ahead; and continued in this position until a fresh breeze sprang up from the northward.

It is necessary here to give some further particulars of this current, in order that it may not be supposed that the whole body of water between the two great continents was setting into the Polar Sea at so considerable a rate. By sinking the patent log first five fathoms, and then three fathoms, and allowing it to remain in the first instance six hours, and in the latter twelve hours, it was clearly ascertained that there was no current at either of those depths; but at the distance of nine feet from the surface the motion of the water was nearly equal to that at the top. Hence we must conclude that the current was superficial, and confined to a depth between nine and twelve feet.

By the freshness of the water alongside there is every reason to believe that the current was occasioned by the many rivers which, at this time of the year, empty themselves into the sea in different parts of the coast, beginning with Schismareff Inlet. The specific gravity of the sea off that place was 1.02502, from which it gradually decreased, and at our station off the point was 1.0173, the temperature at each being 58°. On the other hand, the strength of the stream had gradually increased from half a mile an hour to three miles, which was its greatest rapidity. So far there is nothing extraor-

dinary in the fact; but why this body of water
should continually press to the northward in preference to taking any other direction, or gradually
expending itself in the sea, is a question of considerable interest.

In the afternoon the barge was discovered at
anchor, close in-shore, and being favoured with a
breeze the ship was brought close to the point.
This enabled me to land, accompanied by Mr. Collie,
who, while I was occupied with my theodolite, went
toward the huts, which at first appeared to be deserted; but as he was examining them several old
women and children made their appearance, and
gave him a friendly reception. He brought them
to me, and we underwent the full delights of an
affectionate Esquimaux salutation.

The persons of our new acquaintance were extremely diminutive, dirty, and forbidding. Some
were blind, others decrepit; and, dressed in greasy
worn-out clothes, they looked perfectly wretched.
Their hospitality, however, was even greater than
we could desire; and we were dragged away by the
wrists to their hovels, on approaching which we
passed between heaps of filth and ruined habitations, filled with stinking water, to a part of the
village which was in better repair. We were then
seated upon some skins placed for the purpose; and
bowls of blubber, walrus, and unicorn flesh *(monodon
monoceros)*, with various other delicacies of the same
kind, were successively offered as temptations to our
appetite, which, nevertheless, we felt no inclination
to indulge.

After some few exchanges, the advantage of
which was on the side of our acquaintances, who

had nothing curious to part with, an old man produced a tambourine, and seating himself upon the
roof of one of the miserable hovels, threw his legs
across, and commenced a song, accompanying it
with the tambourine, with as much apparent happiness as if fortune had imparted to him every luxury
of life. The vivacity and humour of the musician
inspired two of the old hags, who joined chorus, and
threw themselves into a variety of attitudes, twisting their bodies, snapping their fingers, and smirk-
ing from behind their seal-skin hoods, with as much
shrewd meaning as if they had been half a century
younger. Several little chubby girls, roused by the
music, came blinking at the daylight through the
greasy roofs of the subterranean abodes, and joined
the performance; and we had the satisfaction of
seeing a set of people happy who did not appear to
possess a single comfort upon earth.

The village consisted of a number of "yourts"
excavated in a ridge of mud and gravel, which had
been heaped up in a parallel line with the beach.
Their construction more nearly approached to the
habitations of the Tschutschi than those of the
Esquimaux of Greenland. They consisted of two
pits about eight feet deep, communicating by a
door at the bottom. The inner one had a dome-
shaped roof, made with dry wood or bones; it was
covered with turf, and rose about four feet above
the surface of the earth. In the centre of this there
was a circular hole or window, covered with a piece
of skin (part of the intestine of the whale), which
gave, however, but very little light. The outer pit
had a flat roof, and was entered by a square hole,
over which there was a shed to protect it from the

snow and the inclemency of the weather. A rude
ladder led to a floor of loose boards, beneath which
our noses as well as our eyes were greeted by a pool
of dirty green water. The inner chamber was the
sleeping and cooking room.

Another yourt, to which a store of provision was
attached, by a low subterraneous passage, was examined by Lieutenant Belcher the ensuing year:
it was in other respects very similar, and needs no
particular description. Of these yourts, one was
of much larger dimensions than the others, which,
it was intimated by the natives, was constructed
for the purpose of dancing and amusing themselves. Mr. Belcher was particularly struck with
the cleanliness of the boards and sleeping places
in the interior of the yourt he examined; whereas the passage and entrance were allowed to remain in a very filthy condition. The air was too
oppressive to continue in them for any length
of time. Every yourt had its rafters for placing
sledges, skins of oil, or other articles upon in the
winter time, to prevent their being buried in the
snow. The number of these frames, some bearing
sledges, and others the skeletons of boats, formed a
complete wood, and had attracted our notice at
the distance of six or seven miles. Of the many
yourts which composed the village, very few were
occupied; the others had their entrances blocked up
with logs of drift-wood and the ribs of whales.
From this circumstance, and the infirm condition of
almost all who remained at the village, it was evident that the inhabitants had gone on sealing excursions, to provide a supply of food for the winter.
The natives, when we were about to take our leave,

accompanied us to the boat, and as we pushed off they each picked up a few pebbles and carried them away with them, but for what purpose we could not guess, nor had we ever seen the custom before.

The point upon which this village stands projects almost sixteen miles from the general line of coast; it is intersected by several lakes and small creeks, the entrances of which are on the north side. There is a bar across the mouth of the opening, consisting of pebbles and mud, which has every appearance of being on the increase; but when the water is smooth a boat may enter, and she will find very excellent security within from all winds. It is remarkable that both Cook and Clerke, who passed within a very short distance of this point, mistook the projection for ice that had been driven against the land, and omitted to mark it in their chart.

The next morning we communicated with the barge, and found she had been visited daily by the natives, who were very friendly. The current inshore was more rapid than in the offing, and the water more fresh. After replenishing her provisions, we steered to the northward, and endeavoured' to get in with the land on the northern side of Point Hope; but the wind was so light that we could not hold our ground against the current, and were drifted away slowly to the northward. In the morning, the wind being still unfavourable for this purpose, we steered for the farthest land in sight to the northward, which answered to Cape Lisburn of Captain Cook. As we approached it, the current slackened, and the depth and specific gravity of the sea both increased. We landed here, and ascended the mountain to obtain a fair view of the coast, which we found turned to the eastward, nearly at a right angle, and then to the north-eastward, as far as the eye could trace. Our height was 850 feet above the sea, and at so short a distance from it on one side, that it was fearful to look down upon the beach below. We ascended by a valley which collected the tributary streams of the mountain, and poured them in a cascade upon the beach. The basis of the mountain was flint of the purest kind, and limestone, abounding in fossil shells, enchinites, and marine animals.

There was very little soil in the valley; the stones were covered with a thick swampy moss, which we traversed with great difficulty, and were soon wet through by it. Vegetation was, however, as luxuriant as in Kotzebue Sound, more than a hundred miles to the southward, or, what is of more consequence, more than that distance farther from the great barrier of ice. Several reindeer were feeding on this luxuriant pasture; the cliffs were covered with birds; and the swamps generated myriads of moskitos, which were more persevering, if possible, than those at Chamisso Island.

After depositing a bottle at this place, and leaving proper directions upon the cliff for finding it, we pursued our course to the eastward, accompanied by the barge. The wind was light, and we made so little progress that on the 9th Cape Lisburn was still in sight. Before it was entirely lost I landed at a small cape, which I named Cape Beaufort, in compliment to Captain Beaufort, the present hydrographer to the Admiralty. The land northward was low and swampy, covered with moss and long grass, which produced all the plants we had met with to the southward, and two or three besides. Cape Beaufort is composed of sandstone, enclosing bits of petrified wood and rushes, and is traversed by narrow veins of coal lying in an E. N. E. and W. S. W. direction. That at the surface was dry and bad, but some pieces which had been thrown up by the burrowing of a small animal, probably the ermine, burned very well.

As this is a part of the coast hitherto unexplored, I may stand excused for being a little more particular in my description. Cape Beaufort is situated in the depth of a great bay, formed between Cape Lisburn and Icy Cape, and is the last point where the hills come close down to the sea, by reason of the coast line curving to the northward, while the range of hills continues its former direction. From the rugged mountains of limestone and flint at Cape Lisburn, there is an uniform descent to the rounded hills of sandstone at Cape Beaufort just described. The range is, however, broken by extensive valleys, intersected by lakes and rivers. Some of these lakes border upon the sea, and in the summer months are accessible to baidars, or even large boats; but as soon as the current from the beds of thawing snow inland ceases, the sea throws up a bar across the mouths of them, and they cannot be entered. The beach, at the places where we landed was shingle and mud, the country mossy and swampy, and infested with moskitos. We noticed recent tracks of wolves, and of some cloven-footed animals, and saw several ptarmigans, ortolans, and a lark. Very little drift wood had found its way upon this part of the coast.

We reached the ship just after a thick fog came

in, from seaward, and only a short time before the
increasing breeze obliged her to quit the coast.
During my absence the boats had been sent to examine a large floating mass which excited a good deal
of curiosity at the time, and found it to be the carcass of a dead whale. It had an Esquimaux harpoon
in it, and a drag attached, made of an inflated seal-
skin, which had no doubt worried the animal to
death. Thus, with knowledge just proportioned to
their wants, do these untutored barbarians, with
their slender boats and limited means, contrive to
take the largest animal of the creation. In the present instance, certainly, their victim had eluded their
efforts, but the carcass was not yet " too high" for
an Esquimaux palate, and would, no doubt, ere
long, be either washed upon the shore, or discovered
by some of the many wandering baidars along the
coast.

Some very extensive flocks of eider ducks had
also been seen from the ship. They consisted entirely of females and young ones, the greater part of
which could not fly, but they nevertheless contrived
to evade pursuit by diving.

On the morning of the 10th we were under treble-
reefed topsails and foresail, with a short head sea, in
which we pitched away the jib-boom. We had a
thick fog, with the wind at N. N. E. A little after
noon, being in lat. 70° 09′ N., and 165° 10′ W., we
had twenty-four fathoms hard bottom: we then
stood toward the shore, and again changed the bottom to mud, the depth of water gradually decreasing.

On the 11th it was calm; by the observations at
noon there had been a current to the S. W., but this
had now ceased, as upon trial it ran west one-third

of a mile per hour, and three hours afterward N. E. five-eighths per hour, which appeared to be the regular tide. In the evening the wind again blew from the northward, and brought a thick fog with it. We stood off and on, guided by the soundings.

In the morning of the 12th we saw a great many birds, walrusses, and small white whales; from which I concluded that we were near a stream of ice, but only one piece was seen in the evening aground. We tacked not far from it in ten fathoms. As we stood in-shore, the temperature of the sea always decreased; the effect, probably, of the rivers of melting snow mingling with it.

As it was impossible to determine the continuity of coast, with the weather so thick, farther than by the gradual decrease of the soundings, I stood to the northward to ascertain the position of the ice, the wind having changed to E. N. E. and become favourable for the purpose. At eight o'clock in the morning of the 13th, the fog cleared off, and exhibited the main body of ice extending from N. 79° E. to S. 29 W. (true). At nine we tacked amongst the *brash*, in twenty-three fathoms water, in lat. 71° 08 N., long. 163° 40′ W. The wind was blowing along the ice, and the outer part of the *pack* was in streams, some of which the ship might have entered, and perhaps have proceeded up them two or three miles; but as this would have served no useful purpose, and would have occasioned unnecessary delay, I again stood in for the land, which at eight o'clock at night was seen in a low unbroken line, extending to the westward as far as Icy Cape, and to the eastward as far as the state of the weather would permit. We tacked at nine, in five fathoms water, within two miles of the shore; and Lieutenant Bel-

cher was despatched in the cutter to examine some
posts that were erected upon it, thinking they might
possibly have been placed there by the land expedition. The boat found a heavy surf breaking upon
a sand bank at a little distance from the beach,
which prevented her landing, and a fog coming on,
she was recalled before the attempt could be made
in another place. There was a thick wetting fog
during the night. The next morning a boat was
again sent on shore, with Lieutenant Belcher,
Messrs. Collie and Wolfe, to make observations,
collect plants, and erect a mark for Captain Franklin. They had nearly the same difficulty in reaching the beach, on account of shoals, as at the former
place, but there was less swell.

Shortly after noon I landed myself, and found
that at the back of the beach there was a lake two
miles long, in the direction of the coast; it had a
shallow entrance at its south-west end, sufficiently
deep for baidars only. The main land at the back
of it presented a range of low earth cliffs, behind
which there were some hills, about two hundred
feet high. Near the entrance to the lake there were
two yourts, inhabited by some Esquimaux, who sold
us two swans and four hundred pounds of venison,
which being divided amongst the crew, formed a
most acceptable meal. These swans were without
their feet, which had been converted into bags, after
the practice of the eastern Esquimaux; and it is
remarkable, that although so far from Kamtschatka
and the usual track of vessels, these people expressed
no surprise at the appearance either of the ship or of
the boat, and that they were provided both with
knives and iron kettles.

In our way to the huts we saw several human

bones scattered about, and a skull which had the teeth worn down nearly to the gums. There appeared to be no place of interment near, and the body had probably decayed where the bones were lying. So little did the natives care for these mouldering remains, that springs for catching birds were set amongst them. The beach upon which we landed was shingle and sand, interspersed with pieces of coal, sandstone, flint, and porphyritic granite. Vegetation was rather luxuriant, and supplied Mr. Collie with three new species. The drift wood was here more abundant than at any place we before visited : it was forced high upon the beach, probably by the pressure of the ice when driven against the coast.

It was high water at this station at noon. The tide fell three feet and a half in four hours, and ebbed to the south-west.

A post was here put up for the land expedition, and a bottle buried near it. We then embarked and got on board, just as a thick fog obscured every thing, and obliged the ship to stand off the coast. In the course of the afternoon the dredge was put over, and supplied us with some specimens of shells of the arca, murex, venus, and buccinum genus, and several lumps of coal. We stood to the N. W., and at midnight tacked amongst the loose ice at the edge of the pack in so thick a fog that we could not see a hundred yards around us.

At half past five in the morning a partial dispersion of the fog discovered to us the land bearing N. 86° E. extending in a N. E. direction as far as we could see. At six we tacked in eleven fathoms within three miles of it, and not far from an open-

ing into a spacious lake which appeared to be the
estuary of a considerable river.   There was a shoal
across the mouth connected with the land on the
northern side, but with a channel for boats in the
opposite direction.   A large piece of ice was aground
near it.   The country around was low, covered with
a brown moss, and intersected by water-courses. To
the northward of the entrance of the lake the coast
became higher, and presented an extensive range of
mud cliffs terminating in a cape, which, as it afterwards proved the most distant land seen from the
ship, I named after Captain Franklin, R. N. under
whose command I had the pleasure to serve on the
first Polar expedition: but as this cape was afterwards found to be a little way inland I transferred
the name to the nearest conspicuous point of the
coast.

The natives taking advantage of this elevated
ground had constructed their winter residences in
it; they were very numerous, and extended some
way along the coast.   The season, however, was not
yet arrived at which the Esquimaux take up their
abode in their subterranean habitations, and they
occupied skin tents upon a low point at the entrance
of the lake.   We had not been long off here before
three baidars from the village paddled alongside and
bartered their articles as usual.   Some of the crew
ascended the side of the ship without any invitation,
and showed not the least surprise at any thing they
beheld; which I could not help particularly remarking, as we were not conscious of any other vessel
having been upon the coast since Kotzebue s voyage,
and he did not reach within two hundred miles of
the residence of these people.   There was nothing

in our visiters different to what we had seen before, except that they were better dressed. One of them, pointing to the shore, drew his hand round the northern horizon as far as the south-west, by which he no doubt intended to instruct us that the ice occupied that space. It would, however, have answered equally well for the land, supposing the coast beyond what we saw to have taken a circuitous direction. With the view of having this explained, I took him to the side of the ship on which the land was, and intimated a desire that he would delineate the coast; but he evidently did not understand me, as he and his companions licked their hands, stroked their breasts, and then went into their boats and paddled on shore.

The apparently good-natured disposition of these people, and indeed of the whole of their tribe upon the coast to this advanced position, was a source of the highest gratification to us all as it regarded Captain Franklin's welfare; for it was natural to conclude that the whole race, which we had reason to think extended a considerable distance to the eastward, would partake of the same friendly feeling, and what was by many considered a material obstacle to his success would thereby be removed. At this place in particular, where the natives appeared to be so numerous that they could have overpowered his party in a minute, it was gratifying to find them so well disposed.

After the natives were gone we stood to the north-westward in the hope that the wind, which had been a long time in the north-eastern quarter, would remain steady until we ascertained the point of conjunction of the ice and the land, which,

from its position when seen in the morning, there
was much reason to suppose would be near the extreme point of land in view from the mast-head. Unfortunately, while we were doing this, the wind fell
light, and gradually drew round to the north-westward; and apprehending it might get so far in that
direction as to embay the ship between the land and
the ice, it became my duty to consider the propriety
of awaiting the result of such a change; knowing
the necessity of keeping the ship in open water, and
at all times, as far as could be done, free from risk,
in order to insure her return to the rendezvous in
Kotzebue Sound. There was at this time no ice in
sight from the ship except a berg that was aground
in-shore of her; and though a blink round the
northern horizon indicated ice in that direction, yet
the prospect was so flattering that a general regret
was entertained that an attempt to effect the north-
eastern passage did not form the object of the expedition. We all felt the greatest desire to advance,
but considering what would be the consequences of
any accident befalling the ship, which might either
oblige her to quit these seas at once, or prevent her
returning to them a second year, it was evident that
by her being kept in open water was paramount to
every other consideration; particularly as she had
been furnished with a decked launch, well adapted
by her size to prosecute a service of this nature. It
was one of those critical situations in which an
officer is sometimes unavoidably placed, and had
further discovery depended upon the Blossom alone,
it is probable I should have proceeded at all hazards.
My orders, however, being positive to avoid the
chance of being beset in the ship, I considered only

how I could most beneficially employ both vessels, and, at the same time, comply with the spirit of my instructions. Thus circumstanced, I determined to get hold of the barge as soon as possible, and to despatch her along the coast, both with a view of rendering Captain Franklin's party the earliest possible assistance, and of ascertaining how far it was possible for a boat to go. Not a moment was to be lost in putting this project in execution, as the middle of August was arrived, and we could not calculate on a continuance of the fine weather with which we had hitherto been favoured. We accordingly returned towards Icy Cape, in order to join the barge which was surveying in that direction.

We passed along the land in about eight fathoms water until near Icy Cape, when we came rather suddenly into three fathoms and three quarters, but immediately deepened the soundings again to seven: the next cast, however, was four fathoms; and not knowing how soon we might have less, the ship was immediately brought to an anchor. Upon examination with the boats, several successive banks were found at about three quarters of a mile apart, lying parallel with the coast line. Upon the outer ones, there were only three and a half or four fathoms, and upon the inner bank, which had hitherto escaped notice from being under the sun, so little water that the sea broke constantly over it. Between the shoals there were nine and ten fathoms, with very irregular casts. These shoals lie immediately off Icy Cape where the land takes an abrupt turn to the eastward, and are probably the effect of a large river, which here empties itself into the sea; though they may be occasioned by heavy ice grounding off the point, and being fixed to the bottom, as we found our anchor had so firm a hold, that in attempting to weigh it the chain cable broke, after enduring a very heavy strain.

This cape, the farthest point reached by Captain Cook, was at the time of its discovery very much encumbered with ice, whence it received its name; none, however, was now visible. The cape is very low, and has a large lake at the back of it, which receives the water of a considerable river, and communicates with the sea through a narrow channel much encumbered with shoals. There are several winter habitations of the Esquimaux upon the cape, which were afterwards visited by Lieutenant Belcher. The main land on both sides of Icy Cape, from Wainwright Inlet on one side to Cape Beaufort on the other, is flat, and covered with swampy moss. It presents a line of low mud cliffs, between which and a shingly beach that every where forms the coast-line there is a succession of narrow lakes capable of being navigated by baidars or small boats. Off here we saw a great many black whales — more than I remember ever to have seen, even in Baffin's Bay.

After the boats had examined the shoals outside the ship, we attempted to weigh the anchor; but in so doing we broke first the messenger, and afterwards the chain, by which the anchor was lost, as I before mentioned, and the buoy rope having been carried away in letting it go, it was never recovered.

We passed over two shoals in three and four fathoms, deepening the water to ten and eleven fathoms between them, and then held our ground for the night. A thick fog came on towards morning, which lasted until noon, when it cleared away, and we had the satisfaction to be joined by the barge.

Since our separation, Mr. Elson had kept close along the beach, and ascertained the continuity of the land from the spot where the ship quitted the coast to this place, thereby removing all doubts on that head, and proving that Captain Franklin would not find a passage south of the cape to which I had given his name. The soundings were every where regular, and the natives always friendly, though not numerous. Their habitations were invariably upon low strips of sand bordering upon some brackish lakes, which extended along the coast in such a manner, that in case the ice was driven against it, a good inland navigation might be performed, by transporting a small boat across the narrow necks that separate them.

Drift-wood was every where abundant, though least so on such parts of the coast as had a western aspect, but without any apparent reason for this difference. After supplying the barge with water, we beat to the northward together, but found so strong a south-westerly current running round Icy Cape, that, the ship being light, we could gain nothing to windward ; and observing that the barge had the advantage of us by keeping in-shore, and that we were only a hindrance to her, I made her signal to close us, and prepared her for the interesting service in view. My intentions were no sooner made known than I had urgent applications for the command of the barge from the superior officers of the ship, who, with the ardour natural to their profession when any enterprise is in view, came forward in the readiest manner, and volunteered their ser-

vices ; but Mr. Elson, the master, who had hitherto
commanded the boat, had acquitted himself so much
to my satisfaction, that I could not in justice remove
him ; more especially at a moment when the service
to be performed was inseparable from risk.   Mr.
Smyth the senior mate of the ship, who executed
the greater part of the drawings which illustrate this
work, was placed with Mr. Elson, who had besides
under his command a crew of six seamen and two
marines.

My instructions to Mr. Elson were to trace the
shore to the north-eastward as far as it was possible
for a boat to navigate, with a view to render the
earliest possible assistance to Captain Franklin, and
to obtain what information he could of the trending
of the coast and of the position of the ice.   He was
also directed to possess himself of facts which, in the
event of the failure of the other expedition, would
enable us to form a judgment of the probable success
which might attend an attempt to effect a north-
eastern passage in this quarter : and further, he was
to avoid being beset in the ice, by returning immediately the wind should get to the north-west or
westward, and not to prolong his absence from the
ship beyond the first week in September.   He was
at the same time ordered to place landmarks and
directions in conspicuous places for Captain Franklin's guidance ; and if possible, on his return, to
examine the shoals off Icy Cape.

We steered together to the northward with foggy
weather until midnight on the 17th, when I made
Mr. Elson's signal to part company, and he commenced his interesting expedition with the good
wishes of all on board.   We continued our course

to the northward until four o'clock in the morning of the 18th, when the fog, as is usual in the neighbourhood of the ice, cleared away, and we saw the main body in latitude 71° 07′ N. nearly in the same position we had left it some days before. It was loose at the edge, but close within, and consisted of heavy floes. We tacked near it, and found it trending from E. to S. W. (mag.) There were no living things near it, except a few tern and kittiwakes; which was rather remarkable, as the edge of the ice is usually frequented by herds of amphibious animals. As we receded from the ice, the fog again thickened, and latterly turned to small snow. The temperature was about the freezing point. At noon the sun broke through, and we found ourselves in latitude 70° 18′ N., and by the soundings about twelve miles from the land, which was not seen. By this we discovered that instead of gaining twenty miles to the eastward, we had lost four: by which it was evident that a current had been running S. 58° W. a mile an hour; off this place, however, it was found upon trial to run S. 60° W. only half a mile per hour. The fog afterwards came on very thick, and remained so during the day.

Finding this inconvenience from the current off Icy Cape, I steered to the westward to ascertain how near the ice approached the coast in that direction, and on the 20th, I stood in for the land about midway between Cape Beaufort and Icy Cape, to verify some points of the survey. About this time immense flocks of ducks, consisting entirely of young ones and females, were seen migrating to the southward. The young birds could not fly; and not having the instinct to avoid the ship in time, one immense flock was run completely over by her. They, however, were more wary when the boats were lowered, and successfully avoided our attempts to shoot them, by diving. At the place where we landed, there was a long lake between us and the main land; and our walk was confined to a strip of shingle and sand, about 150 yards wide, and about six feet above the level of the sea. In the sheltered parts of it there were a few flowers, but no new species. The lake was connected with the sea at high tide, and was consequently salt; but we obtained some water sufficiently fresh to drink by digging at a distance of less than a yard from its margin, a resource of which the natives appeared to be well aware.

An abundance of drift wood was heaped upon the upper part of the shingle. The trees were torn up by the roots, and some were worm-eaten; but the greater part appeared to have been only a short time at sea, and all of it, that I examined, was pine.

From the desolate appearance of the coast where we landed, I scarcely expected to find a human being, but we had no sooner put our foot ashore than a baidar full of people landed a short distance from us. Her crew consisted of three grown-up males and four females, besides two infants. They were as ready as their neighbours to part with what they had in exchange for trifles; esteeming our old brass buttons above all other articles, excepting knives. There was a blear-eyed old hag of the party, who separated from her companions, and seated herself upon a piece of drift wood at a little distance from the baidar, and continued there, muttering an unintelligible language, and apparently believing herself to be holding communion with that invisible world to which she was fast approaching. Though in her dotage, her opinion was often consulted, and on more than one occasion in a mysterious manner. We afterwards witnessed several instances of extremely old women exercising great influence over the younger part of the community. On this occasion I purchased a bow and quiver of arrows for a brooch. The man who sold them referred the bargain to the old woman above-mentioned, who apparently disapproved of it, as the brooch was returned, and the bow and arrows re-demanded.

The males of this party were all provided with lip ornaments; and we noticed a gradation in the size, corresponding to the ages of the party who wore them, as well as a distinction in the nature of them. Two young lads had the orifices in their lips quite raw: they were about the size of a crow-quill, and were distended with small cylindrical pieces of ivory, with a round knob at one end to prevent their falling out. For some time after the operation has been performed, it is necessary to turn the cylinders frequently, that they may not adhere to the festering flesh: in time this action becomes as habitual with some of them as that of twirling the mustachios is with a Mussulman. In the early stage it is attended with great pain, the blood sometimes flowing, and I have seen tears come into the boys' eyes while doing it. Lip ornaments, with the males, appear to correspond with the tattooing of the chins of the females; a mark which is universally borne by the women throughout both the eastern and western Esquimaux tribes: the custom of wearing the labrets, however, does not extend much beyond the

Mackenzie River. The children we saw to-day had none of these marks; a girl, about eleven, had one line only; and a young woman, about twenty-three years of age, the mother of the infants, had the three perfect. One of her children was rolling in the bottom of the baidar, with a large piece of seal-blubber in its mouth, sucking it as an European child would a coral. The mother was rather pretty, and allowed her portrait to be taken. At first she made no objection to being gazed at as stedfastly as was necessary for an indifferent artist to accomplish his purpose; but latterly she shrunk from the scrutiny with a bashfulness that would have done credit to a more civilized female; and on my attempting to uncover her head, she cast a look of inquiry at her husband, who vociferated " naga," when she very properly refused to comply. The young men were very importunate and curious, even to annoyance; and there is little doubt that if any persons in our dress had fallen in with a powerful party of these savages, they would very soon have been made to exchange their suit of broad cloth for the more humble dress of furs. Their honesty was not more conspicuous than their moderation, as they appropriated to themselves several articles belonging to Mr. Collie.

During three hours that we were on shore, the tide fell one foot; it had subsided eighteen inches from its greatest height when we first landed, and when we put off was still ebbing to the S. S. W. at the rate of half a mile an hour. Four hours afterwards, when by our observations on shore it must have changed, it ran N. $\frac{1}{2}$ E. at the same rate, and afforded another instance of the flood coming from the southward.

VOL. I. 2 C

A thick fog came on after we returned on board. The next morning we closed with the land near Cape Beaufort, with a view of trying the veins of coal in its neighbourhood, as we were very short of that article; but the wind veered round to the N. N. W., and by making it a lee shore prevented the boats landing, and rendered it expedient for the ship, which was very light, and hardly capable of beating off, to get an offing. The day was fine, and afforded an opportunity of verifying some of our points, which we had the satisfaction to find quite correct. The next day the wind veered to the S. S. W. and then to the westward. Throughout the 23d, 24th, and part of the 25th, it blew hard, with a short head sea, thick weather, and latterly with snow showers, which obliged the ship to keep at so great a distance that the land expedition would have passed her unobserved, had they been in progress along the coast. With these winds we kept off the coast. The night of the 25th was clear and cold, with about four hours' darkness, during which we beheld a brilliant display of the aurora borealis, which was the first time that phenomenon had been exhibited to us in this part of the world. It first appeared in an arch extending from W. by N. to N. E. mag. (by the north), passing through Benetnasch, β. γ. Ursæ Maj. and β. Aurigæ, decidedly dimming their lustre. The arch, shortly after it was formed, broke up; but united again, threw out a few coruscations, and then entirely disappeared. Soon after, a new display began in the direction of the western foot of the first arch, preceded by a bright flame, from which emanated coruscations of a pale straw-colour. An almost simul-

taneous movement occurred at both extremities of
the arch, until a complete segment was formed of
wavering perpendicular radii. As soon as the arch
was complete, the light became greatly increased ;
and the prismatic colours, which had before been
faint, now shone forth in a very brilliant manner.
The strongest colours, which were also the outside
ones, were pink and green ; the centre colour was
yellow, and the intermediate ones on the pink side
purple and green ; on the green side purple and
pink, all of which were as imperceptibly blended
as in the rainbow. The green was the colour
nearest the zenith. This magnificent display lasted
a few minutes ; and the light had nearly vanished,
when the N. E. quarter sent forth a vigorous display,
and nearly at the same time a corresponding corus-
cation emanated from the opposite extremity. The
western foot of the arch then disengaged itself from
the horizon, crooked to the northward, and the
whole retired to the N. E. quarter, where a bright
spot blazed for a moment, and all was darkness. I
have been thus particular in my description, because
the appearance was unusually brilliant, and because
very few observations on this phenomenon have been
made in this part of the world. There was no noise
audible during any part of our observations, nor
were the compasses perceptibly affected. The night
was afterwards squally, with cumuli and nimbi, which
deposited showers of sleet and snow as they passed
over us, the wind being rather fresh throughout.

On the 26th the weather was moderate, and
being off Point Hope, on which there were several
lakes and a great abundance of driftwood, the boats
were sent to endeavour to procure a supply of fuel

and water. We had completed only one turn, and buried a bottle for Captain Franklin, when the wind freshened from the S. W. and prevented a second landing. During the afternoon we turned to windward, with the wind blowing fresh from the westward.

From the time of our passing Beering's Strait up to the 23d instant, we enjoyed an almost uninterrupted series of fine weather; during which we had fortunately surveyed the whole of the coast from Cape Prince of Wales as far to the northward as I deemed it proper to go, consistent with the necessity of keeping the ship, at all times, in open water and in safety Now, however, there appeared to be a break up, and a commencement of westerly winds, which made the whole of this coast a lee shore, and together with several hours of darkness rendered it necessary to keep the ship at a distance from the land. In doing this the chances were equal that the land expedition, in the event of its success, would pass her. I therefore determined to repair to the rendezvous in Kotzebue Sound, and, as nothing further was to be done at sea, to await there the arrival of our boat and of Captain Franklin's expedition. Accordingly on the 27th we made Cape Krusenstern, and on the following evening anchored at Chamisso Island nearly in our former situation.

Directly the ship was secured, two boats were despatched to the islands to examine the state of the rivulets, and ascertain whether the cask of flour, that had been buried for Captain Franklin's use, had been molested; our suspicion of its safety having been excited by observing six baidars upon the beach opposite the anchorage, none of which ventured off to the ship as was usual. On the return of the boat from Chamisso Island we learned that there was not a drop of water to be had, in consequence of the streams at which we had formerly filled our casks being derived from beds of thawing ice and snow which were now entirely dissolved.

By the other boat, we found, as we expected, that the cask of flour had been dug up and broken open, that the hoops had been taken away, and that the flour had been strewed about the ground, partly in a kneaded state. Suspicion immediately fell upon the natives encamped upon the peninsula, which was strengthened by the manner in which they came off the next morning, dancing and playing a tambourine in the boats, a conciliatory conduct with which we had never before been favoured. When they came alongside, they were shown a handful of flour, and were referred to the island upon which the cask had been buried. Their guilty looks showed that they perfectly understood our meaning ; but they strongly protested their innocence, and as a proof that they could not possibly have committed the theft, they put their fingers to their tongues, and spit into the sea with disgust, to show us how much they disliked the taste of the material, little considering that the fact of their knowing it to be nauseous was a proof of their having tasted it : but no further notice was taken of the matter, as I wished as much as possible to conciliate their friendship on account of the land expedition.

The baidars of these people were better made than any we had seen, excepting those of the St. Lawrence islanders, which they resembled in having a flap made of walrus skin attached to the gunwale for the purpose of keeping their bows and arrows dry. The natives had a great variety of articles for sale, all of which they readily parted with, except their bows, arrows, and spears, and these they would on no account sell. Several old men were among their party, all of whom sat in the stern of the boat, a deference which, as I have already said, we every-where observed to be paid to age by the younger part of this tribe. When they had sold all they intended to part with, and had satisfied their curiosity, they paddled on shore, well satisfied, no doubt, at having escaped detection.

The next morning the boats were sent to find water and to dig wells upon Chamisso Island, as we had but nine days' supply on board at very reduced allowance. In the mean time I paid a visit to the Esquimaux, who were on their travels towards home with cargoes of dried salmon, oil, blubber, and skins, which they had collected in their summer excursion along the coast. When they perceived our boat approaching the shore, they despatched a baidar to invite us to their encampment; and as we rowed toward the place together, observing with what facility they passed our boat, they applied their strength to their paddles, and, exulting on the advantage they possessed, left us far behind. It was perfectly smooth and calm, or this would not have been the case, as their boats have no hold of the water, and are easily thrown back by a wave; and when the wind is on the side, they have the greatest difficulty in keeping them in the right direction.

The shallowness of the water obliged our boat to land a short distance from the village; and the natives, who by this time had hauled up their baidar, walked down to meet us with their arms drawn in from their sleeves, and tucked up inside their frocks. They were also very particular that every one of them should salute us, which they did by licking their hands, and drawing them first over their own faces and bodies, and then over ours. This was considered the most friendly manner in which they could receive us, and they were officiously desirous of ingratiating themselves with us; but they would on no account suffer us to approach their tents; and, when we urged it, seemed determined to resist, even with their weapons, which were carefully laid out upon a low piece of ground near them. They were resolved, nevertheless, that we should partake of their hospitality, and seating us upon a rising ground, placed before us strips of blubber in wooden bowls, and whortle berries mashed up with fat and oil, or some such heterogeneous substance, for we did not taste it. Seeing we would not partake of their fare, they commenced a brisk traffic with dried salmon, of which we procured a great quantity. Generally speaking, they were honest in their dealings, leaving their goods with us, when they were in doubt about a bargain, until they had referred it to a second person, or more commonly to some of the old women. If they approved of it, our offer was accepted; if not, they took back their goods. On several occasions, however, they tried to impose upon us with fish-skins, ingeniously put together to represent a whole fish, though entirely deprived of their original contents; but this artifice succeeded only once: the natives, when detected in other attempts, laughed heartily, and treated the matter as a fair practical joke. Their

cunning and invention were further exhibited in
the great pains which they took to make us understand, before we parted, that the flour had been
stolen by a party who had absconded on seeing the
ship. Their gestures clearly intimated to us that
the attention of this party had been attracted to the
spot by the newly turned earth, though we had replaced it very carefully ; on which, it appears, they
began to dig, and, to their great surprise and joy
no doubt, they soon discovered the cask. They
knocked off the hoops with a large stone, and then
tasted the contents, which they intimated were very
nauseous. The thieves then packed up the hoops,
and carried them over the hills to another part of
the country.

We patiently heard the whole of this circumstantial account, which we had afterwards great reason
to believe was an invention of their own, and that
they had some of the flour secreted in their tents,
which, no doubt, was the reason of their dislike to
our approaching them.

In the forenoon one of our seamen found a piece
of board upon Chamisso Island, upon which was
written, in Russian characters, " Rurick, July 28th,
1816," and underneath it " Blaganome erinoy,
1820." The former was of course cut by Kotzebue
when he visited the island ; and the latter, I
suppose, by Captain Von Basilief Schismareff, his
lieutenant, who paid this island a second visit in
1820.

Upon the low point of this island there was another party of Esquimaux, who differed in several
particulars from those upon the peninsula. I was
about to pay them a visit, but early in the morning

our peninsular friends came off to say they were
going away ; and as I wished to see a little more of
them before they left us, I deferred going there
until the next day, by which I lost the opportunity
of seeing those upon Chamisso, as they decamped
in the evening unobserved. They were, however,
visited by several of the officers. Like the party on
the peninsula, they were on their return to winter-
quarters, with large heaps of dried fish, seals' flesh,
oil, skins, and all the necessary appurtenances to an
Esquimaux residence. They had four tents and
several baidars, which were turned over upon their
nets and fishing-tackle for protection. In one of
their tool-chests was found a part of an elephant's
tooth, of the same species as those which were afterwards collected in Escholtz Bay. They had the
same aversion to our officers approaching their habitations as the party before described on the peninsula, and in all probability it proceeded from the
same cause, as Mr. Osmer detected a young girl
eating some of our flour mixed up with oil and
berries. On seeing him she ran hastily into her
tent, and in so doing spilt some of the mixture,
which led to the discovery.

The women of this party differed from the females
we had hitherto seen, in having the septum of the nose
pierced, and a large blue bead strung upon a strip
of whalebone passed through the orifice, the bead
hanging as low as the opening of the mouth. One
of them, on receiving a large stocking-needle, thrust
it into the orifice, or, as some of the seamen said,
" spritsail-yarded her nose." A youth of the party
who had not yet had his lips perforated wore his
hair in bunches on each side of the head, after the

fashion of the women, which I notice as being the only instance of the kind we met with, and which I trust does not indicate a nearer resemblance to a class of individuals mentioned by Langsdorff as existing in Oonalashka under the denomination of Schopans.

Red and blue beads, buttons, knives, and hatchets were as usual the medium through which every thing they would part with was purchased. The men were more excited than usual by a looking-glass, which, after beholding their own features in it, and admiring alternately the reflection of their head and lip ornaments, they very inconsiderately carried to one of their party who was perfectly blind, and held before his face. As this was done rather seriously, certainly without any appearance of derision, it is possible that they imagined it might produce some effect upon his sight.

On landing at the encampment on the peninsula, I was received in a more friendly manner even than the day before. Each of the natives selected a friend from among our party, and, like the Gambier island-ers, locked their arms in ours, and led us to a small piece of rising ground near their tents, where we sat down upon broad planks and deer-skins. A dried fish was then presented to each of us, and a bowl of cranberries mashed up with sorrel and rancid train-oil was passed round, after the manner of the Krai-kees on the Asiatic shore; but, however palatable this mixture might have been to our hosts, it was very much the reverse to us, and none of our party could be induced to partake of it, except Mr. Osmer, who did so to oblige me at the expense of his appe-tite for the rest of the day. The Esquimaux were surprised at our refusal of this offer, and ridiculed our squeamishness; and by way of convincing us what bad judges we were of good cheer, five of them fell to at the bowl, and with their two fore-fingers very expeditiously transferred the contents to their own mouths; and cleansing their fingers upon the earth, gave the vessel to one of the women.

The whole village then assembled, better dressed than they had been on our first visit, and ranged themselves in a semicircle in front of us, prepara-tory to an exhibition of one of their dances, which merits a description, as it was the best of the kind we saw. A double ring was formed in front of us by men seated upon the grass, and by women and children in the background, who composed the orchestra. The music at the beginning was little better than a buzz of "Ungnā-ăyā, Amnā-ăyā!" —words which always constitute the burthen of an Esquimaux song. The leader of the party, a strong athletic man, jumped into the ring and threw him-self into various attitudes, which would have better become a pugilist than a performer on the light fantastic toe! As his motions became violent, he manifested his inspiration by loud exclamations of Ah! Ah! until he became exhausted and with-drew, amidst shouts of approbation from all pre-sent, and the signal was given for new performers. Five younger men then leaped into the area, and again exhibited feats of activity, which, considering the heavy clothing that encumbered their limbs, were very fair. A simple little girl about eight years of age, dressed for the occasion, joined the jumpers, but did not imitate their actions. Her part consisted in waving her arms and inclining her body from side to side. The poor little thing was so abashed that she did not even lift her head or open her eyes during the whole of her performance, and seemed glad when it was over, though she was not unmindful of the praise bestowed upon her exertions.

The violent action of the male performers required that they should occasionally take breath, during which time the music was lowered; but as soon as the ring was re-furnished it again became loud and animated. A grown-up female now formed one of the party, and appeared to be the prize of contention among several young men, who repeatedly endeavoured to ingratiate themselves with her, but she as often rejected their offers and waved them away. At last an old man, all but naked, jumped into the ring, and was beginning some indecent gesticulations, when his appearance not meeting with our approbation, he withdrew, and the performance having been wrought to its highest pitch of noise and animation, ceased.

Such is the rude dance of these people, in which, as may be seen from the above description, there was neither elegance nor grace; but on the contrary it was noisy, violent, and as barbarous as themselves. The dancers were dressed for the occasion in their best clothes, which they considered indispensable, as they would not sell them to us until the performance was over. In addition to their usual costume, some had a kind of tippet of ermine and sable skins thrown over their shoulders, and others wore a band on their heads, with strips of skin suspended to it at every two inches, to the end of which were attached the nails of seals.

When the dance was over, they presented us with dried salmon, and each person brought his bag of goods, which produced a brisk barter, with great fairness on all sides, and with a more than ordinary sense of propriety on theirs, in never raising or lowering their prices; and by their testifying their disapprobation of it by a groan, when it was attempted by one of our party. But though so strict in this particular, they were not exempt from that failing so unaccountably innate in all uncivilized people, which they endeavoured to gratify in various ways, by engaging our attention at a moment when some of our trinkets were exposed to them for the purpose of selection. Suspecting their designs, however, we generally detected their thefts, and immediately received back our goods, with a hearty laugh in addition. They understood making a good bargain quite as well as ourselves, and were very wary how they received our knives and hatchets, putting their metal to the test by hacking at them with their own. If they stood the blow, they were accepted; but if, on the contrary, they were notched, they were refused. A singular method of deciding a bargain was resorted to by one of their party, almost equivalent to that of tossing up a coin. We had offered an adze for a bundle of skins; but the owner, who at first seemed satisfied with the bargain, upon reflection became doubtful whether he would not be the loser by it; and to decide the doubtful point he caught a small beetle, and set it at liberty upon the palm of his hand, anxiously watching which direction the insect should take. Finding it run towards him, he concluded the bargain to be disadvantageous to him, and took back his goods.

On this day they admitted us to their habitations, and all restrictions were removed, except that upon writing in our remark books, to which they had such an objection, that they refused us any information while they were open, and with great good-nature closed them, or if we persisted, they dodged their heads and made off.

Our new acquaintances, amounting to twenty-five in number, had five tents, constructed with skins of sea-animals, strained upon poles; and for floors they had some broad planks two feet in the clear. I was anxious to learn where they obtained these, knowing that they had themselves no means of reducing a tree to the form of a plank, but I could get no information on this point: in all probability they had been purchased from the Tschutschi, or the Russians. Each tent had its baidar, and there were two to spare, which were turned upside down, and afforded a convenient house for several dogs, resembling those of Baffin's Bay, which were strapped to logs of wood to prevent their straying away. In front of these baidars there were heaps of skins filled with oil and blubber, &c., and near them some very strong nets full of dried salmon, suspended to frames made of drift wood: these frames also contained, upon stretchers, the intestines of whales, which are used for a variety of purposes, particularly for the kamlaikas, a sort of shirt which is put over their skin dresses in wet weather.

More provident than the inhabitants of Melville Peninsula, these people had collected an immense store of provision, if intended only for the number of persons we saw. Besides a great many skins of oil, blubber, and blood, they had about three thousand pounds of dried fish.

On the first visit to this party, they constructed a chart of the coast upon the sand, of which I took very little notice at the time. To-day, however, they renewed their labour, and performed their work upon the sandy beach in a very ingenious and intelligible manner. The coast line was first marked out with a stick, and the distances regulated by the days' journeys. The hills and ranges of mountains were next shown by elevations of sand or stone, and the islands represented by heaps of pebbles, their proportions being duly attended to. As the work proceeded, some of the bystanders occasionally suggested alterations, and I removed one of the Diomede Islands which was misplaced: this was at first objected to by the hydrographer; but one of the party recollecting that the islands were seen *in one* from Cape Prince of Wales confirmed its new position, and made the mistake quite evident to the others, who seemed much surprised that we should have any knowledge of such things. When the mountains and islands were erected, the villages and fishing stations were marked by a number of sticks placed upright, in imitation of those which are put up on the coast wherever these people fix their abode. In time, we had a complete topographical plan of the coast from Point Darby to Cape Krusenstern. In this extent of coast line they exhibited a harbour and a large river situated to the southward of Cape Prince of Wales, of neither of which we had any previous knowledge. The harbour communicated with an inner basin, named Imaurook, which was very spacious, and where the water was fresh. The entrance to the outer one was so narrow, that two baidars could not paddle abreast of each other. This they explained by means of two pieces of wood, placed together, and motioning with their hands that they were paddling. They then drew them along till they came to the channel, when they were obliged to follow one another, and, when through, they took up their position, as before. The river was between this harbour and the cape, and by their description it wound among lofty mountains, and between high rocky cliffs, and extended further than any of the party had been able to trace in their baidars. Its name was Youp-nut, and its course must lie between the ranges of mountains at the back of Cape Prince of Wales. At this last mentioned cape, they placed a village, called Iden-noo; and a little way inland another, named King-a-ghee, which was their own winter residence. Beyond Imau-rook there was a bay, of which we have no knowledge, named I-art-so-rook. A point beyond this, which I took to be the entrance to Norton Sound, was the extent of their geographical knowledge in that direction.

To the Diomede Islands they gave the names of Noo-nar-boak, Ignarlook, and Oo-ghe-eyak; King's Island, Oo-ghe-a-book; and Sledge Island, Ayak. It is singular that this island, which was named Sledge Island by Captain Cook, from the circumstance of one of these implements being found upon it, should be called by a word signifying the same thing in the Esquimaux language. For East Cape they had no name, and they had no knowledge of any other part of the Asiatic coast. Neither Schismareff Bay nor the inlet in the Bay of Good Hope was delineated by them, though they were not ignorant of the former when it was pointed out to them. It has been supposed that these two inlets

communicate, and that the Esquimaux, who intimated to Kotzebue that a boat could proceed nine
days up the latter and would then find the sea,
alluded to this junction; but our rude hydrogra-
phers knew of no such communication; which I
think they certainly would, had it existed, as by
pursuing that course they would have avoided a
passage by sea round Cape Espenburg, which in
deep-laden boats is attended with risk, from the
chance of their not being able to land upon the
coast. They would, at all events, have preferred an
inland navigation had it not been very circuitous.

We passed the greater part of the day with these
intelligent people, who amused us the whole time in
some way or other. The chief, previous to our
embarkation, examined every part of our boat, and
was highly pleased with the workmanship, but he
seemed to regret that so much iron had been expended where thongs would have served as well.
He was more astonished at the weight of a sounding lead than at any thing in the boat, never having
felt any metal so heavy before; iron pyrites being
the heaviest mineral among this tribe.

When we were about to embark, all the village
assembled and took leave of us in the usual manner
of the Esquimaux tribes; and as it was probable we
should never meet again, the parting, much to our
annoyance, was very affectionate. A middle-aged
man, who had taken the lead throughout, and who
was probably their *neakoa* (or head-man) recommended us to depart from these regions; but I signified my intention of waiting some time longer,
and sleeping at least twenty nights where we were;
on which he shivered, and drew his arms in from his

sleeves to apprise us of the approaching cold. I thanked him for his advice, and making them each a parting present we took our leave. The next morning they embarked every thing, and paddled over to Escholtz Bay. After they were gone, we found some of our flour where the tents had stood, and a quantity of it secreted in a bush near the place; so that their cautious behaviour with regard to our approaching their tents the first day was no doubt occasioned by fear of this discovery; and they afterwards secreted their plunder in a manner probably not likely to meet detection.

Among this party there was a man so crippled that he went upon all fours; how it occurred we could not learn, but it was probably in some hunting excursion, as several of his companions had deep scars which they intimated had been inflicted by walrusses, which in the following year we found in great numbers off the coast. In this party we detected a difference of dialect from what we had heard in general, which made their objection to our writing in our books the more provoking, as it prevented us recording any of the variations, except in regard to the negative particle *no*, which with other parties was *naga*, and with these, *aun-ga*. The females were provided with broad iron bracelets, which we had not seen before; and by their having four or five of them upon each wrist, it appeared that this metal, so precious with the tribes to the northward, was with them less rare: nevertheless it is very probable that they intended to appropriate to this purpose the iron hoops they had stolen from us.

I have said nothing of the dress or features of these people, as, with the exception of two of them, they so nearly resemble those already described as to render it unnecessary. These two persons, in the tattooing of the face, and in features, which more nearly resembled those of the Tschutschi, seemed to be allied to the tribes on the Asiatic coast, with whom they no doubt have an occasional intercourse.

On the first of September our sportsmen succeeded in bagging several braces of ptarmigan and wild ducks; but game was not so plentiful as might have been expected at this season of the year, in a country so abundantly provided with berries and so scantily inhabited. It was a pleasure to find that we could now pursue this and other occupations free from the annoyance of moskitos; a nuisance which, whatever it may appear at first, is in reality not trifling. Dr. Richardson fixes the departure of these insects from Fort Franklin on the 11th of September: here, however, it takes place at least a fortnight earlier.

On the 5th I visited the northern side of Escholtz Bay, and found the country almost impassable from swamp, notwithstanding the season was so far advanced. It seemed as if the peaty nature of the covering obstructed the drainage of the water, which the power of the sun had let loose during the summer, and that the frozen state of the ground beneath prevented its escape in that direction. The power of the sun's rays upon the surface was still great, and large stones and fragments of rock that had been split by the frost were momentarily relinquishing their hold and falling down upon the beach. A thermometer exposed upon a piece of black cloth rose to 112°, and in the shade stood at 62°. On the side of the hill that sloped to the southward the

willow and birch grew to the height of eighteen
feet, and formed so dense a wood that we could not
penetrate it.   The trees bordering upon the beach
were quite dead, apparently in consequence of their
bark having been rubbed through by the ice, which
had been forced about nine feet above high water
mark, and had left there a steep ridge of sand and
shingle.   The berries were at this time in great perfection and abundance, and proved a most agreeable
addition to the salt diet of the seamen, who were
occasionally permitted to land and collect them.

The cliffs on this side of Choris Peninsula were
composed of a green-coloured mica slate, in which
the mica predominated, and contained garnets, veins
of feldspar, enclosing crystals of schorl, and had its
fissures filled with quartz ; but I shall avoid saying
any thing on geological subjects here.

On the 6th our curiosity was excited by the appearance of two small boats under sail, which, when
first seen through a light fog, were so different from
the sails of the Esquimaux, that our imagination,
which had latterly converted every unusual appearance in the horizon into the boats of Captain Franklin, really led us to conclude he had at length arrived ; but as they rounded the point, we clearly
distinguished them to be two native baidars.   We
watched their landing, and were astonished at the
rapidity with which they pitched their tents, settled
themselves, and transferred to their new habitation the contents of the baidars, which they drew
out of the sea and turned bottom upwards.   On
visiting their abode an hour after they landed, every
thing was in as complete order as if they had been
established there a month, and scarcely any thing

was wanting to render their situation comfortable. No better idea could have been conveyed to us of the truly independent manner in which this tribe wander about from place to place, transporting their houses, and every thing necessary to their comfort, than that which was afforded on this occasion. Nor were we less struck with the number of articles which their ingenuity finds the means of disposing in their boats, and which, had we not seen them disembarked, we should have doubted the possibility of their having been crammed into them. From two of these they landed fourteen persons, eight tent poles, forty deer skins, two kyacks, many hundred weight of fish, numerous skins of oil, earthen jars for cooking, two living foxes, ten large dogs, bundles of lances, harpoons, bows and arrows, a quantity of whalebone, skins full of clothing, some immense nets made of hide for taking small whales and porpoises, eight broad planks, masts, sails, paddles, &c., besides sea-horse hides and teeth, and a variety of nameless articles always to be found among the Esquimaux.

They received us in the most friendly and open manner, and their conduct throughout was so different from that of their predecessors, that had we had no proof of the latter being guilty of the theft on our flour, this difference of conduct would have afforded a strong presumption against them. The party consisted of two families, each of which had its distinct property, tents, baidar, &c. They were in feature and language nearly connected with the King-a-ghee party, and from what they told us resided near them; but to judge from their dresses and establishment they were of much lower condi-

tion.   However, the women had the same kind of
beads in their ears, and sewn upon their dresses, and
had evidently been to the same market.   We remarked, however, in two of the young ladies a custom which, when first discovered, created considerable laughter.   When they moved, several bells
were set ringing, and on examining their persons, we
discovered that they had each three or four of these
instruments under their clothes, suspended to their
waists, hips, and one even lower down, which was
about the size of a dustman's bell, but without a
clapper.   Whether they had disposed of them in
this manner as charms, or through fear, it was impossible to say ; but by their polished surface, and
the manner in which they were suspended, they appeared to have long occupied these places.   They
were certainly not hung there for convenience, as
the large one in particular must have materially incommoded the ladies in their walking.   One of our
party suggested that this large bell might, perhaps,
be appropriated to the performance of a ceremony
mentioned by Muller, in his " Voyages from Asia
to America," &c., p. 28., where he states that the
bond of friendship or enmity depends upon a guest
rinsing his mouth with the contents of the cup,
which formed an indispensable part of a very singular custom among the Tschutschi, the people of
Cashemir, and some other countries.*

Among other things, this party had small bags of
resin, which appeared to be the natural exudation of
the pine.   From their constantly chewing it, it did
not seem difficult to be had ; and as no trees of this
nature, that we were acquainted with, grew upon the

* M. Paulus Venetus, Witsen, and Trigaut.

coast, we were anxious to learn whence they had procured it, but we could not make our acquaintances understand our wishes.

An old lady, who was the mother of the two girls with the bells, invited me into her tent, where I found her daughters seated amidst a variety of pots and pans, containing the most unsavoury messes, highly repugnant to both the sight and smell of a European, though not at all so to the Esquimaux. These people are in the habit of collecting certain fluids for the purposes of tanning; and that, judging from what took place in the tent, in the most open manner, in the presence of all the family.

The old matron was extremely good-natured, lively, and loquacious; and took great pleasure in telling us the name of every thing, by which she proved more useful than any of our former visiters; and had she but allowed us time to write down one word before she furnished another, we should have greatly extended our vocabularies; but it appeared to her, no doubt, that we could write as fast as she could dictate, and that the greater number of words she supplied, the more thankful we should be. So far from this party having any objection to our books, to which the former one had manifested the greatest repugnance, they took pleasure in seeing them, and were very attentive to the manner in which every thing was committed to paper.

The daughters were fat good-looking girls; the eldest, about thirteen years of age, was marked upon the chin with a single blue line; but the other, about ten, was without any tattooing. I made a sketch of the eldest girl, very much to the satisfaction of the mother, who was so interested in having her daughter's picture, and so impatient to see it finished, that she snatched away the paper several times to observe the progress I was making. The father entered the tent while this was going forward, and observing what I was about, called to his son to bring him a piece of board that was lying outside the tent, and to scrape it clean, which indeed was very necessary. Having procured a piece of plumbago from his wife, he seated himself upon a heap of skins, threw his legs across, and very good-humouredly commenced a portrait of me, aping my manner and tracing every feature with the most affected care, whimsically applying his finger to the point of his pencil instead of a penknife, to the great diversion of his wife and daughters. By the time I had finished my sketch, he had executed his, but with the omission of the hat, which, as he never wore one himself, he had entirely forgotten; and he was extremely puzzled to know how to place it upon the head he had drawn.

On meeting with the Esquimaux, after the first salutation is over an exchange of goods invariably ensues, if the party have any thing to sell, which is almost always the case; and we were no sooner seated in the tent than the old lady produced several bags, from which she drew forth various skins, ornamental parts of the dress of her tribe, and small ivory dolls, allowing us to purchase whatever we liked. Our articles of barter were necklaces of blue beads, brooches, and cutlery, which no sooner came into the possession of our hostess than they were transferred to a stone vessel half filled with train-oil, where they underwent an Esquimaux purification.

We found amongst this party a small Russian

coin of the Empress Catherine, and the head of a
halberd, which had been converted into a knife;
both of which were evidence of the communication
that must exist between their tribe and those of the
Asiatic coasts opposite.

We returned on board with a boat full of dried
salmon, and the next day the party visited the ship.
Notwithstanding the friendly treatment they had
experienced the day before, it required much persuasion to induce them to come upon deck; and
even when some of them were prevailed upon to do
so, they took the precaution of leaving with their
comrades in the boat whatever valuable articles they
had about their persons. They were shown every
thing in the ship most likely to interest them, but
very few objects engaged them long, and they passed
by some that were of the greatest interest, to bestow
their attention upon others which to us were of
none, thus showing the necessity of fully understanding the nature of any thing before the mind
can properly appreciate its value. The sail-maker
sewing a canvass bag, and the chain cable, were two
of the objects which most engaged their attention;
the former from its being an occupation they had
themselves often been engaged in; and the latter as
exhibiting to them the result of prodigious labour,
as they would naturally conclude that our chains—
though so much larger and of so much harder a material than their own—were made in the same manner. The industry and ingenuity of the Esquimaux
are, however, displayed in nothing more than in the
fabrication of chains, two or three of which we met
with cut out of a solid piece of ivory. On showing
these people the plates of natural history in Rees's

Cyclopædia, they were far more intelligent than might have been expected from the difficulty that naturally occurs to uncivilized people in divesting their minds of the comparative size of the living animal and its picture. But the Esquimaux are very superior in this respect to the South Sea Islanders, and immediately recognised every animal they were acquainted with that happened to be in the book, and supplied me with the following list of them :—

English Names.	Esquimaux Names.	English Names.	Esquimaux Names.
Squirrel	*Tsēy-kĕrĕck.*	Porpoise	*Agh-bĕĕ-zēēak.*
Fox	*Kiŏck-tōōt.*	Dog	*Koo-nēak.*
Musk rat	*Paōōna.*	Owl	*Ignă-zĕĕ-wyŭck.*
Rein-deer	*Tootōōt.*	Falcon	*Kje-gōō-ŭt.*
*Musk ox	*Mĭgn-ŭgne.*	Grouse	*A-hăg-ghī-ŭck.*
White bear	*Tsŭ-nark.*	Snipe	*Nŭck-tŏo-ō-lĭt.*
Walrus	*Ei-bwŏ-āk.*	Vulture	*Keegli-āght.*
Seal	*Kasi-gōō-ăk.*	Swan	*Tădi-drācht.*
Otter	*Te-ghĕ-āk-bŏŏk.*	Duck	*Ew-ŭck.*
Porcupine	*Igla-koo-sŏk.*	Puffin	*Kŏŏli-nŏckt.*
Mouse	*Kŏŏblă-ōōk,*	Plover	*Tud-glĭct.*
Beaver	*Ka-boo-ek.*	Pelican	*Pĕĕbli-ark-tōōk.*
Hare	*Oŏ-gōōd-lĭgh.*	Salmon	*Ish-allōōk.*
Goat	*Ip-nū-ŭck.*	Flounder	*Ek-anēē-luk.*
Sheep	*Ok-shŭlk.*	Guard fish	*Iz-nēē-a-ŏŏk.*
*Bull (musk?)	*Mōōng-măk.*	Crab	*Edlŏŏ-azrēy-ŭk.*
White horse	*Izŏŏ-kār-ŭck.*	Shrimp	*Nōwd-lĕnnŏk.*
Narwhal	*Tse-dōō-ăk.*	Lobster	*Pōō-cœ-ō-tuk.*
Whale	*Ah-ōw-lŏŏk.*	Butterfly	*Tăr-dlĕ-ōōt-zŭk.*

Among which there are three animals—the goat, the sheep, and the horse—hitherto unknown upon this coast: probably the sheep may refer to the

---

* See Observations on these names attached to the Vocabulary in the Appendix.

argali, which has been seen near Cook's River. By
the time I had collected these names, our visiters
had become impatient to join their comrades, who
in like manner, finding them a long time absent,
had become equally anxious on their account, and
had quitted the boat in search of them, and both
parties met upon deck to their mutual satisfaction.
Previous to their going away we made them several
useful presents of axes, knives, combs, &c. for which
they seemed thankful, and offered in return a few
skins, pointing at the same time to the south side
of the sound, where their habitations probably were,
intimating that if we went there they would give
us more.   They then pushed off their baidars, rested
on their paddles for a minute, and made off as fast
as they could, to give us an idea of the swiftness of
their boat, which seems to be a favourite practice.

Next day we revisited their abode, and found
that the price of every article had been raised several
hundred per cent., and that nothing of reasonable
value would induce them to part with either bows
or arrows; so that our generosity of the preceding
day had not left any durable impression.

Every visit to these parties furnished some new
insight into their manners, though it was but trifling: on this occasion we witnessed a smoking
party in which the women and children partook
equally with the men.   The pipe used on this occasion was small, and would contain no more tobacco
than could be consumed at a whiff.   To these instruments there were attached a pricker and a strip of
dog's skin, from the last of which they tore off a few
hairs, and placed them at the bottom of the bowl of
the pipe to prevent the tobacco, which was chopped

up very fine, being drawn into the mouth with the smoke. The tobacco which they used had pieces of wood cut up fine with it, a custom which is no doubt derived from the Tschutschi, who use the bark of the birch-tree in this manner, and imagine it improves the quality of the herb.* The pipe being charged with about a pinch of this material, the senior person present took his whiff and passed the empty pipe to the next, who replenished it and passed it on, each person in his turn inflating himself to the fullest extent, and gradually dissipating the fumes through the nostrils. The pungency of the smoke, and the time necessary to hold the breath, occasioned considerable coughing with some of the party, but they nevertheless appeared greatly to enjoy the feast.

On the 8th, Spafarief Bay, which had been but little explored by Captain Kotzebue, underwent a satisfactory examination, and was found to terminate in a small creek navigable a very short distance, and that by boats only. Its whole extent inland is about three miles, when it separates into a number of small branches communicating with several lakes, which, in the spring, no doubt, discharge a large quantity of fresh water into the sound, though at this dry season of the year they were of inconsiderable size. A little to the northward of the creek there is a pointed hill just 640 feet high by measurement, from whence we surveyed the surrounding country, and found that this side of the sound also was covered with a deep swampy moss. The summit of this hill, and indeed of all the others that were ascended in the sound, was the only part destitute of this covering.

* Dobell's Travels in Siberia.

The beach was strewed with a great quantity of drift wood, some of which was in a very perfect state, and appeared to have been recently split with wedges by the natives, who had carried away large portions of the trunks to make their bows, arrows, and fishing implements. They were all pine-trees except one, which by the bark appeared to be a silver birch.

On the 10th we had the satisfaction to see the barge coming down to us under a press of canvass, and the most lively expectations were formed until she approached near enough to discover that the appointed signal of success was wanting at her masthead. Though unfortunate in accomplishing what we most anxiously desired, her voyage was attended with advantage. We had the satisfaction to learn from her commander when he came on board that he had discovered a large extent of coast beyond the extreme cape which we had seen from the masthead of the ship on the 15th ultimo, and which I had named after Captain Franklin; and had proceeded to the latitude of 71° 23′ 31″ N. and to 156° 21′ 30″ W., where the coast formed a low narrow neck beyond which it was impossible to proceed to the eastward, in consequence of the ice being attached to the land, and extending along the horizon to the northward.

The boat had not been at this point many hours, before the wind changed to south-west, and set the whole body of ice in motion toward the land. This was a case in which Mr. Elson had received strict orders to return immediately, and he accordingly began to retrace his route; but in so doing he found that, in addition to the disadvantage of a contrary wind, he had to contend with a current running to the north-east at the rate of three miles and a half an hour, and with large pieces of floating ice which he found it very difficult to avoid, until he was at last obliged to anchor to prevent being carried back. It was not long before he was so closely beset in the ice, that no clear water could be seen in any direction from the hills ; and the ice continuing to press against the shore, his vessel was driven upon the beach, and there left upon her broadside in a most helpless condition ; and to add to his cheerless prospect, the disposition of the natives, whom he had found to increase in numbers as he advanced to the northward, was of very doubtful character.   At Point Barrow, where they were extremely numerous, their overbearing behaviour, and the thefts they openly practised, left no doubt of what would be the fate of his little crew in the event of its falling into their power.   They were in this dilemma several days, during which every endeavour was made to extricate the vessel, but without effect ; and Mr. Elson contemplated sinking her secretly in a lake that was near, to prevent her falling into the hands of the Esquimaux, and then making his way along the coast in a baidar, which he had no doubt he should be able to purchase from the natives.   At length, however, a change of wind loosened the ice ; and after considerable labour and toil, in which the personal strength of the officers' was united to that of the seamen, our shipmates fortunately succeeded in effecting their escape.

The farthest tongue of land which they reached is conspicuous as being the most northerly point yet discovered on the continent of America ; and I named it Point Barrow to mark the progress of northern discovery on each side the American continent which has been so perseveringly advocated by that distinguished member of our naval administration. It lies 126 miles to the north-east of Icy Cape, and is only 146 miles from the extreme of Captain Franklin's discoveries in his progress westward from the Mackenzie River. The bay which appeared to be formed to the eastward of this point I named Elson's Bay, in compliment to the officer in command of the barge; and the extreme point of our discoveries after Captain Franklin, the commander of the land expedition. I could have wished that this point had been marked by some conspicuous headland worthy of the name bestowed upon it; but my hope is that the officer who may be so fortunate as to extend our discoveries will do him the justice to transfer his name to the first object beyond it more deserving of the honour. To the nearest conspicuous object to the southward of Point Barrow I attached the name of Smyth, in compliment to the second officer of the barge, and to the points and inlets to the southward I with pleasure affixed the names of the officers of the ship, whose merits entitled them to this distinction.

I will no longer anticipate the journal of these interesting proceedings, in which are recorded several particulars relating to the natives, the currents, and the geography of these regions; and by which it is evident that the officers and crew acquitted themselves in the most persevering and zealous manner, equally honourable to themselves and to their country. I shall merely remark upon the facts which the journal sets forth, that it was fortunate the ship did not continue near the ice, as she would have been unable to beat successfully against the current, and the violence of the gale would probably have either entangled her amongst the ice, or have driven her on shore.

The narrative was kept by Mr. Smyth under the superintendence of his commander, whose more important duties of surveying prevented his recording more than the necessary detail of a log-book. In publishing it, I have given the most important parts of it in Mr. Smyth's own words, and have only compressed the matter where it could be done with propriety and advantage.

CHAPTER XII.

*Narrative of the Proceedings of the Barge of H. M.
Ship Blossom in quest of Captain Franklin, and to
explore the Coast N. E. of Icy Cape.*

AFTER the signal was made by the Blossom on
the night of the 17th of August, to carry orders
into execution, the barge stood in-shore, and the
next morning was off Icy Cape. Having a contrary
wind, she beat up along the land to the N. E., and
shortly after noon the officers landed opposite a village of yourts, which was found to be deserted, and
the houses to be closed up for the summer. These
habitations closely resembled those of the Esqui-

CHAP.
XII.

Narrative
of the
Barge.

maux, which have been already described. The country here was covered with a thick peat, which retained the water and made it very swampy and almost impassable. Upon the beach there was found an abundance of coal and drift-wood. Working to the north-eastward from this village, they discovered a shoal with only eight feet water upon it lying about 150 yards from the beach, which having deep water within it, offered a security against the ice in the event of its closing the shore, and they did not fail to bear in mind the advantage it might afford in a moment of necessity. About midnight they were visited by four baidars containing about sixty persons, from whom they expected to obtain a supply of venison, as this kind of provision is, generally speaking, abundant to the northward of Cape Lisburn; but being disappointed, they continued their progress along the land. On the morning of the 20th there was a fall of snow, and the weather turned very cold. They found themselves off a village, and were visited by several baidars, the crews of which were very anxious to get alongside the barge, and in so doing one of the baidars was upset. An Esquimaux dress is very ill adapted to aquatic exercises, and persons acquainted with it would think there was considerable danger in being plunged into the sea thus habited; but the natives in the other baidars did not seem to reflect upon these consequences, and laughed most immoderately at the accident: they, however, went to the assistance of their friends, and rescued them all. It must have been a cold dip for these people, as the rigging and masts were partially covered with ice.

About noon they landed to procure observations, and found the latitude of this part of the coast to be 70° 43′ 47″ N., and longitude, from the bearings of Wainwright Inlet, 159° 46′ W. Here a post was erected for Captain Franklin, on which the following inscription was painted: "Blossom's tender, Sunday, August 20th, latitude 70° 43′ N., bound along the coast to the N. E. If Captain Franklin should pass this place, he will probably leave some memorandum." The coast was here low, and more dry than that in the vicinity of Wainwright Inlet, with a beach of sand and gravel mixed, upon which there was an abundance of coal and drift-wood. In the evening they passed several yourts, but saw no inhabitants until nine o'clock, when several came off and annoyed the crew with their importunities and disorderly conduct. The coast was here more populous than any where to the southward, which their visiters probably thought a good protection against the small force of our boat, and they were not easily driven away.

On the 21st they arrived off a chain of sandy islands lying some distance from the main land, which I have distinguished by the name of the Sea Horse Islands. As the wind was light and baffling, they landed upon several of these for observations; and tracking the boat along the shore, at eight in the evening they arrived at the point to which I transferred the name of Cape Franklin, from the cliff on the main land to which I had originally given that name, as I found by the discoveries of Mr. Elson that the cliff was not actually the coast line.* From

* See the Chart.

2 E 2

Cape Franklin, the coast, still consisting of a chain of sandy islands lying off the main land, turned to the south-east and united with the main land, forming a bay on which I bestowed the name of my first lieutenant, Mr. Peard. Two posts were found erected on Cape Franklin, upon which another notice was painted. The surface of the beach was a fine sand, but by digging a few inches down it was mixed with coal; there was here also, as at their former station, a great quantity of drift-wood. Off these islands they were visited by several baidars, the people in which behaved in a very disorderly manner, attempted several depredations, and even cut a piece out of one of the sails of the boat while it was lying upon the gunwale. Finding the natives inclined to part with one of their baidars, she was purchased for two hatchets, under the impression that she might be useful to the boat hereafter. Having run twenty-nine miles along the coast to the N.E., they again landed and obtained some lunar observations. The coast here assumed a different aspect, and consisted of clay cliffs about fifty feet high, and presented an ice formation resembling that which has been described in Escholtz Bay. The interior of the country was flat, and only partially covered with snow. A short distance to the northward of them a river discharged itself into a lake within the shingly beach, which was about twenty yards wide, and the water being perfectly fresh, they obtained a supply, and pursued their course to the north-east. Their latitude was 70° 58' 43" N.; and no ice had as yet been seen, even from the hills. This excited the greatest hopes in our adventurous shipmates, who advanced quite elated at the prospect; but they had not proceeded many miles further before some bergs were seen in the offing nearly in the same parallel in which the margin of the ice had been found by the ship; and from the number of bergs increasing as they advanced, the sanguine expectations in which they had indulged gradually diminished. These bergs were seen off a point of land to which I gave the name of Smyth, in compliment to the officer who accompanied the boat expedition, and very deservedly obtained his promotion for that service. In the course of their run they passed a village, where the inhabitants, seeing them so near, came out of their yourts, and men, women, children, and dogs set up a loud hallooing until they were gone. Upon Cape Smyth there was also a village, the inhabitants of which accosted them with the same hooting noises as before.

Advancing to the northward with the wind off the land, they saw the main body of ice about seven miles distant to the westward, and were much encumbered by the icebergs, which they could only avoid by repeatedly altering the course. The land from Cape Smyth, which was about forty-five feet in height, sloped gradually to the northward, and terminated in a low point which has been named Point Barrow. From the rapidity with which the boat passed the land, there appears to have been a current setting to the north-east. The water, about half a mile from the cape, was between six and seven fathoms deep.

Wednesday, 23rd Aug. " Arriving about two A. M. off the low point, we found it much encumbered with ice, and the current setting N.W. (mag.) between three and four miles an hour. Opening

the prospect on its eastern side, the view was obstructed by a barrier of ice which appeared to join
with the land. This barrier seemed high; but as
there was much refraction, in this we might possibly
have been deceived. The weather assuming a very
unsettled appearance in the offing, (and the S.E.
breeze dying away,) we had every reason to expect
the wind from the westward; and knowing the
ice to extend as far south as 71°, the consequences
that would attend such a shift were so evident, that
we judged it prudent not to attempt penetrating
any farther, especially in this advanced state of the
season. Accordingly we anchored within the eighth
of a mile of the point, under shelter of an iceberg
about fourteen feet high, and from fifty to sixty
feet in length, that had grounded in four fathoms
water. On the eastern side of the point there was
a village, larger than any we had before seen, consisting entirely of yourts. The natives, on seeing
us anchor, came down opposite the boat in great
numbers, but seemed very doubtful whether to treat
us as friends or enemies. We made signs of friendship to them; and a couple of baidars reluctantly
ventured off and accepted a few beads and some tobacco, which on their return to the shore induced
several others to visit us. These people were clothed
like the Esquimaux we had seen on the other parts
of the coast: their implements were also the same,
except that we thought they were more particular
in constructing the bow, the spring of which was
strengthened with whalebone.

Many of the men wore, as lip ornaments, slabs of
bone and stone in an oblong shape, about three inches
in length and one in breadth. They were much

more daring than any people we had before seen, and
attempted many thefts in the most open manner.
Tobacco was the most marketable article; but, excepting their implements, ornaments, or dress, they
had nothing worth purchasing. They were exceedingly difficult to please, and not at all satisfied with
what was given in exchange, insisting, after a bargain had been transacted, on having more for their
articles. One of them who came alongside in a *cai-
ack*, having obtained some tobacco that was offered
for a lance, was resolute in not delivering up either;
and Mr. Elson, considering that if such conduct was
tamely submitted to they would be still more inclined
to impose, endeavoured by threats to regain the tobacco, but without effect. More boats coming off,
and proving by their audacity equally troublesome,
we thought it would be most advantageous to keep
the barge under sail, which in all probability would
prevent any thing serious occurring. Before weighing, the baidar was broken up, as her weight would
materially impede our progress in working to windward on our return; the hides were taken as a covering for the deck, and the frame-work destroyed for
fire-wood. During the time we were at anchor, the
wind shifted to S.W., and we stood to the N.W.
with a light breeze; but finding ourselves drifting
rapidly to the northward by the current, we were
again obliged to anchor, Point Barrow bearing S. by
E.$\frac{1}{2}$E. two and a half miles. Here we remained
till eight o'clock. This point is the termination to
a spit of land, which on examination from the boat's
mast-head seemed to jut out several miles from the
more regular coast line. The width of the neck did
not exceed a mile and a half, and apparently in some

places less. The extremity was broader than any other part, had several small lakes of water on it, which were frozen over, and the village before spoken of is situated on its eastern shore. The eastern side of this neck trended in a S. S. W. (mag.) direction until it became lost to the eye by being joined with a body of ice that encircled the horizon in the N. E. This union scarcely left us room to hazard an opinion which direction it afterwards took; but from the circumstance of the current setting at the rate of three miles and a half an hour N. E. (true), and the ice all drifting to that quarter, we were induced to conjecture that its continuation led well to the eastward.

It was our original intention to have remained at the point till noon, landed, and obtained if possible all the necessary observations, besides depositing instructions for Captain Franklin; but the character of the natives entirely frustrated our plans, and obliged us, to avoid an open rupture, to quit the anchorage — a circumstance we greatly regretted, as we had anticipated gathering much information respecting the coast to the eastward, and on other points of importance. The nights had hitherto been beautifully clear and fine, and we were very sanguine of obtaining a number of lunar distances with the sun, being the only means we had of ascertaining correctly our farthest easting, as the patent log, we knew, from the strength of current, could in no way be depended on. At nine we weighed, and, stemming the current, stood in for the low point, off which there was an iceberg aground, on which we resolved to wait till noon for the latitude. On our way thither we passed another extensive iceberg

Pub.d by H.Colburn & R.Bentley, 1831.

aground in six fathoms water, and not more than
eight or ten feet above the surface. At noon we
were favoured with a clear sun, and determined our
latitude to be 71° 24′ 59″ N., Lunar anchorage bearing from the place of observation one mile north
(true), and the north-eastern part of Point Barrow
S. E. ¾ E. (mag.) 1½′. From which the position of
Point Barrow, the most northern part known of the
American continent, is latitude 71° 23′ 31″ N., longitude 156° 21′ 30″ W. The azimuth sights made the
variation 41° east.

The breeze still continuing light from the S. E.
(although the clouds were approaching from the
westward), we made all sail to the southward, and
with great reluctance left this remarkable point
without being able to leave any traces of our having
visited it for Captain Franklin. The wind about
one P. M. began gradually to fall, and at two it was
perfectly calm. Unfortunately we were now in too
much water to anchor, and were, without the possibility of helping it, being set to the N. E. by the
current at the rate of three miles and a half an hour.
By four o'clock we had lost all we had made during
the day, with a prospect, if it continued calm, of
being drifted quite off the land—an accident that,
had it occurred, would have placed our little vessel
in a very serious situation. We were not, however,
long in this state of suspense ; for an air came again
from the eastward, which strengthening a little, and
with the boat ahead towing, we made good progress
towards the land, where, if it once more fell calm, we
could retain our position with the anchor. When
we had by towing and pulling got within a mile of
the beach (and about two miles west of the point),

nineteen of the natives came down opposite us, armed with bows, arrows, and spears, and imagining that it was our intention to land motioned us to keep off, and seemed quite prepared for hostilities. Some of them were stripped almost naked. They preserved a greater silence than we found customary among them, one only speaking at a time, and apparently interrogating us. Notwithstanding this show of resistance, we still advanced nearer to the shore, as being more out of the current and favourable to our views, at the same time having the arms in readiness in case of an attack.

When within about thirty yards of the beach, we lost the wind, and continued pulling and towing along shore, the natives walking abreast of us upon the beach. At eight P. M. we passed a village of eight tents and four boats, but saw neither women nor children. Whilst approaching this village, we perceived the men hauling their baidars higher up on the beach, fearful, as we supposed, that we should molest them. Their dogs, as usual, set up a most abominable yelling. About eleven our pedestrians began to lag, and shortly after made a general halt, watched us for a little while, and then turned back. At midnight we reached Cape Smyth, and considering ourselves tolerably well secure from the ice (not having seen any until our arrival off this point on the evening before), and the crew being much tired, we anchored, hoping that a few hours would bring a breeze—not caring from which quarter, as we felt confident that, before the ice could approach near enough to block us, we should be able to reach the Sea Horse Islands, where we made certain of being clear. The night dark and cloudy.

Thursday, 24th August.   At two A. M., a fine
breeze rising at E. S. E., we weighed, but found the
current so strong against us that we lost ground and
anchored again : the current setting north (mag.)
three miles and a half an hour.   At three we were
alarmed at the sudden appearance of the ice, which
was drifting fast down on us.   No time was to be
lost.   The crew were instantly sent on shore with a
warp.   We got up the anchor, and hauled within
eight or ten yards of the beach, it being steep
enough to admit our proceeding thus close.   We
now began tracking the boat along, and proceeded
for a short time without much difficulty ; but the
ice increasing fast, and the pieces getting larger,
she received some violent blows.   The main body
nearing the shore to the distance of about 100 yards
left this space less incumbered, and occasioned an
increase in the rapidity of the current one knot an
hour.   To add to our perplexities, at five the wind
freshened up at south (directly against us), and we
also had the mortification to observe the ice speedily
connecting with the beach, scarcely leaving an open
space visible.   Nothing now but the greatest exertion could extricate the boat ; and the crew, willing
to make the most of every trifling advantage, gave
a hearty cheer, and forced her through thick and
heavy ice until we rounded a projecting point that
had hitherto obstructed our view.   This, however,
could only be accomplished with considerable labour
and risk ; for here, as in many other places, we had
to take the track-line up cliffs, frequently covered
with hard snow and ice, which, hanging a considerable distance over the water, prevented the possibility of getting round beneath.   The rope was

then obliged to be thrown down, and the upper end held fast, until the crew hauled themselves up one by one; and in this manner we continued along the cliff until the beach again made its appearance. But here even we found it no easy task to walk, on account of small loose shingle, in which we often sunk to the knees; and having the weight of the boat at the same time, it became excessively fatiguing.

On opening the prospect south of this point, our spirits were greatly enlivened at perceiving the channel clear for a long way, and hoping that by constant tracking we should do much towards getting clear of the ice, we divided the crew into two parties, gave each man a dram, and sent one division on board to rest, whilst the other laboured at the line. About eight A. M. the wind freshened so heavily against us, that we contemplated whether or not it would be advantageous to make a trial with the canvass, particularly as the main body of ice was a little more distant from the shore; but remembering our position at two P. M. on the preceding day, we agreed that the current was too strong, and that if we should get encircled by the ice we must inevitably be separated from the shore, carried back with the stream, or forced to sea. The difficulty of drawing the boat against so strong a wind and current became now very great, and we began to seek a place where she might be laid free of the ice. But the straight line of coast offered us no prospect of such an asylum: we therefore determined to prosecute our first intention of persevering in our endeavours *as long as possible.* By eleven A. M. we reached a village of nine tents, and trusted through the influence of tobacco, beads, &c.

Drawn by William Smyth.

TRACKING THE BARGE ROUND CAPE SMYTH.

Pub.ᵈ by H.Colburn & R.Bentley, 183

to receive some assistance from the inhabitants. Two of them approached us at first with some diffidence; but Mr. Elson throwing the presents on shore, and myself going to meet them, after much gesticulation denoting peaceable intentions, we joined company. The ratification of rubbing noses and cheeks being over, a leaf of tobacco given to each soon gained their confidence. One of them, an old man, seemed very thankful for his present, offering me any part of his garment as a reimbursement, which I declined accepting. Seeing so friendly an interview, several more ventured towards us; and learning from their companions the treasures I possessed, were very eager to obtain some. By a few signs I easily made them understand that their assistance at the track-line would be amply rewarded. Six or seven directly took hold of the rope; and our people relaxing a little in their exertions, though continuing at the line, we proceeded along gaily; but I was frequently obliged to have recourse to the presents to keep them pulling. We had not passed the tents more than half a mile when a new and a very serious difficulty presented itself— the mouth of a river into which the current set with great velocity, carrying with it large masses of ice. After many attempts we succeeded in getting a line across; but had no sooner accomplished it, than it broke, and our repeated trials for a long time were unsuccessful. Eventually we managed to overcome this obstacle, and had just got the boat to the opposite shore when she grounded; and the current setting strong against her, all our exertions to get her afloat were ineffectual. A few minutes before this accident, Mr. Elson, who was on board, hailed me, saying that the channel after crossing the river looked more favourable than ever. Cheered by this report, we worked harder; but so quick was the ice in its movement, that in a few moments we were enclosed on all sides. Nothing more towards freeing the boat could now be done, therefore we carried out her anchors to the shore and secured her, contemplating a retreat by land should we not be so fortunate as to get clear. On looking to the southward, we found the ice perfectly compact, and connected with the shore, not leaving visible a space of water three yards in diameter. The crew now enjoyed a little rest; and Mr. Elson decided that we should remain by the boat until the 1st of September, on which day, should no chance appear of liberating her, we were to start by land for Kotzebue Sound.

Some large ice gounding to windward partially sheltered the boat; but as her situation was on the southern bank of the entrance to the river, the current swept with force round, bringing occasionally some heavy ice in contact with the boat, the violence of which hove her into a foot and a half less water than she drew; and the sand soon formed a bank on the outside, leaving her quite bedded. At six P. M. the current had almost subsided. A most cheerless prospect presented itself, the whole sea being covered with ice sufficiently compact to walk upon; and the clouds becoming heavy and flying swiftly from the S. W., offered not the smallest hope of our escape. The water had likewise fallen a foot and a half, leaving the boat nearly dry. Our feelings now were indescribable, as it appeared very evident that we should be obliged to abandon our

little vessel, and perform the journey to Chamisso
Island on foot — an undertaking we were by no
means adequate to, and which the advanced state of
the season would render extremely fatiguing.   At
eight we ascended a hill, but saw not the slightest
chance of an opening, the ice to the southward being very compact as far as the eye could reach, and
varying in its height from twelve to two feet above
the level of the sea.   At midnight the weather was
cold, dark, and foggy, and seemed to indicate a S. W.
gale.

Friday, 25th Aug.   At four this morning the
current appeared to resume something of its former
rapidity, causing the ice to move to the northward,
and leaving small openings.   This gave us faint
hopes of a release; but the wind springing up as
we had anticipated, soon extinguished them.   After
breakfast we again visited the hill, but with no
better success than before.   The tide returning or
ebbing from the river brought back with it a quantity of the ice, almost every piece of which drifted
athwart the boat; so that we determined on getting
her afloat, and shifting her to a better berth, where
we should be ready to avail ourselves of the smallest prospect of getting clear.   Having laid out an
anchor astern, we with much difficulty got her
through the sand bank that had formed itself round
us; and finding that at her own length farther out a
channel was left for the ice to drive either out or
into the river, we secured her to a large berg that
had grounded and afforded us much shelter.   Towards noon a number of natives visited us, and were
presented with tobacco, &c.   Among them was the
old man spoken of the day before; who, on re-

ceiving his present, offered up what we concluded
to be a prayer, at the same time blowing with his
mouth, as if imploring an east wind and the dispersion of the ice.—In the afternoon the wind had increased to a gale. We went to the hill, and there
observed the line of ice within the horizon, and the
sea breaking very heavily outside: we saw also a
number of large bergs drifting down. At four, fresh
gales, with heavy squalls—the ice around us became
closely wedged, the pieces being forced one over
another, forming a solid mass. The body of ice in
the offing was still drifting to the northward. This
day Mr. Elson determined, if we should be compelled to quit the boat, to take every thing out of her
except the gun, to remove her into the deepest part
of the river, and there sink her, so as to prevent the
natives from destroying or breaking her up to obtain
the iron; from which situation, should we visit this
coast next year, she might with little trouble be
raised. The stores and rigging also we resolved to
bury, and to leave directions where they might be
found. On visiting the village (which was about
half a mile distant), the natives were uncommonly
civil. They resided in tents, the frames of which
were made with poles, and covered with seal-skins:
the bottom or floor was merely a few logs laid side-
wise on the ground: inside there was a second lin-
ing of reindeer skin, which did not reach quite to
the top: this constituted the whole of their dwelling. Their principal food appeared to be reindeer
and seal's flesh; and having procured more than
sufficient of these animals for present use, they had
buried the overplus in the sand, to be kept until
required. They very generously led us to a seal

that had been thus deposited.   The flesh and blubber which had been separated were wrapped in the
skin, and were in a most disgusting oily state.   One
of the natives put in his hand, stirred up the contents,
and offered us some, the sight of which alone was
enough to turn one's stomach.   He seemed to pity
our want of taste, and sucked his fingers with the
greatest relish.   Each of the crew having provided
himself with native boots, &c. for travelling in,
returned to the boat.   During the night the gale
abated and the wind fell almost calm, and it began
to freeze hard.   Wherever there was any opening
before, the water was covered with young ice.   The
tide here rose and fell from eighteen to twenty
inches :—the time of change very irregular, probably
influenced by the ice.

Saturday, 26th Aug.   Our chance of getting clear
seemed more remote now than ever, and we commenced making preparations for the land journey.
The crew were sent on shore to exercise their limbs,
and train themselves for walking.   We traced the
windings of the river for some distance ; the banks
were high on each side.   It seemed deep, and its
turnings frequent and sudden.   The only animal
we saw was a red fox, which avoided our pursuit.
In the evening we returned to the boat—the weather still frosty.

Sunday, 27th Aug.   We had a sharp frost during
the night, attended with frozen particles, which
fell like dust, and covered our clothes.   The wind
light from the S. W., with a thick fog.   The fresh-
water ponds were frozen to the thickness of half
an inch.   After eight A. M. Mr. Elson and myself
walked along the beach to reconnoitre the state of

the ice. We found that, if we could cut the boat
through about a quarter of a mile of ice, we should
get into about double that distance of clear water,
and returned on board with the determination to
accomplish this. Having got the boat afloat, we
began our arduous task of cutting and hauling her
through the ice. The natives, seeing us thus employed, very kindly came (unasked) and lent their
assistance. We persevered in our labours till half
past three, by which time we had moved the boat
a mile and a half south of her former position.
Another and more formidable barrier was now opposed to us, consisting of extensive pieces of ice
aground, closely wedged together by smaller masses,
under which we anchored. After dinner Mr. Elson
and myself again visited the cliffs, and thought we
could perceive a zigzag channel which afforded a
hope of liberation, provided we could force her
through the present obstacle. Immediately we got
on board, we commenced cutting a passage; but
had no sooner made an opening, than it was filled
by the current drifting smaller pieces of ice down.
These we for some time kept cutting and clearing
away; but after two hours and a half of hard work,
we found our exertions endless, and relinquished
the attempt. In the evening the wind veered to the
S. E. and the breeze, though light from this quarter, put some of the smaller pieces of ice in motion
off the land. We remained up till midnight, although fatigued with the toils of the day, and the
wind having increased to a fresh breeze, had the
consolation to witness the moving of several of the
larger pieces. The collision that now took place,
owing to the shift of wind (the ice in the offing still

PACIFIC AND BEERING'S STRAIT. 435

CHAP.
XII.

Narrative
of the
Barge.

holding its former course, whilst that in-shore was opposed to it), occasioned a grinding noise not unlike to that of a heavy roaring surf. Having fully satisfied ourselves of the departure of the ice, if the wind should hold its present direction and force, we retired to rest, anxiously waiting the following morning.

Monday, 28th Aug. Rising early, we had the great satisfaction to see that the formidable barrier which yesterday afternoon had been proof against our attempts, had nearly all drifted to sea, and that the coast, as far as we could discern, was fast clearing of ice. The wind blew strong at S. S. E.; and every preparation being made for weighing, after a hasty breakfast the anchor was got up, and our little vessel again bounded through the waters. Our tacking now was very uncertain, as in some places the ice still remained thick, and obliged us to perform that evolution twice or thrice in the space of a few minutes; and as we made it a rule not to bear up for any thing, we had some close rubs. By two P. M. we could see the southern termination of the main body of ice. There were still a number of large pieces aground, and much drift about us; the current setting to the northward at the rate of a mile and a half an hour. At three the wind fell light. A heavy swell from the S. W. occasioned a furious surf along the beach, and obliged us to keep well out to sea. The ice still extending far to windward made our situation very critical should the wind blow hard from the S. W. It now fell calm, with heavy clouds in the S. W.; and being in want of water, we procured a supply from the bergs that were near us. We watched every cloud with the greatest anxiety, and at eight observed them coming steadily from the westward, bringing with them a thick fog. We then stood to the northward until we reached the ice, when we tacked to the southward, and sailed along its margin. There were several walruses upon it, which at our approach bundled into the water. We had scarcely got clear of this field or body of ice, when it again fell calm — the clouds very heavy, and a thick fog. Finding that the current was again setting us to the northward at the rate of two miles and a half an hour, we anchored, and had no sooner done so, than several large detached bergs were seen driving rapidly down in our hawse. We again got up the anchor, and towed the boat in-shore, where we anchored again, and kept a vigilant look-out.

Tuesday, 29th Aug. In the course of the night the S. W. swell went down, and at one this morning a light air sprang up from the S. E. Weighed and stood in-shore, the wind gradually freshening. In running along the land, passed a quantity of drift ice. At noon, saw another body of ice about two miles distant, extending about eleven miles N. and S.; and as we were not yet far enough south to see Cape Franklin, we were apprehensive the ice might join it, in which case we should be again beset. In the afternoon, with great pleasure, we passed between it and the southern extremity of the ice at the distance of a mile and a half. At three it again fell calm—Cape Franklin, W. S. W. one mile. We were preparing to go on shore to deposit a bottle for Captain Franklin, which we had not done on our way to the northward, when a fresh gale suddenly rising at W. S. W. obliged us to abandon the project, as not a moment was to be lost in getting out of the bight, lest the ice (which experience had now taught us was quick in its motion) might again enclose us. The weather continued very unsettled during the night.

Wednesday, 30th Aug. Having rounded the point, we ran fifty miles on a S. W. course. The wind then suddenly shifted to the S. W., and blew very strong. We shortened sail to the close-reefed mainsail and storm-jib, and stood off and on shore. In the evening we had showers of snow and sleet, and at midnight strong gales with squalls of snow.

Thursday, 31st Aug. At two A. M. a heavy squall came on which split the mainsail, and a little before four the staysail shared the same fate. Towards the morning the weather was more moderate, accompanied with rain. Shortly after eight the wind suddenly veered to W. N. W. and blew strong. Set the close-reefed foresail, and furled the other sails, steering S. S. W. Noon, more moderate. Latitude observed 70° 23′ N. The remainder of the day was fine.

Friday, 1st Sept. Our stock of wood and water being expended, we hauled towards the land and made all sail; but as we drew in, the wind gradually decreased in strength, and before we obtained sight of the land it was almost calm. The breeze, however, again favoured us, and about sunset we reached within a short distance of the shore, on that part where the high land recedes from the coast. The boat was soon despatched to procure what we wanted; but in our thirsty moments we did not perceive that the pool from which we procured the water was brackish; having however filled our casks with it, and obtained some fuel, we again put to sea, with the wind from the southward.

Saturday, 2d Sept. Working along-shore. Noon
calm and fine. Sent the boat on shore to get a supply of better water. Found all the pools near the
beach very brackish; from which we concluded that
the recent westerly gales had thrown the surf so high
that it became mingled with the water of the lakes,
and we determined to have recourse to the first running stream we should come to. About two the
wind again came from the southward, and at four
we had every prospect of a gale from that quarter.
It therefore became necessary to carry a heavy press
of sail all night to obtain an anchorage as near Cape
Lisburn as possible, so that in the event of the wind
shifting to the westward we might be able to get
out of the bay.

On Sunday, as had been anticipated, it blew a
strong gale, but the boat made good weather of it
until eight P. M., when the bowsprit broke, and
obliged us to anchor: Cape Lisburn W.N.W. six
leagues. Strong gales, with heavy gusts of wind off
the land continued until four P.M., at which time
the weather being more moderate, we weighed under
close-reefed sails, and stood towards the cape, Mr.
Elson wishing to be near an entrance to a lake which
was situated a mile or two east of Cape Lisburn, in
which he thought the boat might find shelter, should
it blow hard from the westward. On arriving at
this spot, we found, to our surprise, that the entrance
which Mr. Elson had sounded and examined in the
barge's little boat was quite filled up, and that there
was not the slightest appearance of there ever having
been one. In the evening the wind became light
and variable. Anchored—the cape W.S.W. four
miles.

Monday, 4th Sept. It again blew strong from the southward, and at nine A.M. the wind increased so much as obliged us to let go another anchor to prevent being driven to sea. In the afternoon it again relaxed, but by midnight resumed its former violence.

Tuesday, 5th Sept. The wind somewhat subsiding this morning, completed our wood and water. Whilst thus employed, a native came over the hills and trafficked with us. Afterwards he stole from one of the crew some tobacco, and made off. The theft was not discovered until he was a long way distant and running, being evidently aware of the crime he had committed. At noon a baidar with eleven natives came round the cape and visited us. The wind continued strong from the southward; but being anxious to proceed, as our provisions were beginning to grow short, weighed and stood towards the cape under the foresail and staysail only. At two we got within the influence of the variable winds, occasioned by the steep and high land of the cape. The bubble and violent agitation of the sea exceeded any idea of the kind we had formed, and broke over the boat in every direction. We had no method of extricating her. The gusts of wind that came from every quarter lasting but a moment, left us no prospect of getting clear. We were at this time about two miles from the land. The wind inshore of us blew with astonishing violence; the eddies from the hills making whirlwinds which carried up the spray equal in height to the mountain. However, by four P. M., what with a slight current, and taking advantage of every flaw, we gained an offing of four miles, and, to prevent being set farther to the northward, anchored:—a heavy sea running, but little wind. We had not been more than half an hour in this situation when it blew again from the same point with redoubled violence. With some difficulty we lifted our anchor and made sail in for the land. As we approached it, the gusts came very strong off the hills, notwithstanding which we carried a press of sail to regain an anchorage. For an hour and a half we were literally sailing through a sea of spray. At six, having closed well with the land we anchored and rode out the gale. This evening Mr. Elson put the crew on half an allowance of provisions.

Wednesday, 6th Sept. Early in the morning we observed an alteration in the weather. The clouds collecting fast from the N. W. led us to expect the wind from that quarter. At ten A.M., the wind becoming variable and moderate, weighed, and by three in the afternoon, to our inexpressible joy, got round the windy promontory of Cape Lisburn. The crew were again put on their former allowance; and we made all sail, with an increasing breeze to the southward. Passing the cape, we observed five baidars hauled up and one tent, but saw few of the natives. It had been Mr. Elson's intention to look into the bight on the northern side of Point Hope; but the sea was so high and the weather so threatening that we kept well off, in order to weather the point. We noticed the water, whilst off Marryat Creek, to be of a very muddy colour, as if some river discharged itself there. By nine P. M. we rounded the point and steered S. S. E., to have a good offing in case the wind should again come from the westward.

Thursday, 7th Sept. The weather seemed determined to persecute us to the last. The wind strengthened to a gale, and raised a short, high, dangerous sea. We hauled in for the land as much as it would allow. At nine A. M. it blew extremely hard ; and, considering it dangerous to scud, rounded to on the larboard tack, took in the foresail, and set balance-reefed mainsail and storm-jib. Found the boat behave uncommonly well and continue tolerably dry. At noon our latitude was 67° 19′ N. In the afternoon it moderated, and we made sail in for the land. At four P M. saw Cape Mulgrave on the weather-bow, and altered our course for Kotzebue Sound. The wind dying away left us at midnight becalmed a few miles from Cape Krusenstern.

Friday 8th. After a few hours' calm, a breeze came from the S. E., and we worked along shore. In the forenoon several baidars came off to us. We procured, in exchange for a few beads, a large quantity of salmon, in hopes we should be able to keep enough to supply the ship. While sailing along the land, many more of these boats came off; but on waving them to return, they left us unmolested. We saw immense quantities of fish drying on shore, and concluded that the natives assembled at this inlet to lay in their winter stock.

Saturday, 9th. Owing to the light winds, we made but small progress during the night, and this morning were off Hotham Inlet. At eleven anchored. Sent the boat on shore to obtain wood and water. Noon, the latitude observed (with false horizon) was 66° 58′ N. The spot abreast where we anchored had, when Mr. Elson visited this inlet before, been the site of an Esquimaux village; but there was not a single tent left. In the evening we weighed from here, and the next morning had the

pleasure of seeing the ship at anchor off Chamisso
Island, and the gratification to find all on board of
her well.

(Signed)   WILLIAM SMYTH,
              Mate of H. M. S. Blossom.

Sept.

By this expedition about seventy miles of coast
in addition to those discovered by the Blossom—
making in the whole 126 miles—have been added
to the geography of the polar regions, and the
distance between Captain Franklin's discoveries and
our own has been brought within so small a compass as to leave very little room for further spe-
culation on the northern limits of the continent
of America. The actual distance left unexplored
is thus reduced to 146 miles, and there is much
reason to believe, from the state of the sea about
Point Barrow and along that part of the coast which
was explored by Captain Franklin, that the navigation of the remaining portion of unknown coast in
boats is by no means a hopeless project.

Having now the assistance of the barge, I embarked in her to examine narrowly the shores of
Kotzebue Sound. Proceeding to survey the head of
Escholtz Bay, shallow water obliged the boat to
anchor off Elephant Point, where I left Mr. Collie
with a party to examine again the cliffs in which
the fossils and ice formation had been seen by Kotzebue, and proceeded to the head of the bay in a
small boat. We landed upon a flat muddy beach,
and were obliged to wade a quarter of a mile before
we could reach a cliff for the purpose of having a
view of the surrounding country. Having gained
its summit we were gratified by the discovery of a

large river coming from the southward, and passing between our station and a range of hills. At a few miles distance the river passed between rocky cliffs, whence the land on either side became hilly, and interrupted our further view of its course. The width of the river was about a mile and a half; but this space was broken into narrow and intricate channels by banks—some dry, and others partly so. The stream passed rapidly between them, and at an earlier period of the season a considerable body of water must be poured into the sound; though, from the comparative width of the channels, the current in the latter is not much felt.

The shore around us was flat, broken by several lakes, in which there were a great many wildfowl. The cliff we had ascended was composed of a bluish mud and clay, and was full of deep chasms lying in a direction parallel with the front of the eminence. In appearance this hill was similar to that at Elephant Point, which was said to contain fossils; but there were none seen here, though the earth, in parts, had a disagreeable smell, similar to that which was supposed to proceed from the decayed animal substances in the cliff near Elephant Point.

Returning from this river, we were joined by three caiacs from some tents near us, and four from the river, who were very troublesome, pestering us for *tawack*, and receiving the little we had to give them in the most ungracious manner, without offering any return.

I found Mr. Collie had been successful in his search among the cliffs at Elephant Point, and had discovered several bones and grinders of elephants and other animals in a fossil state, of which a full description and drawings from the remains will be found in the Appendix. Associating these two discoveries, I bestowed the name of Elephant upon the point, to mark its vicinity to the place where the fossils were found; and upon the river that of Buckland, in compliment to Dr. Buckland, the professor of geology at Oxford, to whom I am much indebted for the above mentioned description of the fossils, and for the arrangement of the geological memoranda attached to this work.

The cliff in which these fossils appear to have been imbedded is part of the range in which the ice formation was seen in July. During our absence (a space of five weeks) we found that the edge of the cliff in one place had broken away four feet, and in another two feet and a half, and a further portion of it was on the eve of being precipitated upon the beach. In some places where the icy shields had adhered to the cliff nothing now remained, and frozen earth formed the front of the cliff. By cutting through those parts of the ice which were still attached, the mud in a frozen state presented itself as before, and confirmed our previous opinion of the nature of the cliff. Without putting it to this test, appearances might well have led to the conclusion come to by Kotzebue and M. Escholtz; more especially if it happened to be visited early in the summer, and in a season less favourable than that in which we viewed it. The earth, which is fast falling away from the cliffs—not in this place only, but in all parts of the bay—is carried away by the tide; and throughout the summer there must be a tendency to diminish the depth of the water, which at no very distant period will probably leave it naviga-

ble only by boats.   It is now so shallow off the ice
cliffs, that a bank dries at two miles' distance
from the shore; and it is only at the shingly
points which occur opposite the ravines that a
convenient landing can be effected with small
boats.*

In consequence of this shallow water there was
much difficulty in embarking the fossils, the tusks
in particular, the largest of which weighed 160lbs.,
and it took us the greater part of the night to accomplish it.   In our way on board we met several native
caiacs, and had an exhibition of the skill of one of
the Esquimaux in throwing his dart, which he
placed in a slip, a small wooden instrument about a
foot in length, with a hole cut in the end to receive
the forefinger, and a notch for the thumb. The
stick being thus grasped, the dart was laid along a
groove in the slip, and embraced by the middle
finger and thumb. The man next propelled his
caiac with speed in order to communicate greater
velocity to the dart, and then whirled it through
the air to a considerable distance. As there was no
mark, we could not judge of his skill in taking aim.
His party lived a long distance up Buckland river,
and were acquainted with the musk ox, which I am
the more particular in remarking, as we had never
seen that animal on the coast.

About eight o'clock at night we had a brilliant
display of the aurora borealis, a phenomenon of the

---

* This difficulty of approaching the shore, even in a boat, will,
I trust, convince the reader of the impracticability of trying the
effect of a cannon shot upon the mud cliff with a view of bringing
down some part of its surface, as has been suggested since the
publication of the quarto edition.

heavens so beautiful that it has been justly thought
to surpass all description.

In our return to the ship to deposit the fossils, a
calm obliged us to anchor on the north side of the
bay, where we landed with difficulty, in consequence
of the shallowness of the beach, and of several ridges
of sand thrown up parallel with it, too near the surface for the boat to pass over, and with channels of
water between them too deep to wade through without getting completely wet.   The country abounded
in lakes, in which were many wild ducks, geese, teal,
and widgeon; and was of the same swampy nature
before described: it was covered with moss, and
occasionally by low bushes of juniper, cranberry,
whortleberry, and cloudberry.   Near this spot, two
days before, we saw a herd of eleven reindeer, and
shot a musk rat.

Hence westward, to the neck of Choris Peninsula,
the shore was difficult of access, on account of long
muddy flats extending into the bay, and at low
water drying in some places a quarter of a mile from
the beach.

Bad weather and the duties of the ship prevented
my resuming the examination of the sound until
the 20th, when we ran across in the barge to Spafarief Bay, and explored the coast from thence to the
westward; passing close along the beach, anchoring
at night, and landing occasionally during the day
for observations, and to obtain information of the
nature of the country.

This part of the sound appeared to have so few
temptations to the Esquimaux, that we saw only
two parties upon it; and one of these, by having
their dogs harnessed in the boat, appeared to be only

on an excursion : the other was upon Cape Deceit, a bold promontory, with a conspicuous rock off it, so named by Captain Kotzebue. At two places where we landed there were some deserted yourts, not worthy of description; and at the mouth of two rivers in the first and second bays to the eastward of Cape Deceit, there were several spars and logs of drift-wood placed erect, which showed that the natives had occupied these stations in the summer for the purpose of catching fish, but they were now all deserted. Both these rivers had bars across the entrances, upon which the sea broke, so as to prevent a boat from entering them.

The land on the south side of the sound, as far as the Bay of Good Hope, is higher, more rocky, and of a bolder character than the opposite shore, though it still resembles it in its swampy superficial covering, and in the occurrence of lakes wherever the land is flat. Under water also, it has a bolder character than the northern side, and has generally soundings of four and five fathoms quite close to the promontories. There are two or three places under these headlands which in case of necessity will afford shelter to boats, but each with a particular wind only; and in resorting thither the direction of the wind and the side of the promontory must be taken into consideration.

In a geological point of view this part of the coast is interesting, as being the only place in the sound where volcanic rocks occurred. Near the second promontory to the eastward of Cape Deceit we found slaty limestone, having scales of talc between the layers; and in those parts of the cliff which were most fallen down a talcaceous slate, with thin layers of limestone, and where the rocks were more abrupt, limestone of a more compact nature. In this cliff there was also an alum slate of a dark-bluish colour. We could not land at the next cliff, but on a close view of the rock conjectured it to consist of compact limestone, dipping to the E.N.E. at an angle of 30°. Cape Deceit, the next headland, appeared to be compact limestone also, in large angular blocks devoid of any distinct stratafication. Proceeding on to Gullhead which is a narrow rocky peninsula stretching a mile into the sea, we found it chiefly composed of slaty limestone of a blackish and grayish colour, containing particles of talc in larger or smaller quantities as it was elevated above or on a level with the sea, but without any visible stratafication. A bed of slate to the eastward of the promontory bore strong marks of its having been subjected to the action of fire. The slaty limestone of the cliff on the eastern side of this dips at an angle of about 65° to the eastward. The neck or isthmus is either unstratafied, or its beds are perpendicular ; beyond it the strata dip to the west at nearly a right angle.

Eight miles further along the coast, we landed at the first of a series of low points, with small bays between them, which continue about four miles, beyond which the coast assumes a totally different character. On these low points, as well as upon the shores of the bay, we were surprised to find large blocks of porous vesicular lava, and more compact lava containing portions of olivine. These blocks are accumulated in much larger quantities on the points, and in the bays form reefs off the coast which are dangerous to boats passing close along the shore.

The country here slopes gradually from some hills to the beach, and is so well overgrown that we could not examine its substrata ; but they do not in outward formation exhibit any indication of volcanic agency.

Further on we landed in a small bay formed by a narrow wall of volcanic stones—some wholly above water, others only slightly immersed. These reefs were opposite a low mud cliff, similar in its nature to those in which the fossils were found in Escholtz Bay ; and though they did not furnish any bones, yet it is remarkable that a piece of a tusk was picked up on the beach near them. It must, however, be observed that its edges were rounded off by the surf, to which it had been a long time exposed ; and it might have been either washed up from some other place, or have been left on the beach by the natives.

To the westward of these rocky projections the coast is low, swampy, and intersected by lakes and rivers. The rounded hills which thus far bound the horizon of the sound to the southward here branch off inland, and a distant range of a totally different character rises over the vast plain that extends to Cape Espenburg, and forms the whole of the western side of the sound. In the angle which it makes, we discovered a river, which, we were informed by a few natives who came off to us in a miserable bai-dar, with dogs looking as unhappy as themselves, extended inland five days' journey for their baidars ; but on examination it proved so shallow at the mouth, that even the gig could not enter it. A few miles to the north-westward of this river, we arrived off the inlet which Captain Kotzebue medi-tated to explore in baidars, and was very sanguine that it would lead to some great inland discovery. We consequently approached the spot with interest; and as soon as the low mud capes through which the river has made its way to the ocean opened to our view, bore up, with the intention of sailing into the inlet, which runs in a westerly direction ; but we were here again obliged to desist, in consequence of the shallowness of the water. At two miles and more from the shore, we had less than a fathom water; and we observed the sea breaking heavily upon a bank which extended from shore to shore across the mouth of the inlet. Thinking, however, these breakers might be occasioned by the overfall of the tide, the gig was despatched to endeavour to effect a passage through them ; but the water shoaling gradually, she could not approach within even a cable's length of the breakers. At the top of the tide, probably, when the water is smooth, small boats may enter the inlet; but if the bar is attempted under other circumstances, the crew will probably be subjected to a similar ducking to that which Captain Kotzebue himself experienced in repassing it. Seeing these difficulties, I did not deem any further examination necessary ; and as it could never lead to any useful purpose of navigation, I did not even contemplate a return to it under more favourable circumstances. The inlet occurs in a vast plain of low ground, bounded on the north by Cape Espenburg, on the east by the Bay of Good Hope, on the west by Beering's Strait, and on the south by ranges of mountains. There are also several lakes and creeks in the plain, some of which may probably communicate with the inlet ; or they may all, Schismareff Inlet included, be the mouths of a large river. It is, however, very improbable that there should be any direct communication between these two inlets, as the natives would, in that case, have informed us of it when they drew their chart of the sound.

While we were off here, we noticed a parhelion so bright that it was difficult to distinguish it from the sun; a circumstance the more deserving of remark, in consequence of the naturalist of Kotzebue's expedition having observed that this phenomenon is very rare in these seas, and that a Russian grown old in the Aleutian Islands never saw it more than once. Quitting this inlet, we directed our course along the land toward Cape Espenburg, and found that the bar was not confined to the mouth of the inlet alone, but extended the whole way to the cape, and was not passable in any part; having tried ineffectually in those places which afforded the best prospect of success.

On landing at Cape Espenburg, we found that the sea penetrated to the southward of it, and formed it into a narrow strip of land, upon which were some high sand-hills. The point had a great many poles placed erect upon it, and had evidently been the residence of the Esquimaux; but it was now entirely deserted. Near these poles there were several huts and native burial-places, in which the bodies were disposed in a very different manner to that practised by the eastern Esquimaux. The corpse was here enclosed in a sort of coffin formed of loose planks, and placed upon a platform of drift-wood, covered over with a board and several spars, which were kept in their places by poles driven into the ground in a slanting direction, with their ends cross-

2 G 2

ing each other over the pile. The body was found lying with the head to the westward, and had been interred in a double dress, the under one made of the
skins of eider-drakes, and the upper one of those of
reindeer. It had been exposed a considerable time,
as the skeleton only was left; but enough of the
dress remained to show the manner in which the
body had been clothed.

The beach was in a great measure composed of
dark-coloured volcanic sand, and was strewed with
dead shells of the cardium, Venus, turbo, murex,
solen, trochus, mytilus, mya, lepas, and tellina genera : there were also some large asterias. The
sand-hills were partly covered with elymus grass,
the vaccinium vitis idæa, empetrum nigrum, and
some shrubs, while the carex preferred the hollow
moist places ; the rest of the surface was occupied
by lichens. On the border of the lakes there were
several curlew, sanderlings, and gulls ; while small
flocks of ptarmigan alighted upon those parts which
produced berries. A red fox prowling among the
deserted huts and the graves was the only quadruped seen. Nearly the whole of the day was passed
at this place in making astronomical observations ;
after which we embarked, and were obliged by bad
weather to return to the ship.

The day after my departure, a new cutter, which
had been built of some wood of the porou-tree,
grown upon Otaheite, was completed and launched,
and upon trial found to answer under canvas beyond our expectations, doing great credit to Mr.
Garrett, the carpenter, who built her almost entirely
himself. I placed her under the charge of Lieutenant Belcher, who was afterwards almost daily employed in surveying.

On the 22d the aurora borealis was seen in the W. N. W. ; from which quarter it passed rapidly to the N. E., and formed a splendid arch emitting vivid and brilliantly coloured coruscations.

On the 25th the wind, which had blown strong from the northward the day before, changed to the southward, and had such an effect upon the tide that it ebbed twenty hours without intermission.

In another excursion which I made along the north side of the sound, I landed at a cape which had been named after the ship, and had the satisfaction of examining an ice formation of a similar nature to that in Escholtz Bay, only more extensive, and having a contrary aspect. The ice here, instead of merely forming a shield to the cliff, was imbedded in the indentations along its edge, filling them up nearly even with the front. A quantity of fallen earth was accumulated at the base of the cliff, which uniting with the earthy spaces intervening between the beds of ice, might lead a person to imagine that the ice formed the cliff, and supported a soil two or three feet thick, part of which appeared to have been precipitated over the brow. But on examining it above, the ice was found to be detached from the cliff at the back of it; and in a few instances so much so, that there were deep chasms between the two. These chasms are no doubt widened by the tendency the ice must have towards the edge of the cliff; and I have no doubt the beds of ice are occasionally loosened, and fall upon the beach, where, if they are not carried away by the sea, they become covered with the earthy materials from above, and perhaps remain some time immured. In some places the cliff was undermined, and the surface in general was very rugged; but it was evident in this, as in the former instance, that the ice was lodged in hollow places in the cliff. While we continued here we had an example of the manner in which the face of the cliff might obtain an icy covering similar to that in Escholtz Bay. There had been a sharp frost during the night, which froze a number of small streams that were trickling down the face of the cliff, and cased those parts of it with a sheet of ice, which, if the oozings from the cliff and the freezing process were continued, would without doubt form a thick coating to it.

Upon the beach, under the cliffs, there was an abundance of drift birch and pine wood, among which there was a fir-tree three feet in diameter. This tree, and another, which by the appearance of its bark had been recently torn up by the roots, had been washed up since our visit to this spot in July; but from whence they came we could not even form a conjecture, as we had frequently remarked the absence of floating timber both in the sound and in the strait.

We found some natives at this place laying out their nets for seals, who, perceiving we were about to take up our quarters near them, struck their tents expeditiously, threw every thing into their baidars, to which they harnessed their dogs, and drove off for about half a mile, where they encamped again. We procured from them about two bushels of whortle berries, which they had collected for their own consumption, and learnt that they had been unsuccessful in fishing. We noticed that at their meals they stripped their dried fish of its skin and gave it to the women and children, who ate it very contentedly, while the men regaled themselves upon the flesh.

During the night we had a brilliant display of the

aurora borealis, remarkable for its masses of bright light. It extended from N. E. to W., and at one time formed three arches. As we were taking our departure we were visited by a baidar, from which we procured some fine fresh salmon and trout. The coxswain of this boat wore unusually large labrets, consisting of blue glass beads fixed upon circular pieces of ivory, a full inch in diameter. He drew us a chart of Hotham Inlet, which resembled one that had been traced upon the beach by some natives the day before; both of which represented it as an arm of the sea in the form of an hour-glass, which was not far from the truth. The Esquimaux seem to have a natural talent for such delineations; and though their outlines may serve no essential purpose of navigation, they are still useful in pointing out the nature of a place that has not been visited; an information which may sometimes save a useless journey. It is, however, to be observed, that not unfrequently they appear to trace the route which a boat can pursue, rather than the indentations of the coast, by which rivers and bays not frequented would be overlooked. Such charts are further useful in marking the dwellings and fishing stations of the natives.

From hence we bore away to examine Hotham Inlet, and found it so encumbered with shoals that it was necessary to run seven miles off the land to avoid striking upon them; it had but one small entrance, so very narrow and intricate, that the boats grounded repeatedly in pursuing it. In the middle of the channel there were only five feet water at half-flood; and the tide ebbed so strong through it, that the boat could not stem it; and as there was but a small part of the coast of this inlet that we had not seen, and finding the examination of it would be attended with difficulty, and would occupy a long time, the boats did not ascend it. The shoal which is off the entrance has no good land-marks for it; the bearings from its extremity in two fathoms and a half of water are Cape Blossom, S. 66° 40′ E. (true); Western High Mount, N. 17° 30′ W. (true); and the west extreme, a bluff cape, near Cape Krusenstern, N. 37° 0′ W. (true.) But the best way to avoid it is to go about directly the soundings decrease to six fathoms, as after that depth they shoal so rapidly to two fathoms and a quarter that there is scarcely room to put the ship round.

On the 1st of October we landed upon a sandy point at the western limit of the inlet, and were joined by a few Esquimaux who had their tents not far off to the westward: they had communicated with the boat two months before, and came again in the expectation of getting a few more blue beads and foreign articles for some nets and fish. They immediately recognised such of the officers as they had seen before, and were delighted at meeting them. Some of the beads which they had obtained were now suspended to different parts of their dress, in the same manner as was practised by the Esquimaux of Melville Peninsula, and round their necks, or were made into bracelets. They corroborated the former account of the inlet, the length of which they estimated a long day's paddle: our observations made it thirty-nine miles. At the back of the point where we landed there was another inlet, to the end of which they said their baidars could also

WESTERN ESQUIMAUX MODE OF DISPOSING OF THEIR DEAD.

Published by Henry Colburn and Richard Bentley, New Burlington Street. 1831.

go, notwithstanding we saw a bar across its mouth
so shallow that the gulls waded over from shore to
shore.   Near us there was a burying-ground, which,
in addition to what we had already observed at Cape
Espenburg, furnished several examples of the manner in which this tribe of natives dispose of their
dead.   In some instances a platform was constructed
of drift-wood, raised about two feet and a quarter
from the ground. upon which the body was placed
with its head to the westward, and a double tent
of drift-wood erected over it; the inner one with
spars about seven feet long, and the outer one with
some that were three times that length.   They were
placed close together, and at first no doubt sufficiently so to prevent the depredations of foxes and
wolves; but they had yielded at last; and all the
bodies, and even the hides that covered them, had
suffered by these rapacious animals.

In these tents of the dead there were no coffins
or planks, as at Cape Espenburg; the bodies were
dressed in a frock made of eider-duck skins, with
one of deer-skin over it, and were covered with a sea-
horse hide, such as the natives use for their baidars.
Suspended to the poles, and on the ground near
them, were several Esquimaux implements, consisting of wooden trays, paddles, and a tambourine,
which, we were informed, as well as signs could
convey, were placed there for the use of the deceased,
who, in the next world, (pointing to the western sky),
ate, drank, and sang songs.   Having no interpreter,
this was all the information I could obtain ; but the
custom of placing such implements around the re-
ceptacles of the dead is not unusual, and in all probability the Esquimaux may believe that the soul

has enjoyments in the next world similar to those which constitute happiness in this.

The people whom we saw here were very inquisitive about our fire-arms, and to satisfy one of them I made him fire off a musket, that was loaded with ball, towards a large tree that was lying upon the beach. The explosion and the recoil which succeeded the simple operation of touching the trigger so alarmed him, that he turned pale and put away the gun. As soon as his fear subsided he laughed heartily, as did all his party, and went to examine the wood, which was found to be perforated by the ball, and afforded them a fair specimen of the capability of our arms; but he could not be prevailed upon to repeat the operation.

They had some skins of ravens with them, upon which they placed a high price, though being of no use to us, they did not find a purchaser. On several occasions we had noticed the beaks and claws of these birds attached to ornamental bands for the head and waist, and they were evidently considered valuable. On our return to the ship we fell in with another party of natives, among whom there were two men whose appearance and conduct again led us to conclude that the large blue glass labrets indicated a superiority of rank, and found, as before, that no reasonable offer would induce them to part with these ornaments.

On the 3rd, we reached the ship, and were informed that she had been visited by several baidars in our absence, and had procured from them a quantity of dried salmon, which was afterwards served to the ship's company. These boats were the last that visited the ship, as the season was evidently arrived for commencing their preparations for winter. About this time we had sharp frosts at night; some snow fell; and on the 5th all the lakes on shore were frozen. The hares and ptarmigan were quite white, and all the birds had quitted their abodes in the rocks to seek a milder atmosphere. These unequivocal symptoms of the approach of winter excited great anxiety for the safety of the land expedition.

On the 7th, Mr. Elson went up Escholtz Bay with two boats for the purpose of sounding and obtaining further information of Buckland River, but returned on the 10th, without having been able to effect it, on account of the hostile disposition of the natives, whom he met in the bay When the small boat was detached from the barge, three baidars approached her; and their crews, consisting of between thirty and forty men, drew their knives and attempted to board her, and, on the whole, behaved in so daring and threatening a manner, that Mr. Elson fearing he should be compelled to resort to severe measures, if he proceeded with the examination of the river, desisted, and returned to the ship. This was the first instance of any decidedly hostile conduct of the natives in the sound, whose behaviour in general had left with us a favourable impression of the disposition of their tribe. The barge brought us down a valuable addition to our collection of fossils, the cliff having broken away considerably since the first specimens were obtained.

On the 8th, we had the misfortune to lose one on the marines, by dysentery and general inflammation of the abdomen. On the 10th, having selected a convenient spot for a grave, on the low point of Chamisso Island, his body was interred in the presence of almost all his shipmates, and a stone properly inscribed put up to mark the spot; but the earth was replaced over the grave as evenly as possible, in order that no appearance of excavation might remain to attract the attention of the natives.

We had hitherto remained in the sound, in the expectation of being able to wait till the end of October, the date named in my instructions; but the great change that had recently occurred in the atmosphere, the departure of all the Esquimaux for their winter habitations, the migration of the birds, the frozen state of the lakes, and the gradual cooling down of the sea, were symptoms of approaching winter too apparent to be disregarded, and made it evident that the time was not far distant when it would be necessary to quit the anchorage, to avoid being shut up by the young ice. On every account I was anxious to remain until the above-mentioned period; but as my instructions were peremptory in desiring me not to incur the risk of wintering, it was incumbent upon me seriously to consider how late the ship could remain without encountering that risk. By quitting the rendezvous earlier than had been agreed upon, the lives of Captain Franklin's party might be involved; by remaining too long, those of my own ship's company would be placed in imminent hazard, as but five weeks' provision at full allowance remained in the ship, and the nearest place where we could replenish them was upwards of 2000 miles distant. Thus circumstanced, I was desirous of having the advice of the officers of the ship before I made up my own mind, and accordingly addressed an official letter to them, requesting they would take every circumstance into their considera-

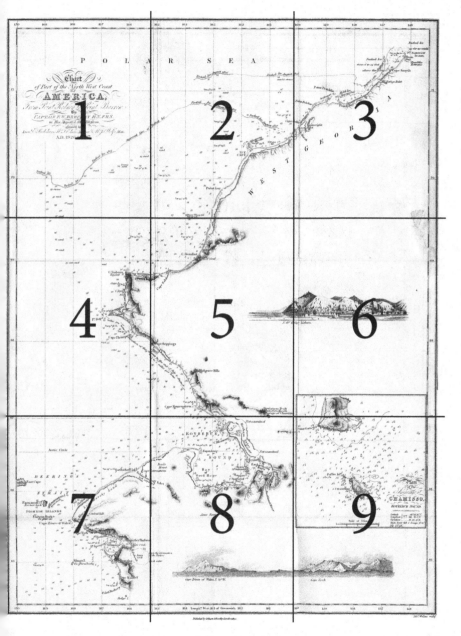

Key to following pages.

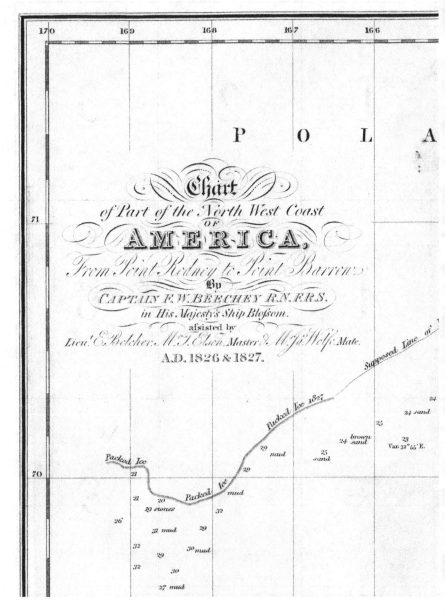

170   169   168   167   166

P O L A

71

Chart
of Part of the North West Coast
OF
AMERICA,
From Point Rodney to Point Barrow
By
CAPTAIN F. W. BEECHEY R.N. F.R.S.
in His Majesty's Ship Blossom.
assisted by
Lieut. E. Belcher, M.T. Elson Master & M. Ja. Wolfe Mate.
A.D. 1826 & 1827.

Supposed Line of

Packed Ice 1827

24
24 sand

25

29   mud   24   brown   29
         sand   Var 32°55'E.

Packed Ice
21   25
           29   sand

70

21   20   Packed Ice   mud   29
19 stones   32
26'
   31 mud   29
32
    29   30 mud
32
      30
  27 mud

1

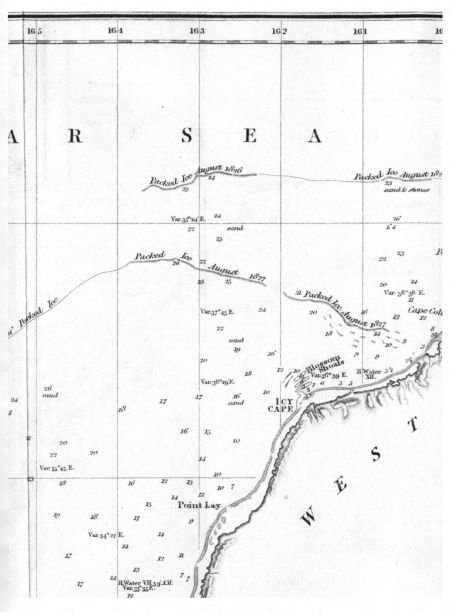

165   164   163   162   161   16

A R   S   E   A

Packed Ice August 1826
24
23

Packed Ice August 182
23
sand & stones

Var: 35°24' E. 24
22   sand
       23

26
l. s

Packed   Ice   22 August 1837
26
25   25

23
22

Var: 37° 43 E.   24

Packed Ice

22
mud
19

20

Var: 36°18' E.

26
mud

24
l

18

17   17

17
16'
sand

ICY
CAPE

16   25   10

20
22   20
Var: 35°45 E.

10
14

18   16   22   13   10 7
12

14
25   Point Lay
19   18   17   9

Var: 34° 27 E.   14
14

17   12

13

24   H.Water VII 3 9 A.M.
27   Var: 33° 35 E.
12

20
18

23

22
20   24
Var: 38° 38 E.

16   Cape Col
13   12

8
3

9

20   Blossom Shoals
7 Var: 36° 39 E. H.Water 3 ½
6   5 3   XII.

W   E   S   T

n W

2

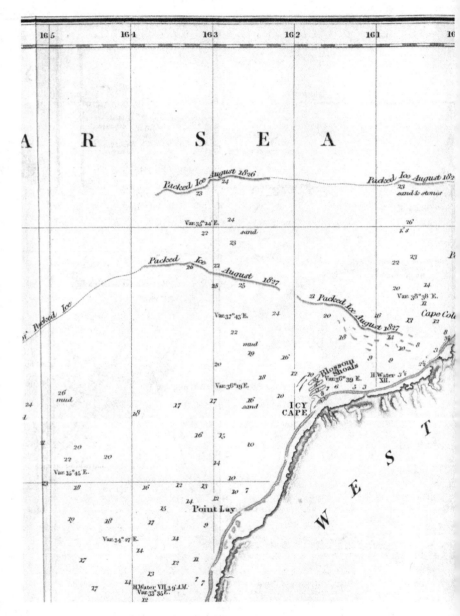

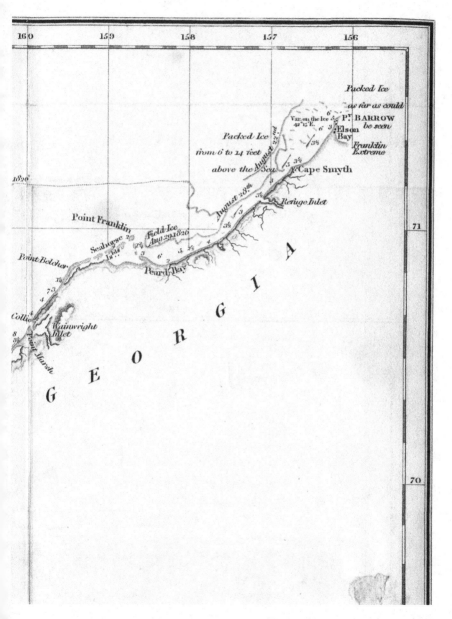

160   159   158   157   156

Packed Ice
as far as could
P.t BARROW   be seen
Var. on the Ice   6   4
42°15'E.   6   3   Elson
Bay   Franklin
Extreme
3½

Packed Ice
from 6 to 14 feet   *Cape Smyth*
above the Sea

1826

Point Franklin
Field Ice
Aug 29 1826
Point Belcher   Seahorse
Is.ts

Peard Bay

G E O R G I A

Collie
Wainwright
Inlet
Marsh

G   E   O   R

Refuge Inlet

August 22.nd

August 28.th

71

70

3

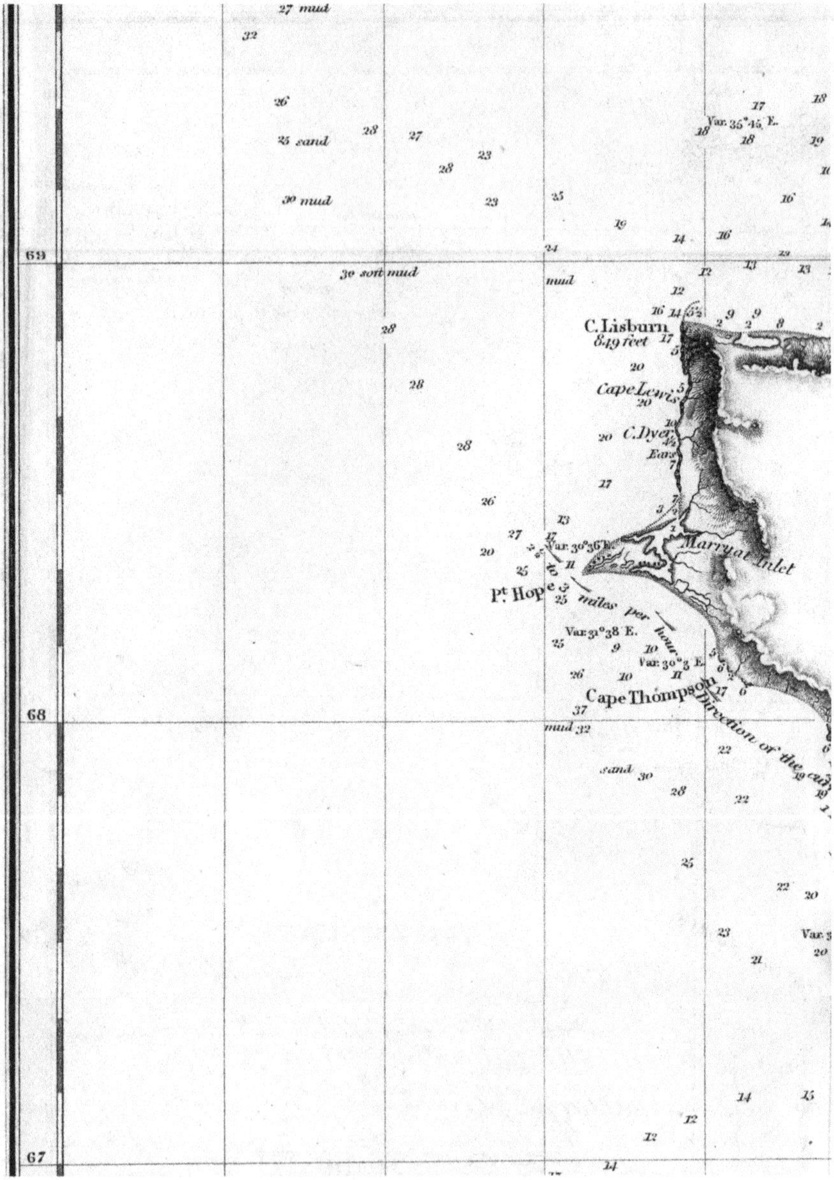

27 mud
32

26
25 sand   28   27
28   23
30 mud   23   25
19   14   16

69   30 soft mud   mud   24   12   13   23
12
26 14   2 9 9 2 8   2
C. Lisburn   17
849 feet   5
28   20
28   Cape Lewis   20
20   C. Dyer
28   Ear
27
26   13   Marryat Inlet
27   Var. 30°36 E.
20   11
25   3
Pt. Hope   25   miles per hour
Var. 31°38 E.   9   10
25   9   10   Var. 30° 3 E.
26   10   11
Cape Thompson   17
68   37   6
mud 32
22
sand   30
28   22

25

22   20

23   Var. 3
21   20

14   25

12   12
67   12
14
14

4

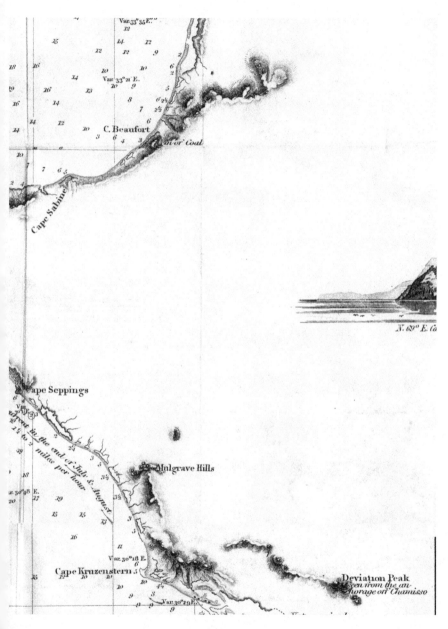

Var.33°35'E.
12
25   14   12
12   12   9
18 16
14   10   10
14   Var.33°31'E.
16 16   13   10
14   12   20   8   6 2½
7   3½
12   10   C.Beaufort   6
3   6
10   6   Seat of Coal
7   6 5
2 4
Cape Sabine

N. 69° E. G.

ape Seppings
Var.
at the end of July & August
18   1½ to 2 miles per hour
39
a.30°48'E.   3½
20   27   19
15   3½
13   15
16
9   II
Var.30°18'E.
6
Cape Kruzenstern 5
13   10   10
10   4½
9   3
Var.30°29'E.

Mulgrave Hills

Deviation Peak
Seen from the an-
chorage off Chamisso

5

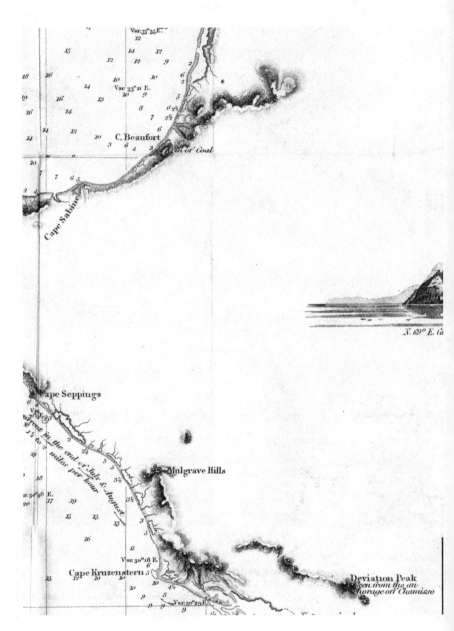

Var. 33° 35' E.

C. Beaufort

Coal or Coal

Cape Sabine

N. 69° E. (a

ape Seppings

reat in the end of July & August
1½ to 2 miles per hour

Mulgrave Hills

Cape Kruzenstern

Deviation Peak
Seen from the an-
chorage off Chamisso

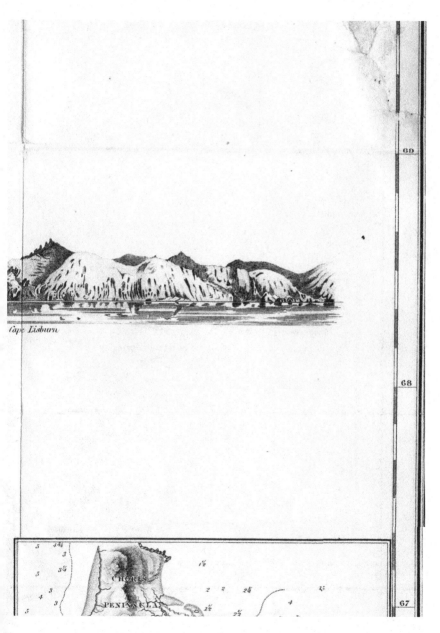

Cape Lisburn

69

68

67

6

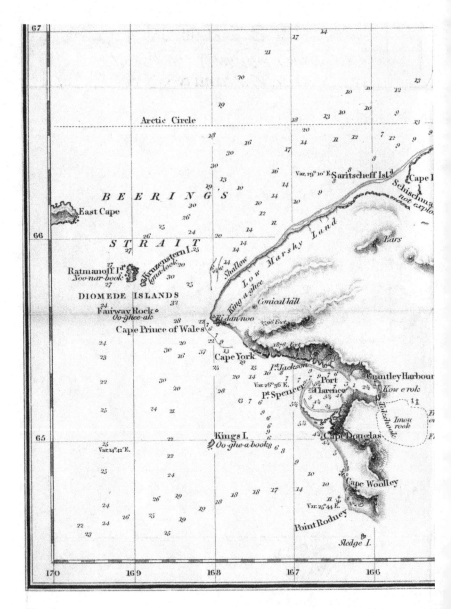

7

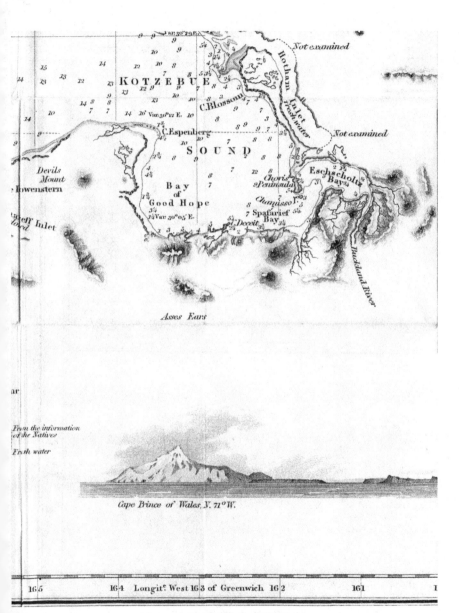

Cape Prince of Wales, N. 71° W.

Published by Colburn & Bentley London 1831.

8

Published by Colburn & Bentley London 1831.

Published by Colburn & Bentley London 1831.

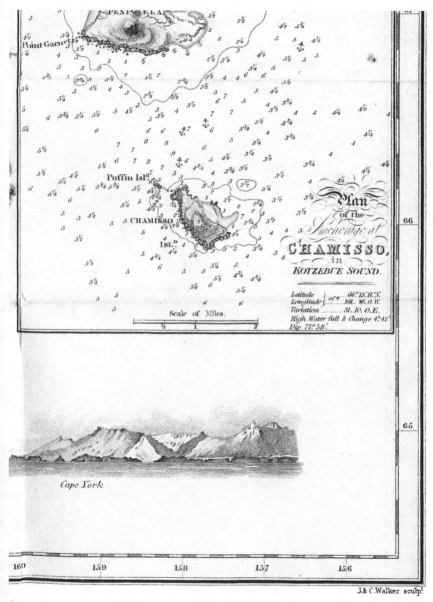

PENINSULA

Point Garnett

Puffin Isl.

CHAMISSO

ISL.^D

Scale of Miles.

Plan
of the
Anchorage at
CHAMISSO,
in
KOTZEBUE SOUND.

Latitude } of + 66°13′.H.N.
Longitude } 161. 16. 0.W.
Variation .... 31. 10, O.E.
High Water full & Change 4°.42′
Dip 77°58′.

66

65

Cape York

160   159   158   157   156

9

J.& C.Walker sculp.^t

tion, and furnish me with their opinion on the propriety of remaining longer in these seas.

Their answer, which I received the next day, conveyed an unanimous opinion that the ship could not continue longer at her present anchorage without incurring the risk of wintering, and suggested her emoval to the entrance of the sound, where the majority of the officers thought she might remain a few days longer; but previous to our taking up our station there, it was considered advisable that the strait should be ascertained to be navigable, lest the ice should have been drifted down from the northward, and the retreat of the ship be cut off. I fully concurred in opinion with them, that if the frost continued the ship could not remain at her anchorage; but as there was a possibility of its yielding, I resolved to wait a day or two longer upon the chance, determined, if it did not give way, to quit the sound; and in the event of Beering's Strait being found clear, to return, as had been proposed, and to wait a few days off Cape Krusenstern, in the hope of meeting the party. Considering, however, the lateness of the season and the long nights, there did not appear to be much chance of the ship being able to maintain an advantageous position at the mouth of the sound; still, as I was unwilling to relinquish the smallest chance of falling in with the party, I purposed making the attempt. In so doing, however, it was necessary to insure our departure by the 23d instant, which, considering our distance from any new supplies, and that at that period there would be but nine weeks' provision remaining at half allowance, was as late as I thought it prudent to continue.

We were now made sensible of the great advantage arising from the economical system that had been adopted at the Society and Sandwich Islands, and also from the reduction of an eleventh part of the ship's company at Portsmouth, without which the provision before this period would have been wholly expended, as the allowance from the time of leaving Chili had been reduced as low as it conveniently could, for a continuance, consistent with the strength of the ship's company, who for several months had been on half allowance.

It now remained for me to consider how Captain Franklin could be most benefited in the event of his party arriving after our departure. It was evident that we could do no more than put him in possession of every information we had obtained, and leave him a temporary supply of provisions and bartering articles, with which he could procure others from the natives. To this end a barrel of flour was buried for him upon the sandy point of Chamisso, a place which, from the nature of the ground, was more likely to escape observation than the former one, where the newly turned turf could not be concealed. A large tin case, containing beads and a letter, was deposited with it, to enable him to purchase provisions from the natives, and to guide his conduct. Ample directions for finding these were both cut and painted on the rock; and to call the attention of the party to the spot, which they might otherwise pass, seeing the ship had departed, her name was painted in very large letters on the cliffs of Puffin Island, accompanied with a notice of her departure, and the period to which she had remained in the sound. Beneath it were written directions

Wm Smyth del^t E. Finden s^c.

Pub^d by H.Colburn & R.Bentley, 1831.

CREW OF THE BLOSSOMS BARGE ERECTING A POST FOR CAPT^N FRANKLIN NEAR BEHRING STRAIT.

for finding the cask of flour, and also a piece of drift-wood which was deposited in a hole in the cliff. This billet had been purposely bored and charged with a letter containing all the useful information I could impart to the party, and then plugged up in such a manner that no traces of its being opened were visible. In fact, nothing was left undone that appeared to me likely to prove useful.

Having thus far performed our duty, we prepared the ship for sea in order that she might start at an hour's notice. On the 13th, the temperature fell to 27°, the lakes on shore had borne two or three days, and the sea had cooled down 8°; in short, there was every appearance of a settled frost. The next day the edges of the sound began to freeze, and it was evident that it needed only calm weather to skin it entirely over. I therefore desired the anchor to be weighed, and having taken on board a large supply of drift-wood, the last thing we procured from the shore, we steered out of the sound.

We passed Cape Krusenstern about midnight, and then shaped a course for the strait. The night, though cold, was fine, and furnished me with eighteen sets of lunar distances, east and west of the moon, which I was very anxious to obtain, in order to fix more accurately the position of Chamisso Island, never having been able to succeed in getting fine weather with the moon to the east of the sun, until his declination was too far south for the lunars to be of any value.

We had no observation at noon the next day, and the land was so refracted that we scarcely re-cognised it; we, however, continued to run for the strait, anxious to reach it before sunset. The breeze increased as we advanced, and before the Diomede Islands came in sight it blew so violently that there was no alternative but to endeavour to push through them before dusk. At this time there was a very thick haze, with a bright setting sun glaring through it, which with the spray around us prevented any thing being seen but the tops of the mountains near Cape Prince of Wales. It was consequently with great pleasure we perceived Fairway Rock, and found the strait quite free from ice.

Having no choice, we passed through it at a rapid rate; and as the night set in dark and thick, with snow showers, we were glad to find ourselves with sea-room around us. A little before midnight the lee-bow port was washed away, and so much water came in that it was necessary to put before the wind to free the ship. In half an hour, however, we resumed our course, and about two o'clock in the morning passed King's Island.

We were now in a situation where, by rounding to, we might have awaited fine weather to return to Cape Krusenstern, and execute the whole of the plan that had been contemplated; but considering that our being able to do so was uncertain, as the barometer, which had fallen to 28,7, afforded no prospect of a change of weather, and that the period I had fixed for my departure might expire before I could repass the straits; together with the state of our provisions, and the improbability of meeting with Captain Franklin after all, it appeared to me that the risks which it involved were greater than the uncertainty of the result justified; and painful as it was to relinquish every hope of this successful issue of our voyage, it became my duty to do so.

In the execution of this necessary resolution, it was some consolation to reflect, that from the nature of Captain Franklin's instructions, it was almost certain that by this time he had either commenced his return or taken up his winter abode. He had been directed to return to his winter quarters on the 15th of August, if he found the prospect of success was not such as to ensure his reaching Icy Cape that season, and if it should prove impracticable to winter at an advanced station on the coast. We were justified, therefore, in supposing that he had already been either compelled to pause or to turn back, as, in the event of the successful prospect anticipated in his instructions, it could hardly happen, considering the open state in which we had found the sea to the northward, that he should not have reached Kotzebue Sound by the time the Blossom left it.

In taking our departure from these seas, some general observations on the country, the natives, the currents, meteorology, and other subjects, naturally present themselves; but as we returned to the same place the following year, and extended our experience, I shall defer them until a future opportunity.

Up to this period of the voyage, my instructions had been a safe guide for my proceedings; but between our departure from these seas, and our return to them the following year, with the exception of touching at the Sandwich Islands, there were no specific directions for my guidance, and it became me seriously to consider how the time could be most usefully employed. It was necessary to repair to some port to refit and caulk the ship, to replenish the provisions and stores, and, what was equally important, to recruit the health of the people, who were much debilitated from their privations ; having been a considerable time on short allowance of salt provision, and in the enjoyment of only seven weeks' fresh meat in the last ten months.

From the favourable account I had heard of Saint Francisco in California, it appeared to be the most desirable place to which a ship under our circumstances could resort ; and as the coast between that port and Cape St. Lucas was very imperfectly known, that the time could not be more usefully employed than in completing the survey of it. I therefore directed our course to that place, and determined to enter the Pacific by the Strait of Oonemak ; which, if not the safest of those formed by the Aleutian Islands, is certainly the best known.

After passing King's Island on the 16th, we saw some very large flocks of ducks migrating to the southward, and fell in with the lummes, which had deserted us more than a month before at Chamisso Island. As we approached St. Lawrence Island, the little crested auks flew around us, and some land birds took refuge in our rigging. We passed to the eastward of this island in very thick weather, and had only a transient view of its eastern extremity, and thence pursued a course to the southward, passing between Gore's Island and Nunevack, an island recently discovered by the Russians, but not known to us at that time. The soundings increased, though not always regularly ; and we had thick misty weather which prevented any thing drying. The barometer fluctuated a little on either side of 28,6. On the 18th, the temperature, which had risen gradually as we advanced to the southward, was twenty degrees higher than it was the day we left Kotzebue Sound—a change which was sensibly felt.

On the 21st we came within sight of the island of St. Paul, the northern island of a small groupe which, though long known to English geographers, has been omitted in some of our most esteemed modern charts. The groupe consists of three islands, named St. George's, St. Paul's, and Sea-otter. We saw only the two latter in this passage, but in the following year passed near to the other, and on the opposite side of St. Paul's to that on which our course was directed at this time. The islands of St. Paul and St. George are both high, with bold shores, and without any port, though there is said to be anchoring ground off both, and soundings in the offing at moderate depths. At a distance of twenty-five miles from Sea-otter Island, in the direction of N. 37° W. (true), and in latitude 59° 22′ N., we had fifty-two fathoms hard ground; after this, proceeding southward, the water deepens. St. Paul's is distinguished by three small peaks, which, one of them in particular, have the appearance of craters; St. George's consists of two hills united by moderately high ground, and is higher than St. Paul's; both were covered with a brown vegetation. Sea-otter Island is very small, and little better than a rock. The Russians have long had settlements upon both the large islands, subordinate to the establishment at Sitka, and annually send thither for peltry, consisting principally of the skins of

amphibious animals, which, from their fine furry
nature, are highly valued by the Chinese and Tartar
nations. I have given the geographical position of
these islands in the Appendix; and for a further
account of them, the reader is referred to Langsdorff's Travels, and to Kotzebue's Voyage.

At sunset we lost sight of St. Paul's Island, and
being at that time ignorant of the position of St.
George's, further than what knowledge was derived
from a rough notice of it in the geological account
of Kotzebue's Voyage, we pursued our course with
some anxiety, as the night was dark and unsettled,
and the morning came without our obtaining a sight
of the island. On approaching the Aleutian Islands,
we found them obscured by a dense white haze
which hung to windward of the land; and the wind
increasing with every appearance of a gale, our
situation became one of great difficulty. Early in
the morning a peak was seen for so short a time
that it only served to show us that we were not far
from the land, without enabling us to determine
which of the islands we were near; and as in this
part of the Aleutian Chain there are several passages so close together, that one may easily be mistaken for the other, an accurate knowledge of the
position of the ship is of the greatest importance.
Under our circumstances, I relied on the accuracy
of Cook's chart, and steered due east, knowing
that if land were seen in that pàrallel, it could
be no other than the island of Oonemak; and that
then, should the fog not clear away, the course
might still be directed along that island to the
southward.

This is a precaution I strongly recommend to any

person who may have to seek a channel through this chain in foggy weather, particularly as these passages are said to be rendered dangerous by the rapid tides which set through them. It was no doubt these tides, added to the prevalence of fogs, that caused many of the misfortunes which befel the early Russian navigators. Shelekoff, in speaking of the strait to the westward of Oonemak, through which we passed, observes that it is free from the danger of rocks and shoals, but is troubled with a strong current. In our passage through it, however, we did not remark that this was the case; but no doubt there are just grounds for the observation.*

After running five miles, breakers were seen upon both bows, and, at the same time, very high cliffs above them. We stood on a little further, and then, satisfied that the land must be that of Oonemak, bore up along it, and passed through the strait. We had no soundings with forty fathoms of line until we were about four miles off the S. W. end of the island; and there we found thirty fathoms on a bank of dark-coloured lava, pebbles, and scoriæ, but immediately lost it again, and had no bottom afterwards. The south-west angle of Oonemak is distinguished by a wedge-shaped cape, with a pointed rock off it. This cape and the island of Coogalga form the narrowest part of the strait, which is nine miles and a half across. Coogalga is about four miles long, and rendered very conspicuous by a peak on its

---

* I afterwards learned from a very respectable master of an American brig, that in passing through the strait to the westward of Oonalaska he experienced a current running to the northward at the rate of six miles an hour, and was unable to stem it.

N. E. extremity. Acouan, the island to the north-ward of this, which also forms part of the strait, is high and remarkable; but on this occasion we did not see it, in consequence of the bright haze that hung over the hills on the northern part of the chain.

Oonemak was the only island upon which snow was observed. Its summit was capped about one-third down, even with a line of clouds which formed a canopy over the northern half of the groupe. The limits of this canopy were so well defined, that in passing through the strait on one side of us there was a dense fog, while on the other the sun was shining bright from a cloud-less sky.

As soon as we had fairly entered the Pacific the wind abated, and we had a fine clear night, as if in passing through the chain that divides the Kams-chatkan Sea from the Pacific we had left behind us the ungenial climate of the former. Shortly after dark flashes were observed in the heavens, in the di-rection of the burning mountain of Alaska, some-times so strong as to be mistaken for sheet light-ning, at others very confined; viewed with a tele-scope, they appeared to consist entirely of bright sparks. They seemed to proceed from different parts of a long narrow cloud elevated 8°, and lying in the direction of the wind. Our distance from the volcano at this time was about seventy miles, and as similar flashes were observed in this place the following year, it is very probable they were caused by an eruption. This mountain, I am informed, has burnt lately with great activity, and has been truncated much lower than is represented in the drawings of it in Captain Cook's Voyage.*

After clearing the Aleutian Chain, we had the winds from the westward, and made rapid progress towards our port. The first part of the passage was remarkable for heavy rolling seas, misty weather, and a low barometer, which varied a little each side of 28,5; in the latter part of the passage we had dry foggy weather, and the barometer was at 30,5.

On the 5th of November we made the high land of New Albion about Bodega, and soon afterwards saw Punta de los Reyes, a remarkable promontory, from which the general line of coast turns abruptly to the eastward, and leads to the port of St. Francisco.

We stood to the southward during the night, and about three o'clock in the morning unexpectedly struck soundings upon a clayey bank in 35 fathoms very near the Farallónes, a dangerous cluster of rocks, which, until better known, ought to be avoided. The ship was put about immediately; but the next cast was 25 fathoms in so stiff a clay that the line was broken. The weather was very misty, and a long swell rolled towards the reefs, which, had there been less wind, would have obliged us to anchor; but we increased our distance from them, and deepened the water. This cluster of rocks is properly divided into two parts, of which the south-eastern is the largest and the highest, and may be seen nine or ten leagues in clear weather. The most dangerous part is apparently towards the north-west.

* See also Kotzebue's Voyage, vol. iii p. 283.

The next evening we passed Púnta de los Reyes, and awaited the return of day off some white cliffs, which, from their being situated so near the parallel of 38° N. are in all probability those which induced Sir Francis Drake to bestow upon this country the name of New Albion. They appear on the eastern side of a bay too exposed to authorize the conjecture of Vancouver, that it is the same in which Sir Francis refitted his vessel.

END OF THE FIRST VOLUME.

LONDON:
PRINTED BY SAMUEL BENTLEY,
Dorset Street, Fleet Street.

Printed in the United States
By Bookmasters